Energy Storage and Management for Electric Vehicles

Energy Storage and Management for Electric Vehicles

Special Issue Editors

James Marco

Quang Truong Dinh

Stefano Longo

MDPI • Basel • Beijing • Wuhan • Barcelona • Belgrade

Special Issue Editors
James Marco
University of Warwick
UK

Quang Truong Dinh
University of Warwick
UK

Stefano Longo
Cranfield University
UK

Editorial Office
MDPI
St. Alban-Anlage 66
4052 Basel, Switzerland

This is a reprint of articles from the Special Issue published online in the open access journal *Energies* (ISSN 1996-1073) in 2019 (available at: https://www.mdpi.com/journal/energies/special_issues/Energy_for_EV)

For citation purposes, cite each article independently as indicated on the article page online and as indicated below:

LastName, A.A.; LastName, B.B.; LastName, C.C. Article Title. *Journal Name* **Year**, *Article Number*, Page Range.

ISBN 978-3-03921-862-2 (Pbk)
ISBN 978-3-03921-863-9 (PDF)

Contents

About the Special Issue Editors . . . . **vii**

Preface to "Energy Storage and Management for Electric Vehicles" . . . . **ix**

Shouguang Yao, Xiaofei Sun, Min Xiao, Jie Cheng and Yaju Shen
Equivalent Circuit Model Construction and Dynamic Flow Optimization Based on Zinc–Nickel Single-Flow Battery
Reprinted from: *Energies* **2019**, *12*, 582, doi:10.3390/en12040582 . . . . **1**

Chaofeng Pan, Yanyan Liang, Long Chen and Liao Chen
Optimal Control for Hybrid Energy Storage Electric Vehicle to Achieve Energy Saving Using Dynamic Programming Approach
Reprinted from: *Energies* **2019**, *12*, 588, doi:10.3390/en12040588 . . . . **18**

Seyed Saeed Madani, Erik Schaltz and Søren Knudsen Kær
Simulation of Thermal Behaviour of a Lithium Titanate Oxide Battery
Reprinted from: *Energies* **2019**, *12*, 679, doi:10.3390/en12040679 . . . . **37**

José Luis Sampietro, Vicenç Puig, and Ramon Costa-Castelló
Optimal Sizing of Storage Elements for a Vehicle Based on Fuel Cells, Supercapacitors, and Batteries
Reprinted from: *Energies* **2019**, *12*, 925, doi:10.3390/en12050925 . . . . **52**

Stefan Englberger, Holger Hesse, Daniel Kucevic and Andreas Jossen
A Techno-Economic Analysis of Vehicle-to-Building: Battery Degradation and Efficiency Analysis in the Context of Coordinated Electric Vehicle Charging
Reprinted from: *Energies* **2019**, *12*, 955, doi:10.3390/en12050955 . . . . **79**

Henry Miniguano, Andrés Barrado, Cristina Fernández, Pablo Zumel and Antonio Lázaro
A General Parameter Identification Procedure Used for the Comparative Study of Supercapacitors Models
Reprinted from: *Energies* **2019**, *12*, 1776, doi:10.3390/en12091776 . . . . **96**

Hongjie Liu, Tao Tang, Jidong Lv and Ming Chai
A Dual-Objective Substation Energy Consumption Optimization Problem in Subway Systems
Reprinted from: *Energies* **2019**, *12*, 1876, doi:10.3390/en12101876 . . . . **116**

Xiangdong Sun, Jingrun Ji, Biying Ren, Chenxue Xie and Dan Yan
Adaptive Forgetting Factor Recursive Least Square Algorithm for Online Identification of Equivalent Circuit Model Parameters of a Lithium-Ion Battery
Reprinted from: *Energies* **2019**, *12*, 2242, doi:10.3390/en12122242 . . . . **144**

Deidre Wolff, Lluc Canals Casals, Gabriela Benveniste, Cristina Corchero and Lluís Trilla
The Effects of Lithium Sulfur Battery Ageing on Second-Life Possibilities and Environmental Life Cycle Assessment Studies
Reprinted from: *Energies* **2019**, *12*, 2440, doi:10.3390/en12122440 . . . . **159**

Bizhong Xia, Yadi Yang, Jie Zhou, Guanghao Chen, Yifan Liu, Huawen Wang, Mingwang Wang and Yongzhi Lai
Using Self Organizing Maps to Achieve Lithium-Ion Battery Cells Multi-Parameter Sorting Based on Principle Components Analysis
Reprinted from: *Energies* **2019**, *12*, 2980, doi:10.3390/en12152980 . . . . **178**

In-Ho Cho, Pyeong-Yeon Lee and Jong-Hoon Kim
Analysis of the Effect of the Variable Charging Current Control Method on Cycle Life of Li-ion Batteries
Reprinted from: *Energies* **2019**, *12*, 3023, doi:10.3390/en12153023 . . . . **195**

Woo-Yong Kim, Pyeong-Yeon Lee, Jonghoon Kim and Kyung-Soo Kim
A Nonlinear-Model-Based Observer for a State-of-Charge Estimation of a Lithium-ion Battery in Electric Vehicles
Reprinted from: *Energies* **2019**, *12*, 3383, doi:10.3390/en12173383 . . . . **206**

About the Special Issue Editors

James Marco is a Chartered Engineer and a Fellow of the Institution of Engineering and Technology (FIET). After graduating with an Engineering Doctorate from Warwick in 2000, James worked for several years within the automotive industry, leading engineering research teams for Ford (North America and Europe), Jaguar Cars, Land Rover, and DaimlerChrysler (Germany). In 2006, James returned to academia, taking a research position at Cranfield University and later joined WMG, University of Warwick, in 2013. Since joining WMG, his research has focused on (1) the design of novel thermal management solutions for energy storage systems, (2) the integration of electric vehicles into a future charging infrastructure—vehicle-to-grid (V2G) operation and optimising fast charging to mitigate battery degradation, (3) the design of new control functions to support second-life energy storage applications, and (4) the design of accelerated energy storage characterisation techniques to enable the broader adoption of circular economy strategies for used vehicle batteries.

Quang Truong Dinh is Assistant Professor in Energy Management and System Control. He was awarded his B.E. mechatronics degree, first-class, from the Mechanical Engineering Department at Ho Chi Minh City University of Technology, Vietnam, in March 2006. He completed his first-class Ph.D. mechatronics degree in the School of Mechanical Engineering at University of Ulsan, South Korea, in early 2010. After obtaining his Ph.D. degree, he worked in the School of Mechanical Engineering (University of Ulsan) as a Postdoctoral Researcher for two years and as a Research Professor for three-and-a-half years. From August 2015, he joined WMG, the University of Warwick. Dr Dinh's research interests are mainly concentrated on the following fields: Mechatronics—advanced mechatronic systems, integrated systems with both sensing, communication and automation technologies, system engineering and power management and control; Control theories and applications—nonlinear control (robust, adaptive, predictive, hybrid), networked control, remote control, fault tolerant control; Renewable energy—wind/wave energy converter, self-performance optimization control to maximize energy harvesting efficiency.

Stefano Longo received his MSc in Control Systems from the Department of Automatic Control and System Engineering at the University of Sheffield in 2007 and completed his PhD in Control Systems in the Department of Mechanical Engineering at the University of Bristol in 2011. His PhD thesis was awarded the prestigious Institution of Engineering and Technology (IET) Control and Automation Prize for significant achievements in the area of control engineering. In November 2010, he was appointed for the position of Research Associate in the Department of Electrical and Electronic Engineering at Imperial College London, where he worked in the intersection of the Control & Power, and Circuit & Systems Research Groups. He retained the position of Honorary Research Associate until 2016. He was appointed Lecturer (Assistant Professor) at Cranfield University in 2012 and promoted to Senior Lecturer (Associate Professor) in 2017. His work and research interests gravitate around the problem of designing and implementing advanced control and estimation algorithms in hardware, where algorithm design and hardware implementation are not seen as two separate and decoupled problems but, rather, as a single interconnected problem. These ideas have been applied to networked control systems (whilst at the University of Bristol), to parallelisable hardware for constrained optimal control and real-time optimisation (whilst at Imperial College London) and, more recently, to alternative powertrain and chassis control.

Preface to "Energy Storage and Management for Electric Vehicles"

One of the main drivers for technological development and innovation within the automotive and road transport sectors in recent years is the need to reduce the fuel consumption and exhaust emissions of vehicles while concurrently exceeding consumer expectations of quality, driveability, refinement, and vehicle range. To meet this challenge, engineers and researchers have worked together to design, integrate, and validate future powertrain technologies for the next generation of hybridised and fully electric vehicles. Within the context of many electrified vehicle applications, the design and management of high-voltage battery systems represents the greatest element of research novelty. The aim of this Special Issue of Energies is to explore research innovation within the battery systems engineering domain that incorporates optimization, mathematical modelling, control engineering, thermal management, mechanical design, and component sizing and packaging.

James Marco, Quang Truong Dinh, Stefano Longo
Special Issue Editors

Article

Equivalent Circuit Model Construction and Dynamic Flow Optimization Based on Zinc–Nickel Single-Flow Battery

Shouguang Yao [1,*], Xiaofei Sun [1], Min Xiao [1], Jie Cheng [2] and Yaju Shen [2]

1 School of Energy and Power Engineering, Jiangsu University of Science and Technology, Zhengjiang 212000, China; ntsunxf@126.com (X.S.); xiaomin_just@126.com (M.X.)

2 Zhangjiagang Zhidian Fanghua Storage Research Institute, Zhangjiagang 215600, China; chengjie_chj@126.com (J.C.); syjee7766@163.com (Y.S.)

* Correspondence: zjyaosg@126.com; Tel.: +86-15051110000

Received: 15 January 2019; Accepted: 11 February 2019; Published: 13 February 2019

Abstract: Based on the zinc–nickel single-flow battery, a generalized electrical simulation model considering the effects of flow rate, self-discharge, and pump power loss is proposed. The results compared with the experiment show that the simulation results considering the effect of self-discharge are closer to the experimental values, and the error range of voltage estimation during charging and discharging is between 0% and 3.85%. In addition, under the rated electrolyte flow rate and different charge–discharge currents, the estimation of Coulomb efficiency by the simulation model is in good agreement with the experimental values. Electrolyte flow rate is one of the parameters that have a great influence on system performance. Designing a suitable flow controller is an effective means to improve system performance. In this paper, the genetic algorithm and the theoretical minimum flow multiplied by different flow factors are used to optimize the variable electrolyte flow rate under dynamic SOC (state of charge). The comparative analysis results show that the flow factor optimization method is a simple means under constant charge–discharge power, while genetic algorithm has better performance in optimizing flow rate under varying (dis-)charge power and state of charge condition in practical engineering.

Keywords: zinc–nickel single-flow battery; equivalent circuit model; self-discharge; dynamic flow rate optimization; genetic algorithm

---

1. Introduction

The shortage of primary energy and environmental problems have led to increased development of renewable energy in all countries of the world. However, renewable energy has the characteristics of discontinuity, instability, and uncontrollability. Large-scale integration of renewable energy into power grids will bring serious impact on the safe and stable operation of power grids, resulting in a large number of abandoned light and wind [1]. Large-scale energy storage technology is one of the effective methods to solve this problem [2–4]. Among them, the liquid flow battery has attracted wide attention in the home and abroad because of its independent capacity, flexible location, safety, and reliability. In view of the problems of ion cross-contamination and high cost of ion exchange membrane in traditional dual-flow batteries, Professor Pletcher of Cape Town University had proposed single-flow lead–acid batteries [5–8] in 2004. Due to the obvious advantages of single-flow batteries over dual-flow batteries, different series of single-flow batteries have been developed at home and abroad, such as zinc–nickel single-flow batteries [9], lead dioxide/copper single-flow batteries [10], and quinone/cadmium [11] single-flow batteries. Among them, zinc–nickel single-flow batteries have attracted wide attention due to their long life, high energy efficiency, safety, and environmental protection [9]. In recent years, the research and development of zinc–nickel single-flow batteries have been mainly based on experiments, including the selection and testing of key materials [12–14], electrolyte composition addition [15–18], and flow structure design [19–22] to improve the performance of zinc–nickel single-flow batteries and promote large-scale zinc–nickel single-flow battery systems (ZNBs) to form an energy storage system for engineering applications [23].

Establishing a general electrical model that can accurately reflect the external characteristics of the stack is the premise of predicting and analyzing the parameters of ZNBs energy storage system and optimizing its operation, and then building an efficient battery stack management system. At present, there are few studies on the electrical model construction of zinc–nickel single-flow battery stacks, and the development of more complete vanadium redox flow batteries can be referred to. Barote et al. [24,25] and Chahwan et al. [26] proposed the basic equivalent circuit model of the vanadium redox flow battery. The model used a controlled current source and a fixed resistance to represent parasitic loss, reaction resistance, and electrode capacitance, and a voltage source to represent stack voltage. However, their models do not take into account the dynamic characteristics of batteries and lack of experimental verification. Recently, Ankur et al. [27] aimed to make vanadium redox flow batteries further oriented to renewable energy sources, and built an equivalent circuit model of vanadium redox flow batteries considering electrolyte flow rate, pump loss, and self-discharge. Accurate estimation of battery stack terminal voltage and dynamic SOC was achieved, and the optimal range of variable electrolyte flow under dynamic SOC was investigated, which provided support for the design of flow controller. On the basis of the above, reference [28] further estimated the parameters of the internal electrical components of the equivalent circuit of the vanadium redox flow battery under different electrolyte flow rates, charge–discharge current densities, and charge states, and coupled the obtained parameters with the simulation model. The comparison with the experimental results showed that the accuracy of the model has been significantly improved. For the zinc–nickel single-flow battery stack studied in this paper, Yao Shou-guang et al. [29,30], based on the working principle of zinc–nickel single-flow batteries, built the PNGV (the Partnership for a New Generation Vehicles) equivalent circuit model, and further obtained the PNGV model parameters by parameter identification based on the experimental data of the pulse discharge of the battery at 100 A. Then, the high-order polynomial and exponential function fitting method was used to obtain the analytical formula of each model parameter. Xiao M. et al. [31] proposed an improved Thevenin equivalent circuit model of the zinc–nickel single-flow battery, based on the principle of parameter identification and the least-squares curve-fitting method to obtain the parameters of the improved model, and then the discrete mathematical model of each parameter in the improved model was obtained by discretization. However, the above equivalent circuit model established for the zinc–nickel single-flow battery does not consider the effects of self-discharge, electrolyte flow, and pump loss.

Based on the preliminary work, a general electrical model considering the factor of flow rate, self-discharge, and pump loss which can accurately reflect the external characteristics of the stack is proposed in the paper. In addition to this, another significant contribution of this paper is to use flow factor multiplied by the theoretical minimum flow and genetic algorithm to determine an optimal flow rate for minimum loss in the ZNBs system, considering both the internal power loss and pump power loss. Such a comprehensive modeling of zinc–nickel single-flow batteries has not been reported in the literature available at home and abroad. The general electrical model is simulated in MATLAB/Simulink and is verified by a zinc–nickel single-flow battery stack composed of 23 single batteries in parallel. The simulation model can support the design of efficient battery management systems for large-scale ZNBs energy storage system.

2. Equivalent Circuit Model

The positive electrode of the zinc–nickel single-flow battery adopts a nickel oxide electrode used in a secondary battery; the negative electrode is an inert metal current collector (nickel-plated steel strip), and 10 mol/L KOH + 5 g/L LiOH + 0.5 mol/L ZnO solution is used as the base electrolyte.

The positive electrode reaction is completed in the porous nickel positive electrode, and the negative electrode reaction is a surface deposition/dissolution reaction. Figure 1 is a schematic diagram of the basic structure of a zinc–nickel single-flow battery stack (300 Ah), which comprises 23 parallel cells, and the electrolyte is driven by a pump to flow through the stack from the bottom during the charge and discharge cycle. Figure 2 is a schematic structural view of a partially parallel single cell, and d_1 is an interval between the positive and negative electrodes. The specific structural parameters of the model are shown in Table 1.

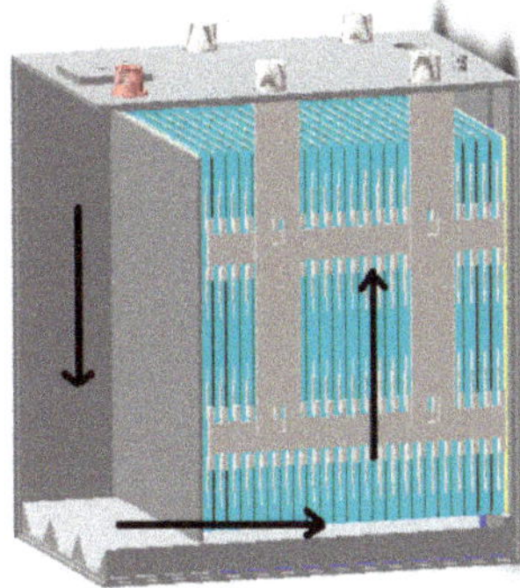

Figure 1. Basic structure of zinc–nickel single-flow battery.

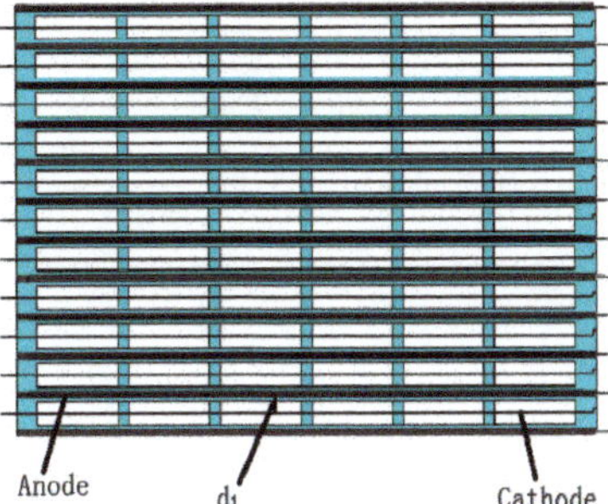

Figure 2. Basic structure of partially parallel single cells.

Table 1. Size parameters of the initial model.

Main Components	Size Parameters
Height of porous nickel electrode (mm)	240
Width of porous nickel electrode (mm)	186
Thickness of porous nickel electrode (mm)	0.64
Height of negative pole (mm)	240
Width of negative pole (mm)	186
Thickness of negative pole (mm)	0.08
Distance between anode and cathode(d_1/mm)	160
Electrolyte density ($kg \cdot m^{-3}$)	1456.1
Electrolyte viscosity ($kg \cdot m^{-1} \cdot s^{-1}$)	0.003139
No. of parallel cells in stack	23
Inner diameter of the pipeline (mm)	15
Length of pipeline (cm)	40
Pipeline import and export height difference (cm)	5
Number of bends	3

The active substance in the nickel oxide electrode undergoes a chemical reaction during charge and discharge. The charge–discharge reaction process is as shown in Equation (1). The zinc negative electrode is accompanied by deposition and dissolution during charge and discharge. The charge–discharge reaction process is as shown in Equation (2). The total reaction in the zinc–nickel single-flow battery is shown in Equation (3).

$$2NiOOH + 2H_2O + 2e^- \rightleftarrows 2Ni(OH)_2 + 2OH^- \qquad E^0 = 0.49\ V \tag{1}$$

$$Zn + 4OH^- \leftrightarrows Zn(OH)_4^{2-} + 2e^- \qquad E^0 = -1.215\ V \tag{2}$$

$$2NiOOH + 2H_2O + Zn \rightleftarrows 2Ni(OH)_2 + Zn(OH)_2 \qquad E^0 = 1.705\ V \tag{3}$$

Taking the above-mentioned zinc–nickel single-flow battery stack (300 Ah) as the research object, the equivalent circuit model considering the flow rate, pump power loss, and self-discharge is built. The final general electrical model of the zinc–nickel single-flow battery stack is shown in Figure 3. The following Sections 2.1–2.5 elaborate on each module of the general electrical simulation model of the zinc–nickel single-flow battery.

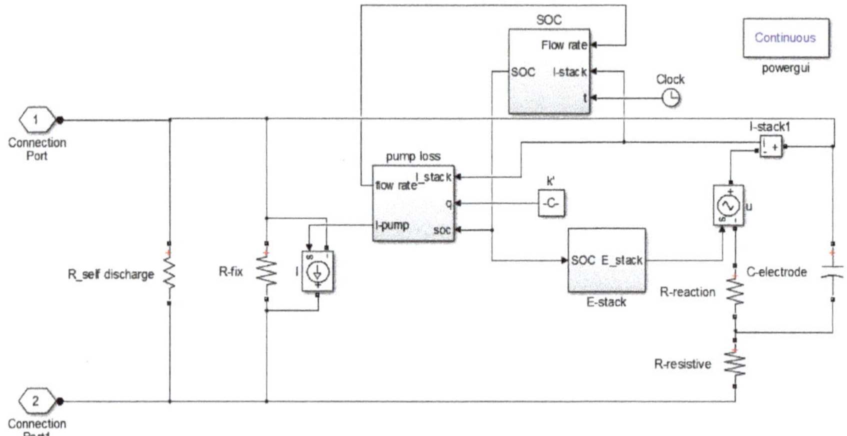

Figure 3. Generalized electrical model of zinc–nickel single-flow battery stack.

2.1. Internal Loss

Experimental tests show that the system efficiency of the zinc–nickel single-flow battery stack (300 Ah) is about 69% when the charge–discharge current is 100 A, and the remaining 31% is internal loss. The actual power inside the stack can be calculated by Equation (4). The internal loss of the stack can be divided into ohmic loss and polarization loss. The effect on the stack can be reflected in the equivalent circuit model as ohmic loss resistance ($R_{resistive}$) and polarization loss resistance ($R_{reaction}$), which can be calculated by Equation (4) [32].

$$P_{stack} = \frac{P_{rate}}{\eta_{system}} \tag{4}$$

$$R = \frac{K \cdot P_{stack}}{I_{max}^2} \tag{5}$$

In Equation (4), P_{rate} is rated power and η_{system} is system efficiency. In Equation (5), K is power loss coefficient, I_{max} is the maximum charge–discharge current of the battery stack, and R is the internal loss resistance (ohmic loss resistance or polarization loss resistance). Equivalent circuit model parameters are calculated under very bad conditions [32], that is, when the charge–discharge current is the maximum current and SOC is 0.2. This paper is based on the function expression of the ohmic loss resistance ($R_{resistive}$) and the polarization loss resistance ($R_{reaction}$) of the zinc–nickel single-flow battery stack (300 Ah) proposed in reference [29]. When the SOC is 0.2, the values of $R_{resistive}$ and $R_{reaction}$ are respectively 0.623 mΩ and 0.2504 mΩ, and then the ohmic loss coefficient ($K_{resistive}$) and polarization loss coefficient ($K_{reaction}$) are calculated by Equation (5) to be 10.8% and 4.35%, respectively, and the parasitic loss is about 15.85% of the total loss.

2.2. Pump Loss Model

The pump loss model of the zinc–nickel single-flow battery is shown in Figure 4. The pump loss is characterized by fixed loss (R_{fix}) and pump current loss (I_{pump}). Fixed loss resistance (R_{fix}) is calculated by Equation (6), in which U_{min} is the minimum voltage of the stack and P_{fix} is the fixed loss power, which is experimentally measured to account for about 2% of P_{stack}.

$$R_{fix} = \frac{U_{min}^2}{P_{fix}} \tag{6}$$

The function relationship between pump loss current (I_{pump}) and pump power (P_{mech}) in the electrical model is shown in Equation (7). The pump loss coefficient (M) is related to pump loss power. Definition of M see Equation (8).

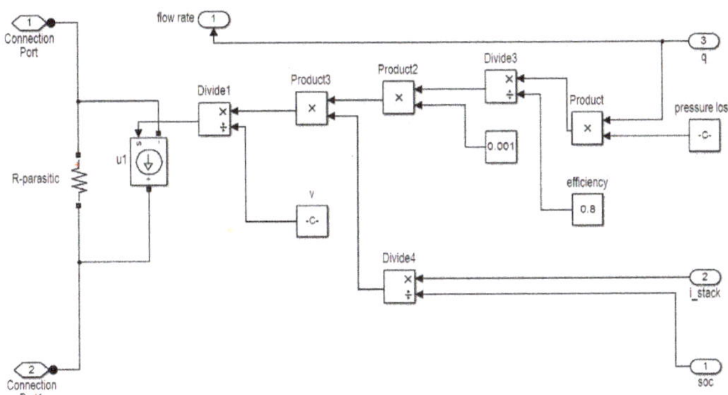

Figure 4. Pump loss model of zinc–nickel single-flow battery stack.

$$I_{pump} = \frac{P_{mech\_loss}}{U_{stack}} = \frac{M \cdot \left(\frac{I_{stack}}{SOC}\right)}{U_{stack}} \tag{7}$$

$$M = \frac{P_{mech} \cdot SOC_{worse}}{I_{max}} \tag{8}$$

The mechanical loss (P_{mech_loss}) includes two parts: the mechanical loss (P_{pipe_loss}) caused by the electrolyte flowing through the pipeline connecting the stack and the external storage tank, and the mechanical loss (P_{stack_loss}) caused by the electrolyte flowing through the stack. The total loss (P_{mech_loss}) is shown in Equation (9).

$$P_{mech\_loss} = P_{stack\_loss} + P_{pipe\_loss} \tag{9}$$

When the electrolyte of the zinc–nickel single-flow battery flows through pipes, valves, and liquid storage tanks, it will cause a certain pressure drop, which is collectively called pipeline pressure drop. The pressure drop equation of the pipeline can be obtained by the Bernoulli equation, which is related to electrolyte flow rate, loss along the pipeline, local loss, and height difference between inlet and outlet of the pipeline. Pipeline pressure drop and mechanical loss can be expressed as Equations (10) and (11). The pressure drop of the tube outside the stack is estimated to be about 65.5 kPa.

$$\Delta P_{pipe} = -\gamma\left(\frac{\Delta V_s^2}{2g} + \Delta Z + h_f + h_m\right) \tag{10}$$

$$P_{pipe} = \Delta P_{pipe} \times Q \tag{11}$$

The pressure drop in the stack is determined by the flow rate of the electrolyte and the resistance of the electrolyte, so the expressions of pressure drop and mechanical loss in the stack are as follows:

$$\Delta P_{stack} = Q \times \widetilde{R} \tag{12}$$

$$P_{stack} = \Delta P_{stack} \times Q \tag{13}$$

In Equation (12), $\widetilde{R}$ is the hydraulic resistance of the stack, and its value can be seen in the previous research work of our group [33]. The formula for calculating P_{stack} is shown in Equation (13).Considering the pump efficiency, the total mechanical loss of the battery system can be defined as Equation (14).

$$P_{mech\_loss} = \frac{P_{pipe\_loss} + P_{stack\_loss}}{\eta_{pump}} \tag{14}$$

2.3. Self-Discharge Loss

The self-discharge of the zinc–nickel single-flow battery is mainly caused by the negative reaction of the negative electrode, which forms a microprimary battery on the surface of the negative electrode, which has a significant influence on the attenuation of the battery capacity. In this paper, the self-discharge effect is equivalent to the loss resistance (R_{self}) in the equivalent circuit model. The calculation formula is shown in Equation (15), where P_{self} is the power loss caused by self-discharge, and its expression is given by Equation (16). For the self-discharge power loss coefficient (f), the calculation formula is shown in Equation (17), where U_1 and U_2 are the changes of battery voltage with time in the charge–discharge process without considering self-discharge effect and considering self-discharge effect, respectively.

$$R_{self} = \frac{U_{min}^2}{P_{self}} \tag{15}$$

$$P_{self} = f \cdot P_{stack} \tag{16}$$

$$f = \frac{\int_{t_1}^{t_2} U_1 I_1 dt - \int_{t_1}^{t_2} U_2 I_2 dt}{\int_{t_1}^{t_2} U_1 I_1 dt} \tag{17}$$

2.4. Voltage Estimation Model

The voltage estimation module of the zinc–nickel single-flow battery stack is shown in Figure 5. The ion activity should be used when calculating the battery electromotive force using the Nernst equation. When the ionic strength is not large, and the valence state of the oxides and the reductants is not high, the battery electromotive force can be directly calculated by using the ion concentration. In the zinc–nickel single-liquid battery, the valence states of the hydroxide ion and zincate ion are −1 and −2, respectively. The active material nickel oxide of the positive electrode is not present in the battery in the form of ions, and its ion activity cannot be further measured. Only the proton concentration of hydrogen can be used to indicate the content of nickel hydroxide. Whether it is theoretical analysis or comparison with experimental results, it is shown that the error caused by the calculation of the voltage of the stack using the ion concentration is small and within an acceptable range. The potentials of the positive and negative electrodes are as follows:

$$\text{Positive electrode potential}: E^{+} = E^{0}_{+} + \frac{RT}{nF} \ln\left(\frac{C_{NiOOH}}{C_{Ni(OH)_2} \cdot C_{OH^-}}\right)^{2} \tag{18}$$

$$\text{Negative electrode potential}: E^{-} = E^{0}_{-} + \frac{RT}{nF} ln\left(\frac{C_{Zn(OH)_4^{2-}}}{C_{OH^{-4}}}\right) \tag{19}$$

E^{+} is the positive equilibrium potential, E^{-} is the negative equilibrium potential, T is the ambient temperature, and n is the electron transfer number in the electrode reaction. The concentration of positive active substance can be replaced by H proton concentration. Equation (18) can be rewritten as follows:

$$\text{Positive electrode potential}: E^{+} = E^{0}_{+} + \frac{RT}{nF} \ln\left(\frac{C^{H}_{max} - C^{H}}{C^{H} \cdot C_{OH^-}}\right)^{2} \tag{20}$$

The battery stack potential is as follows:

$$E_{stack} = E^{0} + \frac{RT}{F} \ln\left(\frac{C^{H}_{max} - C^{H}}{C^{H}} \times \frac{C_{OH^-}}{C_{Zn(OH)_4^{2-}}{}^{1/2}}\right) \tag{21}$$

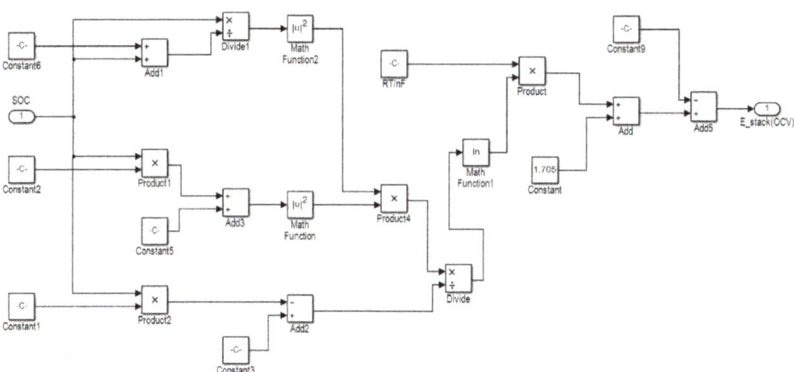

Figure 5. Open-circuit voltage estimation model of zinc–nickel single-flow battery stack.

Based on the above-mentioned calculations in Equations (18)–(21) for the potential of the zinc–nickel single-flow battery stack, combined with the range of concentration of each substance in Table 2, the battery potential can be further expressed by SOC as Equation (22), where E^{0} is 1.705 V. Under different operating conditions, the terminal voltage is affected by internal loss and self-discharge. The terminal voltage is estimated by Equation (23), where "±" indicates the charging process and the discharging process. $E_{self\text{-}discharge}$ is the average voltage drop caused by the self-discharge during charge and discharge, which is 3.65 mV and 6.9 mV, respectively [33].

$$E_{stack} = E^{0} + \frac{RT}{nF} \ln\left(\left(\frac{SOC}{1 - SOC}\right)^{2} \times \frac{(1.4SOC + 9.6)^{2}}{1 - 0.7SOC}\right) \tag{22}$$

$$E_{terminal} = E_{stack(OCV)} \pm I_{stack}(R_{reaction} + R_{resistive}) - E_{self-discharge} \tag{23}$$

Table 2. Range [28].

Parameters	Unit	Range
C^H	$mol\cdot m^{-3}$	0–35,300
C_{OH^-}	$mol\cdot m^{-3}$	9600–11,000
$C_{Zn(OH)_4^{2-}}$	$mol\cdot m^{-3}$	300–1000
C^H_{max}	$mol\cdot m^{-3}$	35,300

2.5. SOC Estimation Model

SOC is used to characterize the state of charge of batteries. Its estimation module is shown in Figure 6. Based on the change of concentration of $Zn(OH)_4^{2-}$, the dynamic SOC value of the zinc–nickel single-flow battery is reflected in Equation (24). "±" indicates the charging and discharging process. The value of $C_{max}^{Zn(OH)_4^{2-}}$ can be obtained as 1 mol/L from Table 2.

$$SOC = 1 - \frac{C_{initial}^{Zn(OH)_4^{2-}} \pm C_{variable}^{Zn(OH)_4^{2-}}}{C_{max}^{Zn(OH)_4^{2-}}} \tag{24}$$

The SOC of the zinc–nickel single-flow battery stack storage system is divided into SOC_{tank} in the tank and SOC_{stack} in the stack. The SOC in the stack is given by Equation (25). To simplify the estimation of the SOC, the formula for calculating the dynamic SOC of the stack is shown in Equation (26). Equation (27) is a formula for calculating the SOC_{stack}. When charging, b takes a value of 1, and when discharged, it is −1. The simulation parameters involved in the model are shown in Table 3.

$$SOC_{stack} = \frac{SOC_{stack\_in} + SOC_{stack\_out}}{2} = \frac{SOC_{tank} + SOC_{tank} + \frac{I_{stack}}{F\times Q\times C}}{2} \tag{25}$$

$$SOC_{stack\_t} = SOC_{tank\_t} + \frac{I_{stack}}{2\times F\times Q\times C} \tag{26}$$

$$SOC_{tank\_t} = SOC_{tank\_initial} + \frac{b\times\int_{t_1}^{t_2} I_{stack\_t}dt}{F\times V\times C} \tag{27}$$

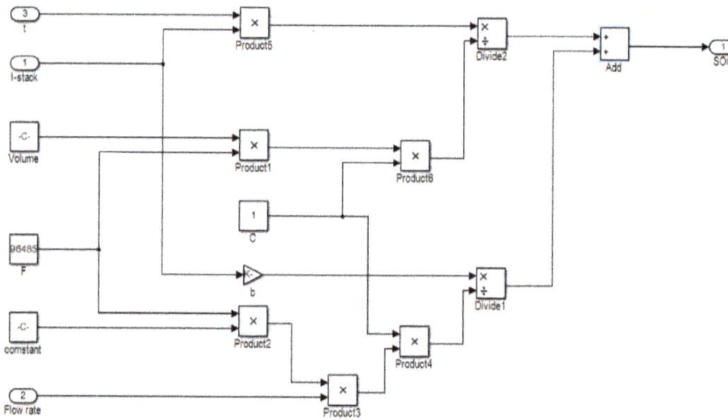

Figure 6. SOC estimation model of zinc–nickel single-flow battery stack.

Table 3. Parameters [29–34].

Parameters	Unit	Value
Rated voltage	V	1.6
I_{max}	A	200
U_{min}	V	1.2
P_{rate}	W	160
Stack capacity	Ah	300
Number of parallel cells	-	23
Operating temperature range	°C	−40~40
Volume of electrolyte	L	8.5
$R_{\text{resistive}}$	Ω	0.00064
$R_{reaction}$	Ω	0.00036
$K_{\text{resistive}}$	-	10.8%
K_{reaction}	-	4.35%
R_{fix}	Ω	0.313
$C_{electrode}$	F	138
R_{self}	Ω	0.16
f	-	0.039
η_{pump}	-	0.8
F	C/mol^{-1}	96485
SOC_{worse}	-	0.2
$\widetilde{R}$	Pa/m^3	14186843
n	-	2
T	K	298
C	mol/L	1

3. Results and Discussions

3.1. Terminal Voltage Estimation and Error Analysis of the Charging

This section compares the voltage values of the zinc–nickel single-flow battery stacks obtained from experimental and simulation models at different charging currents (50 A, 100 A, 150 A). Figure 7a shows the comparison between the terminal voltage value of the stack obtained by the experiment and the voltage of the stack of the equivalent circuit model (considering self-discharge and without considering self-discharge) when the charging current is 100 A. The results show that the simulation results without considering the self-discharge effect have a large error with the experimental values. When the model considers the capacity loss and voltage drop caused by self-discharge, the charging time and voltage value obtained by the simulation are more consistent with the experimental values, so as to avoid the undercharge phenomenon caused by the large voltage estimation error. Figure 7b is a relative error analysis of the model simulation voltage value considering the self-discharge effect and the experimental value, and the error range is between 0.001% and 2.61%.

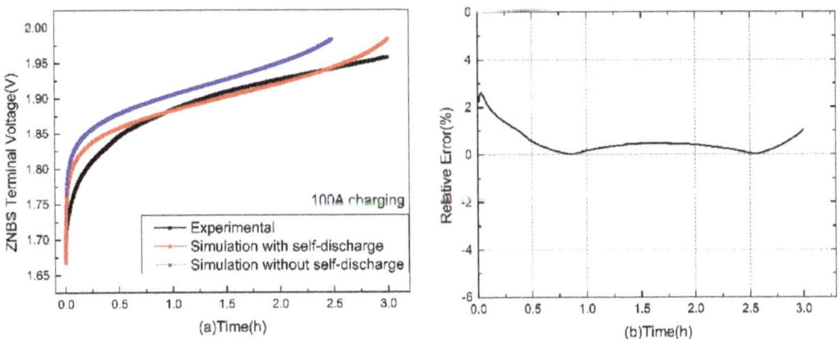

Figure 7. (**a**) Simulation results and experimental verification of ZNBs voltage at 100 A charging current; (**b**) relative error between simulation results considering self-discharge and experimental results.

Figure 8 is a comparison of simulated voltage values obtained from electrical models (considering self-discharge and without considering self-discharge) with experimentally obtained voltage values at 50 A and 150 A. The results show that the simulation results considering the self-discharge are more accurate, and the error analysis is shown in Table 4.

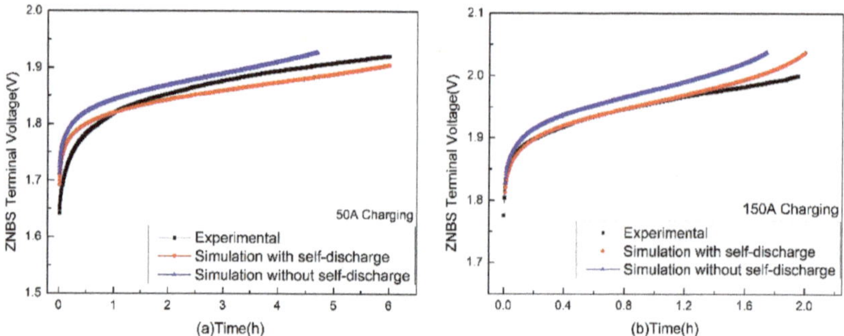

Figure 8. (**a**) Simulation results and experimental verification of the ZNBs terminal voltage when the charging current is 50 A; (**b**) simulation results and experimental verification of the ZNBs terminal voltage when the charging current is 150 A.

Table 4. Voltage error analysis of stack under different charging currents.

Charging Current (A)	Maximum Relative Error (%)	Minimum Relative Error (%)	Charging Completion Time (h)
50	1.1	0.02	6
100	2.61	0.001	3
150	1.44	0	2

3.2. Terminal Voltage Estimation and Error Analysis of the Discharging

Similar to Section 3.1, this section analyzes and validates the simulated voltage values obtained from the equivalent circuit model of the zinc–nickel single-flow battery stack under different discharge currents (50 A, 100 A, 150 A) and the experimentally obtained voltage values. Figure 9a shows the terminal voltage estimation of a zinc–nickel single-flow battery stack under different conditions (experiment, simulation of self-discharge, simulation without self-discharge) when the discharge current is 100 A. The results show that the simulation results without considering the self-discharge effect have a large error with the experimental values. When the model considers the capacity loss and voltage drop caused by self-discharge, the discharge time and voltage value obtained by the simulation are more consistent with the experimental values, so as to avoid the overdischarge phenomenon caused by the large voltage estimation error. Figure 9b is a relative error analysis of the model simulation voltage value considering the self-discharge and the experimental value, and the error range is between 0.004% and 3.75%.

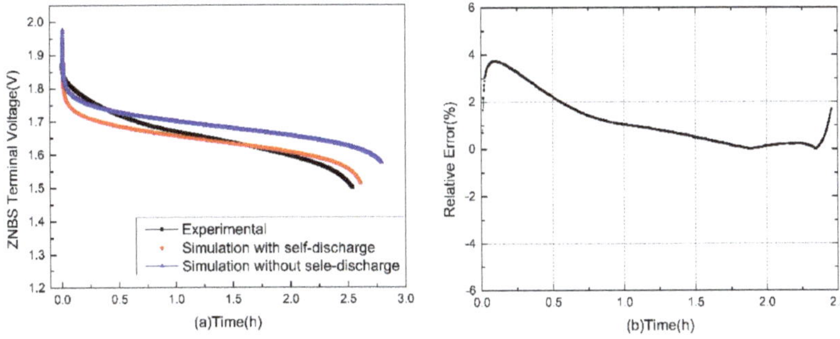

Figure 9. (**a**) Simulation results and experimental verification of ZNBs voltage at 100 A discharging current; (**b**) relative error between simulation results considering self-discharge and experimental results.

Figure 10 is a comparison of simulated voltage values obtained from electrical models (considering self-discharge and without considering self-discharge) with experimentally obtained voltage values at 50 A and 150 A. The results show that the simulation results considering the self-discharge are more accurate, and the error analysis is shown in Table 5.

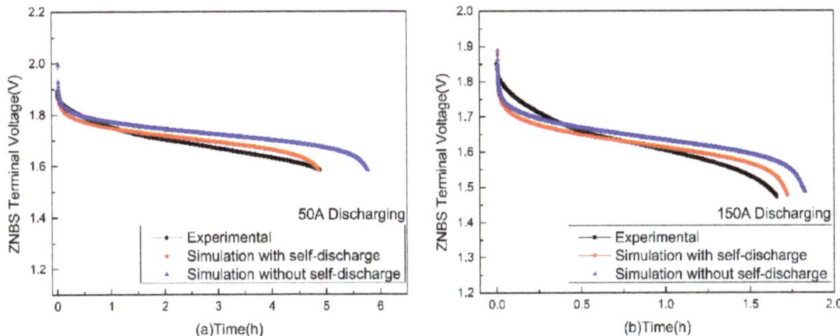

Figure 10. (**a**) Simulation results and experimental verification of the ZNBs terminal voltage when the discharging current is 50 A; (**b**) simulation results and experimental verification of the ZNBs terminal voltage when the discharging current is 150 A.

Table 5. Voltage error analysis of stack under different discharging currents.

Discharging Current (A)	Maximum Relative Error (%)	Minimum Relative Error (%)	Discharging Completion Time (h)
50	1.8	0.002	5.3
100	3.75	0.004	2.48
150	3.85	0.02	1.78

3.3. Coulomb Efficiency Analysis

This section evaluates the Coulomb efficiency of a complete charge–discharge cycle for a zinc–nickel single-flow battery stack. Charging current is 100 A, discharge current is 50 A, 100 A, 150 A, and Coulomb efficiency ($\eta_{coulombic}$) is defined as Equation (28).

$$\eta_{coulombic} = \frac{\int_0^{t_d} i_{dicharge} dt}{\int_0^{t_c} i_{charge} dt} \tag{28}$$

Figure 11 shows the Coulombic efficiency of the ZNBs energy storage system under the same charging current and different discharge currents. Under the operating conditions of 50 A, 100 A, 150 A discharge, the Coulomb efficiency calculated by the experiment is 89%, 89.9%, and 88%, respectively, and the Coulomb efficiency calculated by the simulation model is close to the experimental values, at 88.1%, 90.3%, and 88.6%, respectively. Therefore, the model can be used to estimate the Coulombic efficiency of a zinc–nickel single-flow battery stack under different operating conditions.

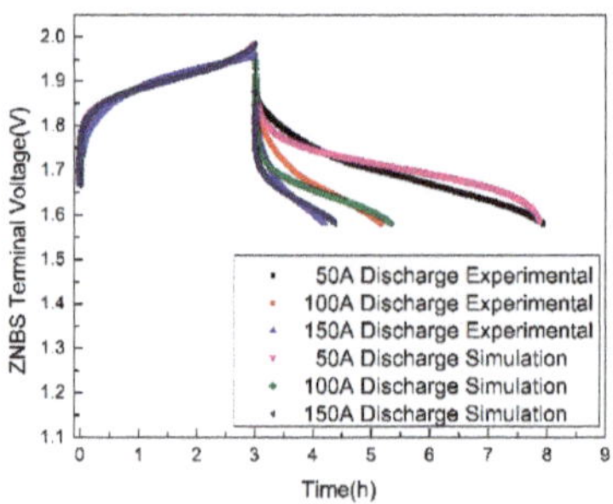

Figure 11. Coulombic efficiency estimation of 300 Ah zinc–nickel single-flow battery stack when charging at 100 A and discharging at three different currents (50 A, 100 A, 150 A).

3.4. *Dynamic Flow Rate Optimization*

Electrolyte flow rate is one of the parameters that have a great influence on the performance of the flow battery stack energy storage system, and is closely related to its internal mass transfer, temperature distribution, and system loss. For the concentration overpotential, Ma X. et al. [35] first proposed the theoretical minimum electrolyte flow rate (Q_{min}) based on Faraday's law; see Equations (29) and (30). On this basis, relevant scholars use the theoretical minimum flow multiplied by different flow factors to optimize the electrolyte flow. Fu et al. [36] found that the minimum flow of the stack system should consider the concentration overpotential and pump power loss; Tang et al. [37] found that the system efficiency is the highest when the electrolyte flow rate is 7.5 times the theoretical minimum flow rate (factor = 7.5). In this paper, with reference to the optimization method proposed by the predecessors, the overall power loss (pump loss and internal loss) of the system is taken as the objective function. Firstly, the theoretical minimum flow multiplied by different flow factor (factor) is used to optimize the flow. The expression of flow rate can be seen in Equation (31).

$$\text{Charge}: \ Q_{min} = \frac{I}{F \times n \times c \times (1 - SOC)} \tag{29}$$

$$\text{Discharge}: \ Q_{min} = \frac{I}{F \times n \times c \times SOC} \tag{30}$$

$$Q = \text{factor} \cdot Q_{min} \tag{31}$$

Figure 12 shows the theoretical minimum electrolyte flow rate of the zinc–nickel single-flow battery stack (300 Ah) as a function of SOC and current. The results show that the theoretical minimum flow rate of the electrolyte is large at the end of charge and at the end of discharge to avoid a large concentration overpotential [34].

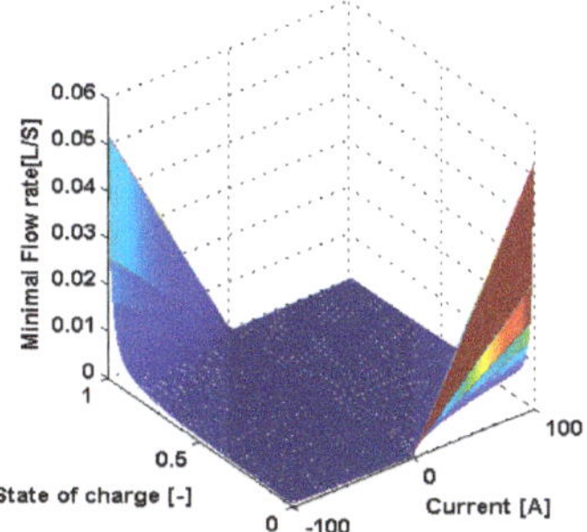

Figure 12. The change of theoretical minimum flow rate with current (I) and charge state (SOC). ("-" denotes charging.)

Figure 13 shows the power consumption and output of the system under different flow factors (factor = 5, factor = 10, factor = 15) and charging and discharging currents of 100 A. The simulation results show that with the increase of the flow factor, the power consumption of the stack system is slightly improved, and the power output of the stack system has a small decrease.

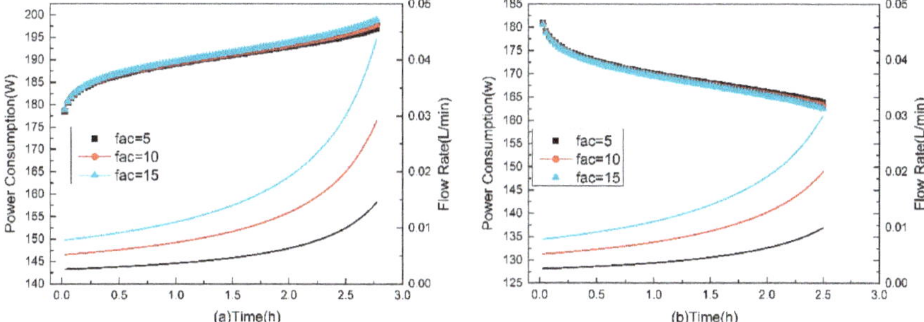

Figure 13. (**a**) The change of power consumption with the flow factor during charging; (**b**) the variation of output power with the flow factor during discharging.

In the face of the phenomenon of charge and discharge "peak and valley" in actual engineering, there is a time-varying optimal flow factor corresponding to different charge and discharge powers under the corresponding state of charge. If a fixed flow factor is used, it may not reach the expected optimization effect. Tao W. et al. [38] combined dichotomy with the flow factor optimization method to realize real-time optimization of electrolyte flow under dynamic charging and discharging power, but this method is only applicable to single-parameter optimization, and the objective function must be a single peak function. Compared with the traditional optimization algorithm, genetic algorithm has good optimization ability for nonlinear problems, and can optimize multiobjective and multiparameter simultaneously. Therefore, the genetic algorithm is introduced as an optimization method in this paper. The constant charge–discharge power condition is taken as an example to optimize the electrolyte flow rate of the zinc–nickel single-flow battery stack in real time, which provides theoretical support for multiparameter and multiobjective optimization under dynamic charge–discharge power.

In this paper, the total loss (internal loss and pump loss) of the zinc–nickel single-flow battery stack (300 Ah) energy storage system during charging and discharging process is taken as objective function, and the electrolyte flow rate was optimized at each time step in the simulation model. Figure 14a,b show the system power under two different flow control strategies (not optimized flow rate 0.09 L/s and genetic algorithm optimized flow) during charging and discharging, respectively. The results show that under the optimized electrolyte flow rate, the power consumption of the charging process is significantly reduced, and the power output of the discharge process is significantly improved.

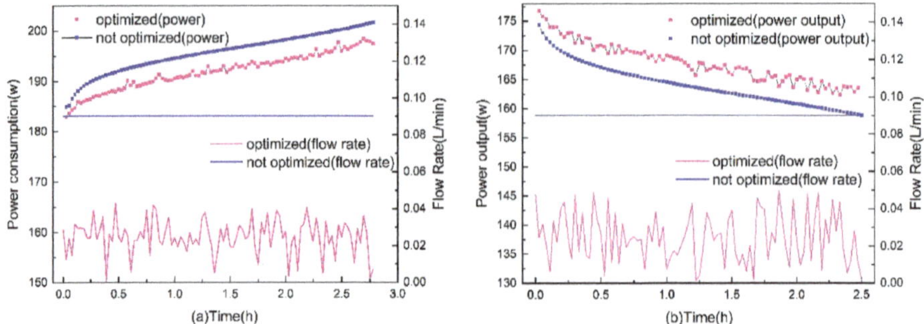

Figure 14. (**a**) System power consumption under different flow control strategies during charging process; (**b**) system output power under different flow control strategies during discharge process.

In addition, the overall performance (Coulomb efficiency, energy efficiency, and system efficiency) of the zinc–nickel single-flow battery stack (300 Ah) under 100 A charge–discharge current and different electrolyte flow control strategies (optimized electrolyte flow rate by genetic algorithm, electrolyte flow rate corresponding to different flow factors, and rated flow rate of 0.09 L/s) are compared and analyzed. The calculation formulas for Coulombic efficiency, energy efficiency, and system efficiency are as follows in Equations (32)–(34).

$$\text{Coulombic efficiency}: \ \eta_{\text{Coulombic}} = \frac{\int_0^{t_d} I_{\text{discharge}} dt}{\int_0^{t_c} I_{\text{charge}} dt} \tag{32}$$

$$\text{Energy efficiency}: \ \eta_{\text{Eenergy}} = \frac{\int_0^{t_d} I_{\text{discharge}} E_{\text{discharge}} dt}{\int_0^{t_c} I_{\text{charge}} E_{\text{charge}} dt} \tag{33}$$

$$\text{System efficiency}: \ \eta_{\text{System}} = \frac{\int_0^{t_d} (P_{\text{stack}} - P_{\text{loss}}) dt}{\int_0^{t_c} (P_{\text{stack}} + P_{\text{loss}}) dt} \tag{34}$$

Figure 15 shows the performance parameters calculated by the complete charge and discharge cycle of the zinc–nickel single-flow battery stack under different flow control strategies with a current of 100 A. The results show that when the flow control strategy is optimized by the genetic algorithm, the system efficiency is the highest, reaching 86.7%. When the theoretical minimum flow is multiplied by different flow factors for flow optimization, it can be found that with the increase of factor, the system efficiency has a small decrease. The theoretical minimum flow multiplied by the different flow factor optimization method can make the system efficiency of the stack energy storage system reach a higher value, for example, when the factor value is 5, the system efficiency is 85.6%. From the trend of change, the factor value is smaller, which may further improve the system efficiency, but the optimization result of the flow factor is only suitable for a specific working condition. With the fluctuation of charging and discharging power, if a fixed flow factor is used, the expected optimization effect may not be achieved. However, the optimization method of genetic algorithm solves this problem well. In practical engineering applications, a superior flow control strategy can be derived by genetic algorithm when in the face of the "peak and valley" phenomenon of charge and discharge power caused by the discontinuous, unstable, and uncontrollable characteristics of renewable energy and uncontrollable changes in user demand.

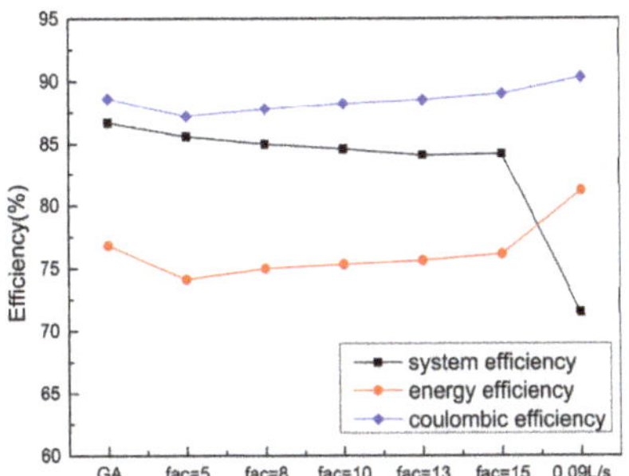

Figure 15. Performance analysis of the stack under different flow control strategies.

4. Conclusions

In this paper, the zinc–nickel single-flow battery stack is taken as the research object, and a general electrical model considering self-discharge, pump loss, and flow is built by using MATLAB/Simulink software. The self-discharge module, pump loss module, SOC, and voltage estimation module in this model are described in detail in Section 2. In order to evaluate the accuracy of the electrical model, the charging and discharging experiments of the zinc–nickel single-flow battery stack (300 Ah) were carried out under different charging and discharging currents (50 A, 100 A, 150 A). The results are compared with the simulation values (considering self-discharging and without considering self-discharging). The results show that the simulation values obtained by the simulation model considering self-discharging are closer to the experimental results. The minimum error of voltage in charging is 0–0.02%, the maximum error is 1.1–2.61%, the minimum error of voltage in discharging is 0.002–0.02%, and the maximum error is 1.8–3.85%. In addition, the Coulombic efficiency of the complete charge and discharge cycle of the simulation model is estimated. Under the operating conditions of rated electrolyte flow rate (0.09 L/S), charging current 100 A, and discharge current 50 A, 100 A, and 150 A, the comparison with experimental data shows that the simulation model has high accuracy in estimating Coulomb efficiency. The flow rate of electrolytes is one of the most influential parameters in the operation of battery stacks. Excessive flow rate of electrolytes will cause high pump loss, and too low a flow rate of electrolyte will increase the internal loss of the battery stack. Therefore, there exists a time-varying optimal electrolyte flow rate to maximize the system efficiency of the zinc–nickel single-flow battery stack corresponding to the dynamic SOC. In this paper, the overall power loss (pump loss, internal loss) of the system is taken as the objective function, and two methods, genetic algorithm and theoretical minimum flow multiplied by different flow factors, are used to optimize the flow rate. The results show that, compared with the rated flow rate (0.09 L/s), the optimized flow rate of electrolytes improves the system efficiency significantly. The results show that under the constant charge and discharge power, the above two optimization methods have significantly improved the system performance, and the flow factor optimization method is more convenient. However, in the face of the "peak-valley" phenomenon of charge and discharge power in actual engineering, the optimization method of fixed flow factor may not achieve the expected effect, and the genetic algorithm can optimize the electrolyte flow in real time to provide better flow control strategy.

Author Contributions: S.Y. provided financial support and put forward research ideas. X.S. completed model building and manuscript writing. M.X. put forward research ideas. J.C. and Y.S. designed the experiment.

Funding: This research was funded by National Natural Science Foundation of China (grant number: 51776092).

Acknowledgments: We would like to thank the School of Energy and Power Engineering of Jiangsu University of Science and Technology for its support. In addition, we would like to thank our colleagues in the laboratory for their technical assistance to the software problems encountered during the project.

Conflicts of Interest: The authors declare no conflict of interest.

Abbreviations

P_{stack}	Stack power	h_m	Localized loss
P_{rate}	Stack power rating	P_{self}	Self-discharge power loss
K	Power loss coefficient	R_{self}	Self-discharge resistance
I_{max}	Maximum charge and discharge current	V_{stack}	Stack terminal voltage
R	Internal loss resistance	E_{stack}	Stack potential
V_{min}	Stack minimum voltage	SOC_{stack}	Stack stage of charge
P_{fix}	Fixed loss power	SOC_{tank}	Tank stage of charge
R_{fix}	Fixed loss resistor	n	No. of electrons transferred per mole
P_{mech}	Mechanical loss	Q	Electrolyte flow rate
P_{stack_loss}	Internal mechanical loss of stack	F	Faraday constant
P_{pipe_loss}	Mechanical loss in the pipe	factor	Flow rate factor
V_s	Velocity of the electrolyte inside the pipe	$\eta_{Coulombic}$	Coulombic efficiency
Z	Height of the pipe	$\eta_{Eenergy}$	Energy efficiency
h_f	Pipeline loss	η_{System}	System efficiency
$\widetilde{R}$	Hydraulic resistance	η_{pump}	Pump efficiency

References

1. Xie, X.; Zheng, Q.; Li, X.; Zhang, H. Current advances in the flow battery technology. *Energy Storage Sci. Technol.* **2017**, *6*, 1050–1057.
2. Ding, M.; Chen, Z.; Su, J.; Chen, Z.; Zhu, C. Optimal Control of Battery Energy Storage Based on Variables Smoothing Time Constant. *Autom. Electr. Syst.* **2013**, *01*, 19–25.
3. Zhao, P.; Zhang, H.; Zhou, H. Research outline of redox flow cells for energy storage in china. *Chin. Battery Ind.* **2005**, *10*, 96–99.
4. Jia, Z.; Song, S.; Wang, B. A critical review on redox flow batteries for electrical energy storage. *Energy Storage Sci. Technol.* **2012**, *01*, 50–57.
5. Hazza, A.; Pletcher, D.; Wills, R. A novel flow battery: A lead acid battery based on an electrolyte with soluble lead (II) Part I. *Chem. Phys.* **2004**, *6*, 1773–1778. [CrossRef]
6. Pletcher, D.; Wills, R. A novel flow battery: A lead acid battery based on an electrolyte with soluble lead (II) Part II. *Chem. Phys.* **2004**, *6*, 1779–1785.
7. Pletcher, D.; Wills, R. A novel flow battery—A lead acid battery based on an electrolyte with soluble lead (II) III. *J. Power Sour.* **2005**, *149*, 96–102. [CrossRef]
8. Hazza, A.; Pletcher, D.; Wills, R. A novel flow battery—A lead acid battery based on an electrolyte with soluble lead (II) IV. *J. Power Sour.* **2005**, *149*, 103–111. [CrossRef]
9. Cheng, J.; Zhang, L.; Yang, Y.; Wen, Y.; Cao, G.; Wang, X. Preliminary study of single flow zinc–Nickel battery. *Electrochem. Commun.* **2007**, *9*, 2639–2642. [CrossRef]
10. Pan, J.; Sun, Y.; Cheng, J.; Wen, Y.; Yang, Y.; Wan, P. Study on a new single flow acid Cu-Pb O2battery. *Electrochem. Commun.* **2008**, *10*, 1226–1229. [CrossRef]
11. Xu, Y.; Wen, Y.; Cheng, J.; Cao, G.; Yang, Y. Study on a single flow acid Cd-chloranil battery. *Electrochem. Commun.* **2009**, *11*, 1422–1424. [CrossRef]
12. Zhang, L.; Cheng, J.; Yang, Y.; Wen, Y.; Xie, Z. Preliminary Study of Single Flow Zinc-Nickel Battery. *Electrochemistry* **2008**, *14*, 248–252.
13. Zhang, L.; Cheng, J.; Yang, Y.; Wen, Y.; Wang, X.; Cao, G. Study of zinc electrodes for single flow zinc/nickel battery application. *J. Power Sour.* **2008**, *179*, 381–387. [CrossRef]
14. Cheng, J.; Wen, Y.; Xu, Y.; Cao, G.; Yang, Y. Effects of Substrates on Deposition of Zinc from flowing Alkaline Zincate Solutions. *Chem. J. Chin. Univ.-Chin.* **2011**, *32*, 2640–2644.
15. Wang, J.; Zhang, L.; Zhang, C.; Xiao, Q.; Zhang, J.; Cao, C. The Influence of $Bi3^+$ and Tetrabutylammonium Bromide on the Dendritic Growth Behavior of Alkaline Rechargeable Zinc Electrode. *Funct. Mater.* **2001**, *32*, 45–47.
16. Wen, Y.; Cheng, J. The inhibition of the spongy electrocry stallization of zinc from dopedflowing alkaline zincate solutions. *J. Power Sour.* **2009**, *193*, 890–894. [CrossRef]

17. Wen, Y.; Wang, T.; Cheng, J.; Pan, J.; Cao, G.; Yang, Y. Lead ion and tetrabutylammonium bromide as inhibitors of the growth of spongy zinc in single flow zinc/nickel batteries. *Electrochim. Acta* **2012**, *59*, 64–68. [CrossRef]
18. Song, S. Study on Electrolyte of Zinc/Nickel Single Flow Battery. Ph.D. Thesis, Beijing University of Chemical Technology, Beijing, China, 2014.
19. Cheng, Y.; Zhang, H. A high power density single flow zinc-nickel battery with three-dimensional porous negative electrode. *J. Power Sour.* **2013**, *241*, 196–202. [CrossRef]
20. Cheng, Y.; Zhang, H. Performance gains in single flow zinc-nickel batteries through novel cell configuration. *Electrochim. Acta* **2013**, *105*, 618–621. [CrossRef]
21. Cheng, Y.; Zhang, H. Effect of temperature on the performances and in situ polarization analysis of zinc-nickel single flow batteries. *J. Power Sour.* **2014**, *249*, 435–439. [CrossRef]
22. Yao, S.; Ji, Y.; Wang, Y.; Song, Y.; Xiao, M.; Cheng, J. Optimization Analysis for the Internal Flow Field of Nickel-zinc Single Flow Energy Storage Battery with 32-Cell. *Oxid. Commun.* **2016**, *39*, 3223–3234.
23. Zhao, P.; Cheng, J.; Xu, Y.; Wen, Y.; He, K.; Cao, G. Pilot Scale Development of Zn/Ni single flow redox battery. In *Summary of the 29th Annual Academic Meeting of the Chinese Chemical Society—Chapter 24: Chemical Power*; Peking University: Beijing, China, 2014.
24. Barote, L.; Marinescu, C.; Georgescu, M. VRB modeling for storage in stand-alone wind energy systems. *IEEE Powertech. Conf.* **2009**. [CrossRef]
25. Barote, L.; Marinescu, C. A new control method for VRB SOC estimation in stand-alone wind energy systems. *Intern. Conf. Clean Electr. Power.* **2009**, 253–257.
26. Chahwan, J.; Abbey, C.; Joos, G. VRB modelling for the study of output terminal voltages, internal losses and performance. *Electr. Power Conf.* **2008**. [CrossRef]
27. Bhattacharjee, A.; Saha, H. Design and experimental validation of a generalised electrical equivalent model of Vanadium Redox Flow Battery for interfacing with renewable energy sources. *J. Energy Storage* **2017**, *13*, 220–232. [CrossRef]
28. Bhattacharjee, A.; Roy, A.; Banerjee, N.; Patra, S.; Saha, H. Precision dynamic equivalent circuit model of a Vanadium Redox Flow Battery and determination of circuit parameters for its optimal performance in renewable energy applications. *J. Power Sour.* **2018**, *396*, 506–518. [CrossRef]
29. Yao, S.; Liao, P.; Xiao, M.; Cheng, J.; He, K. Equivalent circuit modeling and simulation of the zinc-nickel single flow batteries. *AIP Adv.* **2017**, *7*, 1–10. [CrossRef]
30. Xiao, M.; Liao, P.; Yao, S.; Cheng, J. Experimental study on charge/discharge characteristics of zinc-nickel single-flow battery. *J. Renew. Sustain. Energy* **2017**, *9*. [CrossRef]
31. Yao, S.; Liao, P.; Xiao, M.; Cheng, J.; He, K. Modeling and simulation of the zinc-nickel single flow batteries based on MATLAB/Simulink. *AIP Adv.* **2016**, *6*. [CrossRef]
32. Chi, X.; Zhu, M.; Wu, Q. Research on optimal operation control based on the equivalent model of VRFB system. *Energy Storage Sci. Technol.* **2018**, *7*, 530–538.
33. Yao, S.; Liao, P.; Xiao, M.; Cheng, J.; Cai, W. Study on Electrode Potential of Zinc-nickel Single-Flow Battery during Charge. *Energies* **2017**, *10*, 1101. [CrossRef]
34. Blanc, C. Modelling of Vanadium Redox Flow Battery Electricity Storage System. Ph.D. Thesis, Echole Polytechnique Federale De Lausanne, Lausanne, Switzerland, 2017.
35. Ma, X.; Zhang, H.; Sun, C.; Zou, Y.; Zhang, T. An optimal strategy of electrolyte flow rate for vanadium redox flow battery. *J. Power Sour.* **2012**, *203*, 153–158. [CrossRef]
36. Fu, J.; Zheng, M.; Wang, X.; Sun, J.; Wang, T. Flow-Rate Optimization and Economic Analysis of Vanadium Redox Flow Batteries in a Load-Shifting Application. *J. Energy Eng.* **2017**, *143*, 1–13. [CrossRef]
37. Tang, A.; Bao, J.; Skyllaskazacos, M. Studies on pressure losses and flow rate optimization in vanadium redox flow battery. *J. Power Sour.* **2014**, *248*, 154–162. [CrossRef]
38. Wang, T.; Fu, J.; Zheng, M.; Yu, Z. Dynamic control strategy for the electrolyte flow rate of vanadium redox flow batteries. *Appl. Energy* **2018**, *227*, 613–623. [CrossRef]

Article

Optimal Control for Hybrid Energy Storage Electric Vehicle to Achieve Energy Saving Using Dynamic Programming Approach

Chaofeng Pan [1,2,*], Yanyan Liang [2], Long Chen [1,2] and Liao Chen [2]

1 Automotive engineering research institute, Jiangsu University, Zhenjiang 212013, China; chenlong@ujs.edu.cn

2 College of automotive and traffic engineering, Jiangsu University, Zhenjiang 212013, China; 2211604019@stmail.ujs.edu.cn (Y.L.); qinhe@ujs.edu.cn (L.C.)

* Correspondence: chfpan@ujs.edu.cn; Tel.: +86-0511-88782845

Received: 27 December 2018; Accepted: 11 February 2019; Published: 13 February 2019

Abstract: In this paper, the efficiency characteristics of battery, super capacitor (SC), direct current (DC)-DC converter and electric motor in a hybrid power system of an electric vehicle (EV) are analyzed. In addition, the optimal efficiency model of the hybrid power system is proposed based on the hybrid power system component's models. A rule-based strategy is then proposed based on the projection partition of composite power system efficiency, so it has strong adaptive adjustment ability. Additionally. the simulation results under the New European Driving Cycle (NEDC) condition show that the efficiency of rule-based strategy is higher than that of single power system. Furthermore, in order to explore the maximum energy-saving potential of hybrid power electric vehicles, a dynamic programming (DP) optimization method is proposed on the basis of the establishment of the whole hybrid power system, which takes into account various energy consumption factors of the whole system. Compared to the battery-only EV based on simulation results, the hybrid power system controlled by rule-based strategy can decrease energy consumption by 13.4% in line with the NEDC condition, while the power-split strategy derived from the DP approach can reduce energy consumption by 17.6%. The results show that compared with rule-based strategy, the optimized DP strategy has higher system efficiency and lower energy consumption.

Keywords: hybrid power system; electric vehicle; rule-based optimal strategy; dynamic programming approach

---

1. Introduction

Due to the shortcomings of short life and low power density of power battery, if power battery is used as the sole energy source of electric vehicle (EV), the power and economy of vehicles will be greatly limited [1,2]. The utilization of high-power density super capacitor (SC) into the EV power system and the establishment of a battery-super capacitor hybrid power system can achieve complementary advantages to make up for the lack of power battery [3,4]. Yi Hongming simulated the important modules of the SC-battery hybrid power system in MATLAB/Simulink. The results show that the hybrid power system can exert its high energy density and high-power density characteristics, thus improving the vehicle's dynamic performance and energy utilization [5]. Xu and Wang combined high-power SC with traditional batteries, and adopted parallel interleaving technology in DC/DC converter, which changed the topology of the hybrid power supply, greatly improving the overall performance of the composite power system. The fuzzy control method is used to manage the energy storage system [6]. Cezar improved the performance of the combined energy storage unit by introducing SC as auxiliary power supply. This paper presents a complete energy storage system model, including a battery, a SC and a rule-based control strategy. When the power required for energy storage is higher than the threshold, the SC is released, which means that the power of the driver needs to be increased for a period of time [7]. Therefore, a SC with battery hybrid power system is proposed in this paper, which is composed of the battery-super-capacitor hybrids, transmission and the electric motor in this research. Specific efficiency characteristics are displayed by each component of the hybrid power system, which is strongly affected by the power demands according to driving conditions and driver's intentions. EV with battery-super-capacitor hybrids can attain minimum energy consumption through switching different driving modes according to the high efficiency area of the hybrid power system [8].

In addition, the rationality of the energy distribution strategy of the composite power system is also an important factor affecting energy consumption. Special efforts have been devoted to the design and implementation of optimal energy management strategies concerning their importance to urban EV. Essentially, existing approaches may be categorized in rule-based control strategies, optimization and intelligent control strategies [9]. (1) Rule-based methods and analytic methods are usually operation mode dependent. (2) Optimal theory methods can be classified as global optimization and real-time optimization methods, including minimum principle, quadratic programming and dynamic programming (DP) method. (3) Intelligent control methods include neural networks, and model predictive control methods, fuzzy logic, genetic algorithm method, and swarm optimization method [10]. Rule-based methods have difficulty achieving optimal control effect, but they are simple and easily conducted; real-time application of intelligent control methods are limited because of they involve more calculation and are time-consuming [11]. Banvait proposed a rule-based energy management strategy for plug-in hybrid electric vehicle (PHEV), then a PHEV model was built using Advisor software, and the simulation results show that the strategy can significantly reduce fuel consumption [12]. Hemi proposed a rule-based energy management strategy combined with the equivalent consumption minimization strategy (ECMS), which is developed and simulated by using a dynamic model of the vehicle developed in the Matlab/Simulink environment. The simulation results verify the effectiveness of the strategy under various vehicle masses [13]. Previous research about energy management algorithms are concentrated in the field of energy management algorithm based on optimization. DP is a widely-used method that applies search for absolutely optimal controls under a predetermined driving cycle [14]. Optimal power management strategy obtained by DP was employed in parallel hybrid electric vehicle (HEV) to minimize fuel consumption [15]. A finite horizon dynamical optimization problem with constraints of proper energy limits and solved by a DP approach was proposed by Xiaosong Hu, in order to avoid physical damage of the electrical storage system [16]. A driving pattern recognition technique of switching among the control rule employed in the optimal power management strategy for range extended electric vehicle sets extracted from DP results of each representative driving pattern [17]. An optimal solution to the energy management problem in fuel-cell hybrid vehicles with dual storage buffer for fuel economy in a standard driving cycle using multi-dimensional dynamic programming (MDDP) was suggested and turned out to be applicable [18]. An energy management strategy based on stochastic dynamic programming was proposed for a serial hybrid electric tracked vehicle [19]. DP typically focuses on the energy consumed during the driving event as its objective, with the SOC indicating the state of the system, and either the power split ratio or the torque split ratio as the control variable [20]. However, the real-time controller based on DP is effective only for the driving cycle that is used for rule extraction [21]. For the near-optimal rule-based energy split strategy, control rules can also be extracted from the DP results [22]. In summary, rules-based and DP method used in composite power pure electric vehicles, the existing research shows that the main optimization lies in the optimization of motor control and optimized space can be limited; in this paper, the hybrid power system and motor drive system are comprehensively considered, and the optimal efficiency model of the hybrid power system is established to explore the best feasible scheme of energy utilization for pure electric vehicles.

In order to propose a systematic optimal solution for hybrid power system and energy split strategy, high efficiency areas of hybrid power system under single power and hybrid power modes needed to be rationally distributed. Efficiency characteristics analysis of battery, SC, electric motor and DC/DC converter is a vital part of the solution. Based on simulation and experiment methods, the efficiency formulas of the hybrid power system can be summarized under different working conditions, with vehicle acceleration and velocity as independent variables. It is critical to set the status parameters of battery and SC as constraints of optimization problems, because constraints represent the work status of both energy storage units, and work status directly influences the efficiency of the hybrid power system. In this sense, after analyzing the characteristics of each component of the hybrid power system, the efficiency calculation model of the hybrid power system is established. On this basis, a rule-based energy management strategy is proposed, and then the DP method is used to solve the optimization problem of the optimal energy allocation strategy for the hybrid power system of EV.

The structure of the paper is organized as follows: firstly, the materials and methods are provided in Section 2, and the structure and the key components models of hybrid power system in EV are presented in Sections 2.1–2.4, which include description of the hybrid power system, battery model, SC model and electric motor; and the rule-based strategy and DP optimization strategy is described in Sections 2.5 and 2.6. In Section 3, the results and discussions of rule-based energy management strategy and energy allocation optimization strategy based on DP are presented. Finally, conclusions are presented in Section 4.

2. Materials and Methods

2.1. Description of the Hybrid Power System

The vehicle is equipped with a brushless direct current motor (BLDC) connected to a fixed-ratio transmission. BLDC is powered by battery and SC. The bidirectional DC/DC converter is used to interface the SC with the BLDC. Energy management controller plays the role of power split in the hybrid power system. The architecture of the hybrid power system is as shown in Figure 1. The main parameters of the EV used in this study are listed in Table 1.

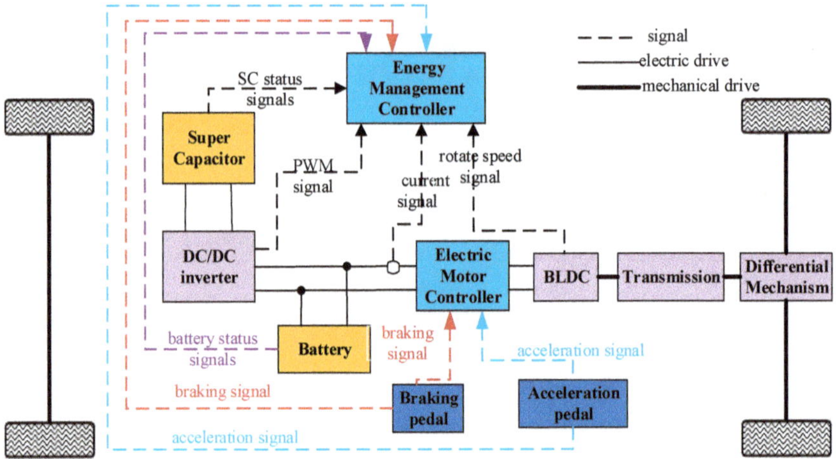

Figure 1. Architecture of the hybrid power system.

Table 1. Basic parameters of the electric vehicle (EV).

Parameter	Value
m, Vehicle mass (kg)	1360
R, Wheel radius (m)	0.277
C_D, Air drag coefficient	0.35
A, Front area (m$^{2)}$	2.3
ρ, Air density (kg/m$^{3)}$	1.29
i_0, Transmission ratio	7.881
η_T, Transmission efficiency (%)	95
η_r, Regenerative braking efficiency (%)	65
η_{DC}, DC/DC converter efficiency (%)	92
DC bus voltage (V)	260–350

2.2. Battery Model

Lithium-ion batteries are preferred in EV applications owing to their high voltage, good safety property, and long cycling life [23]. The main parameters of the Lithium-ion battery used in this study are given in Table 2.

Table 2. Basic parameters of the battery cell.

Parameter	Value
Nominal voltage (V)	3.65
Capacity (Ah)	42
Stored energy (kWh)	21
R_0 (mΩ)	16.8

Rint model is commonly used in describing the characteristics of the lithium battery, which is applied because of its simplicity and little inaccuracy. The battery behavior is represented by the Rint model depicted in Figure 2a.

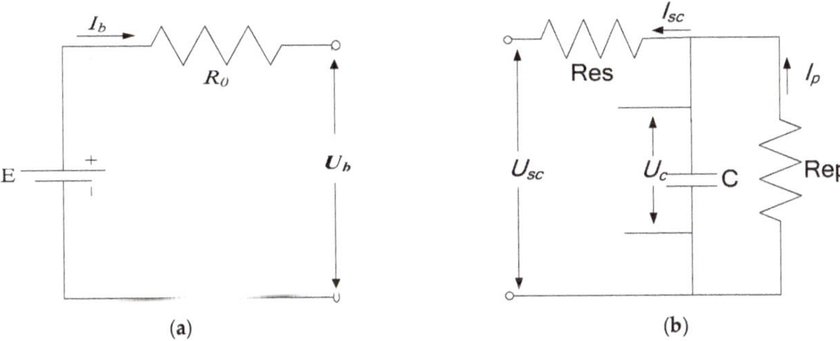

Figure 2. Simplified circuit models of battery and super capacitor (SC). (**a**) Rint model of the battery; (**b**) Transmission line model of SC.

The parameter data of variables is obtained through battery characteristics experiments, and the calculation results of the battery module internal resistance are shown in Table 3. According to the calculation results of the internal resistance of the battery module in Table 3, the relationship between battery SOC and internal resistance can be obtained by polynomial fitting under MATLAB environment (The MathWorks, Inc, Natick, MA, USA), and 6 times polynomial is obtained, such as Equation (1).

$$R_0 = 221.2x^6 - 695.5x^5 + 729x^4 - 358.2x^3 + 73.12x^2 - 6.338x + 17.49 \quad (1)$$

where x represents the SOC value of the battery, R_0 (mΩ) is the internal resistance of the battery. The fitting curve is shown in Figure 3, and a conclusion can be drawn through analyzing the data that the resistance appears smaller when the SOC was between 0.3 and 0.9. Hence, 0.3 to 0.9 is a high-efficiency working region for battery which will be used as boundaries of constraints.

Table 3. The calculation results of battery cell internal resistance.

SOC	**1**	**0.9**	**0.8**	**0.7**	**0.6**
R_0/mΩ	16.81	16.41	16.24	16.24	16.25
SOC	0.5	0.4	0.3	0.2	0.1
R_0/mΩ	16.29	16.35	17.09	17.26	17.26

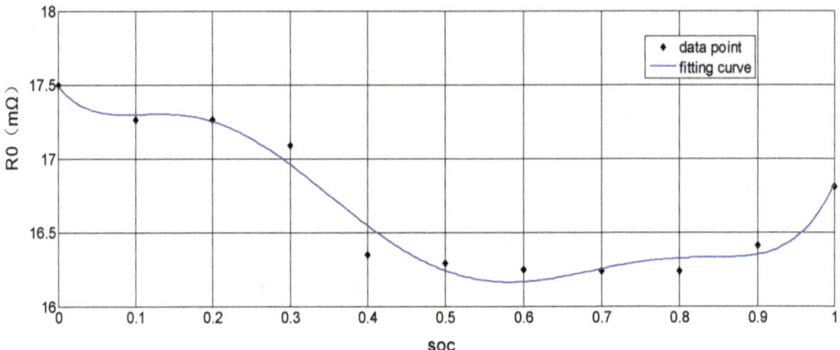

Figure 3. Change curve of internal resistance of battery discharge.

The following two formulas demonstrate the mathematical character of battery based on the equivalent circuit model above.

$$E = U_b + I_b R_0 \tag{2}$$

$$P_{w-bat} = I_b{}^2 R_0 \tag{3}$$

where E is the electromotive force; U_b is the battery terminal voltage; I_b represents the battery current load; R_0 represents internal resistance; P_{w-bat} is the power loss.

2.3. SC Model

The basic parameters of the SC modules used in this paper are listed in Table 4. The transmission line model shown in Figure 2b was adopted to represent the characteristics of the SC with its simplicity and sufficient accuracy. The series/parallel numbers S_{SC}/P_{SC} of the SC module within the SC pack is 40/9 in this study. SC working states are divided into 2 states in this paper beforehand, as shown in Figure 4. When the terminal voltage of SC is between U_H and V_{sc_max}, SC single drive is the priority selection; when the terminal voltage of SC is between U_H and V_{sc_min}, the driving modes are decided by the system efficiency.

Table 4. Basic parameters of the SC module.

Parameter	Value
Maximum voltage (V)	2.7
Capacity (F)	350
Stored energy (Wh)	0.35
Maximum discharge current (A)	170
Resistance (mΩ)	3.2

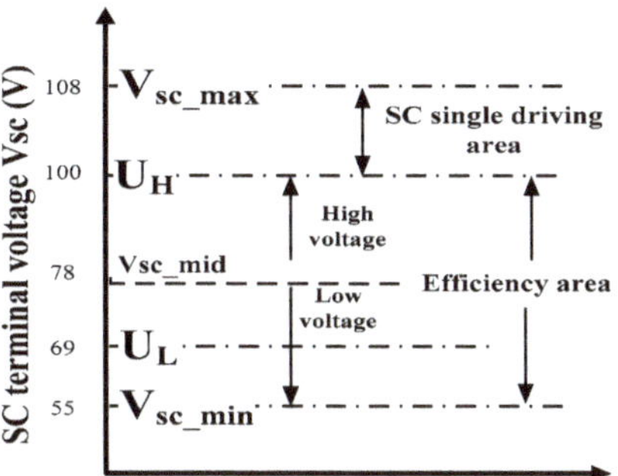

Figure 4. Division of the SC working states.

The series resistance (R_{es}) is relatively steady, and the parallel resistance (R_{ep}) demonstrates the current leakage, which is small enough to be neglected. Hence, the approximate power wastage of SC is mainly caused by R_{es}. The mathematical model of SC is expressed by the equations below.

$$U_{sc} = I_{sc} R_{es} + U_c \tag{4}$$

$$I_{sc} = C\frac{dU_c}{dt} + I_p \tag{5}$$

where U_{SC} represents the terminal voltage of SC; I_{SC} is the outputting current; U_C is the actual voltage of the capacitor in the SC; I_p is the leakage current.

2.4. Electric Motor

The basic parameters of the motor used are listed in Table 5. The motor output power was evaluated through testing the rotor speed and torque, which could be used to distinguish the efficiency region and divide the power split mode, meanwhile the electric motor efficiency under different torque and rotor speed could be obtained. An efficiency map of the motor is presented in Figure 5. The top blue line represents the characteristics curve of the motor: the first half is constant torque phase, and the latter part is constant power stage.

Table 5. Basic parameters of the electric motor.

Type	Nominal Power (kW)	Maximum Power (kW)	Maximum Speed (r/min)
BLDC	29	40	9000

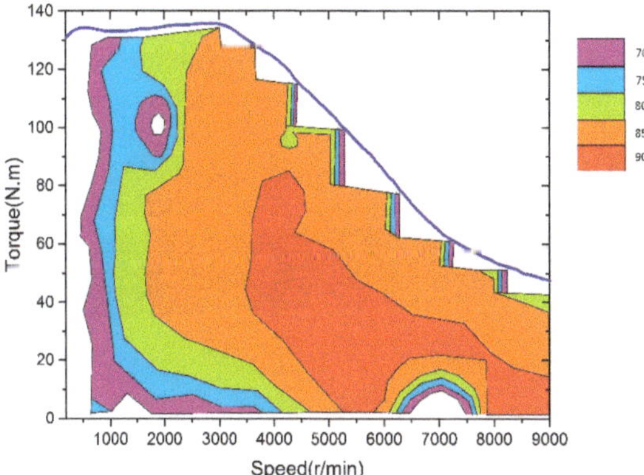

Figure 5. Efficiency map of the electric motor.

2.5. Rule-Based Energy Management Strategy

The aim of optimal control strategy is to maximize the efficiency of the hybrid power system, the efficiency of which is expressed as follows.

$$\eta_{sys(v,a,soc,I)} = \frac{P_{req(v,a)}}{P_{req(v,a)} + P_{w-bat(soc,I_b)} + P_{w-sc(I_{sc},U_{sc})} + P_{w-DC(I_{sc},U_{sc})} + P_{w-mot(I_{bus},n,w)}} \tag{6}$$

where P_{req} is the power requirement of hybrid power system, $P_{w\text{-}bat}$ is the power dissipation of battery. $P_{w\text{-}SC}$ is the power dissipation of SC. $P_{w\text{-}DC}$ represents the power dissipation of DC/DC converter. $P_{w\text{-}MOT}$ represents the power loss in electric motor.

The rolling resistance, the air resistance, the ramp resistance and the acceleration resistance are mainly used in the process of vehicle driving, and the force equation is shown in Equation (7):

$$F_t = F_f + F_w + F_i + F_j \tag{7}$$

where F_f is rolling resistance, F_w is air resistance, F_i is ramp resistance, and F_j is acceleration resistance. The specific expressions of each force are expressed as Equation (8), and the calculation of demand power is as shown in Equation (9).

$$\begin{cases} F_f = mgf \\ F_w = \frac{C_D A v^2}{21.15} \\ F_i = mgi \\ F_j = \delta ma \end{cases} \tag{8}$$

$$P_{req(v,a)} = \frac{F_t v}{3600\eta_T} \tag{9}$$

where m is the EV mass, g is the gravitational acceleration, f is the rolling resistance coefficient, v is the EV velocity, a is the EV acceleration, i is the climbing angle, C_D is the air drag coefficient, A is the front area, δ is the generalized inertia coefficient, η_T is the transmission efficiency. All the variables are formulated as the equations below:

$$\begin{cases} P_{req(v,a)} = \frac{1}{\eta_T}(mgfv/3600 + mgiv/3600 + C_D A v^3/76140 + \delta mva/3600) \\ P_{w-bat(soc,I_b)} = {I_b}^2(R_{0(SOC)} + R_{1(SOC)}) \\ P_{w-sc(I_{sc})} = {I_{sc}}^2 R_{es} \\ P_{w-DC(I_{sc},U_{sc})} = \eta_{DC} I_{sc} U_{sc} \end{cases} \tag{10}$$

where η_{DC} is the efficiency of the DC/DC converter.

According to the above efficiency calculation mathematical model and the characteristics analysis of each component of the composite power system, the efficiency calculation model is built in Simulink, as shown in Figure 6. Vehicle speed and acceleration as the input of the model can be converted into the motor output speed and torque through the conversion module, thus the required driving power of the vehicle can be calculated through the demand power module. Then it is transferred to the logic module, which distributes the direct current bus power to the battery module and the SC module according to the different driving modes of the vehicle. The DC/DC converter is included in the SC module. Finally, the efficiency of the system can be calculated by summing up the power loss of battery, SC, DC/DC converter and motor to the efficiency calculation module.

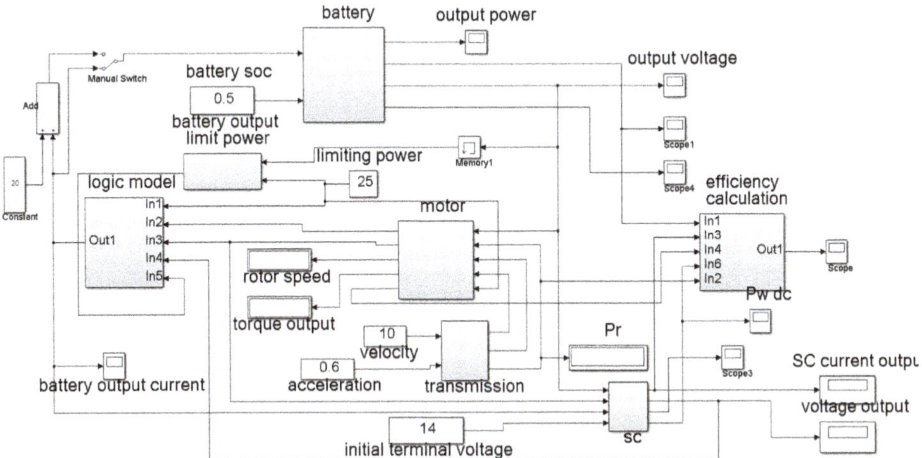

Figure 6. Efficiency model of power system.

System efficiency varies significantly under different conditions concerning different battery and SC states. Depending on the working states of battery and SC, the hybrid power system is classified into four conditions. Based on the relevant tests on battery, a relative low resistance region ($SOC_b <= SOC <= SOC_t$) was obtained as mentioned above, and for SC, high and low terminal voltage work region was determined by a middle voltage line U_m.

$$\begin{cases} SOC \in (SOC_b, SOC_t)\&U_{sc} > U_m & condition1 \\ SOC \notin (SOC_b, SOC_t)\&U_{sc} > U_m & condition2 \\ SOC \in (SOC_b, SOC_t)\&U_{sc} \le U_m & condition3 \\ SOC \notin (SOC_b, SOC_t)\&U_{sc} \le U_m & condition4 \end{cases} \quad (11)$$

(1) Condition 1: Battery stays in high efficiency region, and the initial terminal voltage of SC stays in high value. Adjusting the power limit of battery, and a maximum efficiency value under each couple of vehicle velocity and acceleration is seek out. The simulation results of the efficiency model of the power system under Condition 1 are depicted in Figure 7.

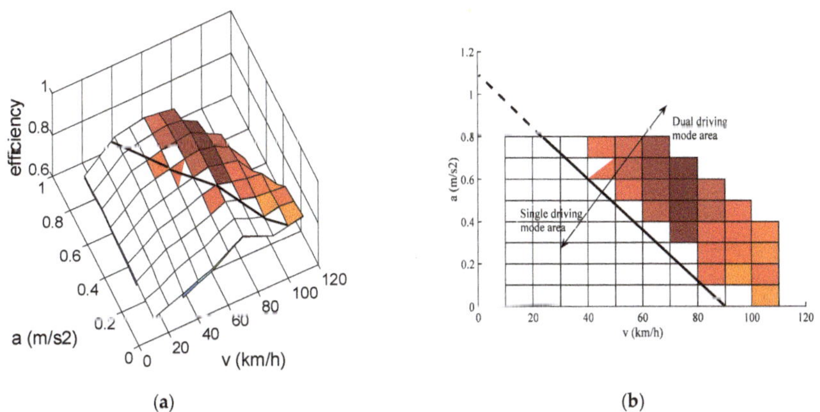

Figure 7. Efficiency comparison and switching rule under Condition 1. (**a**) Efficiency comparison under Condition 1; (**b**) Switching rule under Condition 1.

As the Figure 7 shows (similar to Figures 8–10), the white mesh stands for the efficiency in single driving mode; the colorful one stands for the dual driving mode; meanwhile the switching rule of single driving mode and dual driving mode under Condition 1 can be summarized. It can be seen from Figure 7a (similar to Figures 8a, 9a and 10a) that the efficiency of single driving mode is higher than that of dual driving mode when the speed and acceleration are smaller, and the power system should work in single driving mode at this time. With the increase of the vehicle speed and acceleration, the efficiency of dual driving mode increases slowly and is larger than that of single driving mode, and the power system should work in dual driving mode at this time. In order to accurately find out the switching rules of single driving mode and dual driving mode in power system, the three-dimensional surface graph shown in Figure 7a (similar to Figures 8a, 9a and 10a) is projected to the v-a plane, and the efficiency differentiation curve of single driving mode and dual driving mode can be obtained in Figure 7b (similar to Figures 8b, 9b and 10b). The dividing line between white grid and color grid is the switching rule of single driving mode and dual driving mode: when the speed and acceleration are located at the lower left of the dividing line, the efficiency of single driving mode is higher and the power system should switch to single driving mode; when the speed and acceleration are at the upper right of the dividing line, the efficiency of dual driving mode is higher and the power system switches to dual driving mode.

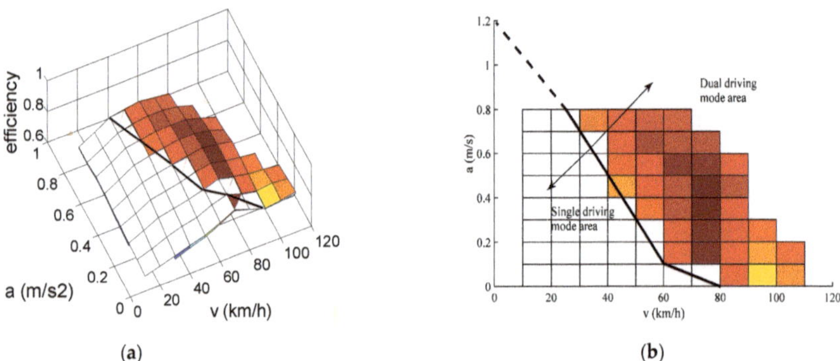

Figure 8. Efficiency comparison and switching rule under Condition 2. (**a**) Efficiency comparison under Condition 2; (**b**) Switching rule under Condition 2.

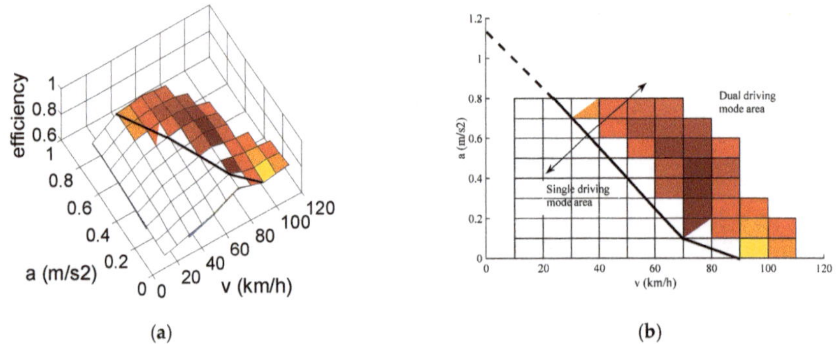

Figure 9. Efficiency comparison and switching rule under Condition 3. (**a**) Efficiency comparison under Condition 3; (**b**) Switching rule under Condition 3.

(2) Condition 2: Battery is in lower efficiency region, and SC is in high voltage. The comparison and switching rule are depicted in Figure 8, and the switching rule of single driving mode and dual

driving mode under Condition 2 can be summarized according to the explain of switching rules under Condition 1.

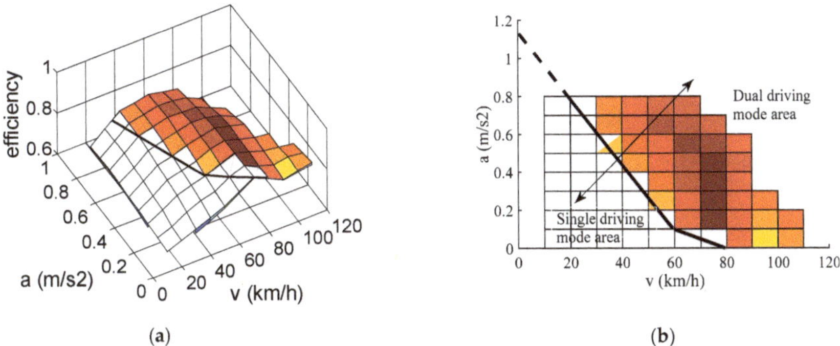

Figure 10. Efficiency comparison and switching rule under Condition 4. (**a**) Efficiency comparison under Condition 4; (**b**) Switching rule under Condition 4.

(3) Condition 3: Battery stays in high-efficiency region, while SC keeps in low voltage. The comparison and switching rule under Condition 3 can be seen in Figure 9. The switching rule of single driving mode and dual driving mode can be summarized similarly according to the explanation of switching rules under Condition 1.

(4) Condition 4: Both battery and SC stay in low-efficiency region. The simulation results of the power system efficiency model are depicted in Figure 10. The switching rule of single driving mode and dual driving mode under Condition 4 can be obtained similarly according to the explain of switching rules under Condition 1.

Based on the switching rule of single driving mode and dual driving mode under four different conditions, the schematic flow-chart of the rule-based strategy can be achieved, as shown in Figure 11. It can be seen from the Figure 11 that when the SC terminal voltage is higher than 100, the power system adopts single driving mode. When the terminal voltage of the SC is less than 100, the power distribution of the SC and the battery is determined according to the switching rules under different conditions.

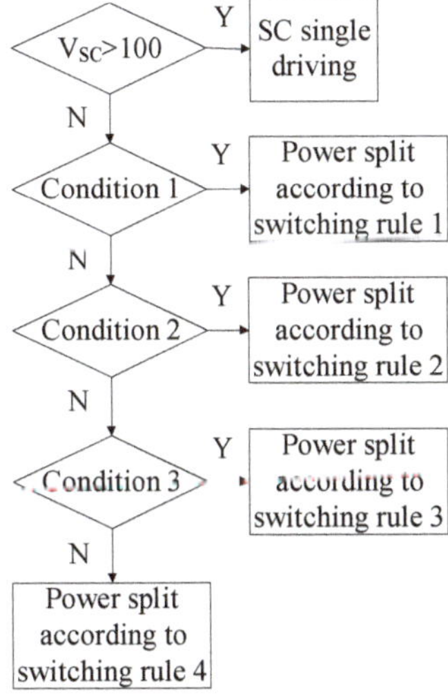

Figure 11. The flow-chart of the rule-based strategy.

2.6. Energy Management Optimal Strategy Derived from DP Approach

The rule-based strategy is proposed on the basis of power source state and efficiency, and the strategy is empirical, but it cannot achieve the best efficiency optimization effect.

The DP algorithm will be utilized to solve the optimal problem in this section, which is a famous principal of optimization which transfers multistage decision problem to single stage, and solves each problem with the relations between different stages. The recursion equation below plays an essential role in the DP sequential algorithm.

$$\begin{cases} f_k(x_{k+1}) = \min\{\dot{v}_k(x_{k+1}, u_k) + f_{k-1}(x_k)\},\ k = 1, 2, \cdots, n-1, n \\ \qquad boundary\ condition : f_0(x_1) = 0 \end{cases} \tag{12}$$

where x_k is the state of the kth stage, the decision variable u_k is the decision of the state at x_{k+1}, the state transition equation is $x_{k+1} = T_k(x_k, u_k)$, the set of allowed decision of k phase is denoted as $D_k(x_k)$, $v_k(x_{k+1}, u_k)$ is as an indicator function.

The DP algorithm aims at minimizing the energy loss of the hybrid power system at every moment under cyclic conditions, so as to achieve the optimization of vehicle efficiency and extend the driving distance. In this paper, the optimal control strategy of energy allocation based on DP algorithm takes the output power of the SC ($P_{SC}(t)$) as the decision variable of global optimization, and time series as the stage sequence with the interval of 1 s. The SOC of the power battery (SOC_b) and the terminal voltage of the SC (V_{SC}) are the state variables of global optimization. The expression is as follows:

$$u = \{P_{SC}(t)\} \tag{13}$$

$$x = \{SOC_b, V_{SC}\} \tag{14}$$

The DP optimization process of the energy allocation of the hybrid power system can be expressed as: (1) dividing the search for the minimum energy consumption path of the hybrid power system into several time series at a time interval of 1 s; (2) searching for the optimal decision variable $u(t)$ during the transition from the initial state $x(0)$ to the final state $x(t)$ according to the vehicle power demand at each time; (3) making the hybrid power system achieve smaller energy loss. Therefore, the objective function of DP optimization for energy allocation of composite power system is as follows:

$$J = \int_0^T f_w(x(t), u(t), t)dt \tag{15}$$

In the above formula, $f_w(x(t),u(t),t)$ is the energy loss of the hybrid power system at t time.

According to the above analysis, the cycle condition is divided into N stages in this paper, and the step size between each stage is Δt (1 s in this paper). Therefore, the discrete global cost function of the hybrid power system is as follows:

$$J = \sum_{k=1}^{N} f(x(k), u(k), k) \cdot k \tag{16}$$

The transfer function of the hybrid power system is as follows in this paper:

$$x(k) = \begin{cases} SOC_b(k+1) = SOC_b(k) - \frac{\eta_b(k) \cdot I(k)}{3600 \cdot C_{ini}} \\ V_{SC}(k+1) = \sqrt{V_{SC}^2(k) - 2\frac{P_{SC}(k)}{C_{SC}}} \end{cases} \tag{17}$$

where $\eta_b(k)$ is the charging and discharging efficiency of battery, $I(k)$ is the current of battery, C_{ini} is the initial capacity of battery, C_{SC} is the rated capacity of SC.

At the same time, the constraints that should be met are as follows:

$$\begin{cases} 0 \leq P_r \leq P_{r\_\max} \\ 0 \leq I_b \leq I_{b\_\max} \\ 0 \leq I_{SC} \leq I_{SC\_\max} \end{cases} \tag{18}$$

where P_{r_max} is the vehicle maximum demand power, I_{b_max} is the battery maximum output current, I_{SC_max} is the maximum current of SC.

According to the basic principles and basic equations of the above DP method, the objective function recursive equation of the hybrid power system at any k time can be expressed as:

$$J_k^*(x(k)) = \min_{u(k)}\{f_w(x(k), u(k), k) + J_{k+1}^*(x(k+1))\} \tag{19}$$

The DP algorithm adopted in this paper takes the SC terminal voltage of as the initial state variable and discretizes it into 140 states from $0.5V_{SC,max}$ to $V_{SC,max}$, as shown in Figure 12. The DP algorithm first carries on the reverse calculation. After the SOC of the battery and the terminal voltage of the SC are given, the optimal decision results of each stage are obtained by the reverse calculation according to the terminal conditions $k = k_{max}$. Taking the decision variables of k time $u(k)$, $u_i(k)$ as an example, the state of k+1 time $x(k+1)$, $x_i(k+1)$ can be obtained according to the system state variables of k-time. Because $x(k+1)$, $x_i(k+1)$ is not necessarily the state value at $k + 1$ time, the energy loss of composite power system corresponding to state $x(k+1)$, $x_i(k+1)$ is obtained by interpolation method. The energy loss value of the composite power system with state variables $u(k)$, $u_i(k)$ at k time is $f(k)$, $f_i(k)$. The output power of the SC corresponding to the smaller one is taken as the optimal decision variable in this stage. All other stages are the same until all stages are reversed. At the end of the reverse recursion, the forward calculation is needed, and the optimal decision variables at each time are obtained by interpolation calculation based on the results of the reverse operation as known variables.

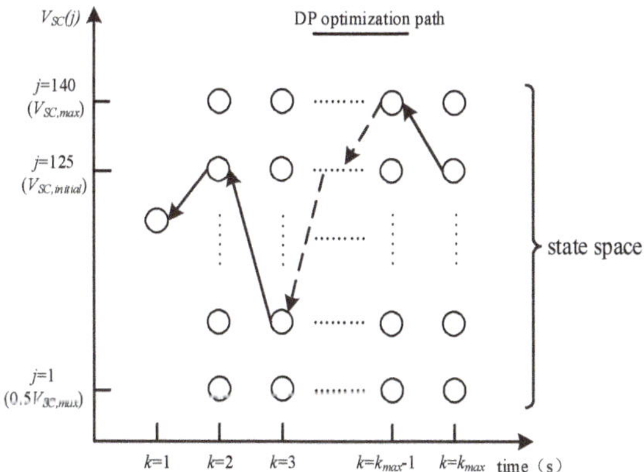

Figure 12. Dynamic programming (DP) approach flowchart.

3. Results and Discussion

3.1. The Results of Rule-based Strategy

According to GB/T 18386-2005 Electric vehicles—Energy consumption and range—Test procedures [24], the New European Driving Cycle (NEDC) is utilized to verify the rule-based strategy proposed in the paper. As shown in Figure 13, period ① is urban driving cycle, and period ③ belongs to period ①, which represents the basic urban driving cycle, and period ② means suburban driving cycle.

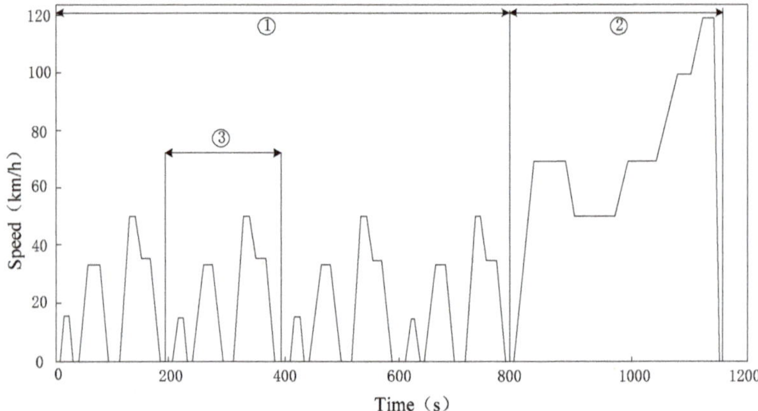

Figure 13. The New European Driving Cycle.

It can be seen from Figure 13 that the whole cycle is composed of conditions such as parking, driving and braking. Concerning optimal control during driving period is the main aim of this paper, the power loss during parking can be neglected, thus the system efficiency is equivalent to 95% under parking conditions. Based on previous research results [25], the recovery of regenerative braking can be assumed to reach 20%.

The hybrid power system can enter the dual driving mode with higher efficiency in time and reasonably, so the efficiency of the power system will be higher theoretically. In order to verify the advantages of the system efficiency under the rule-based strategy, the NEDC cycle condition is still used as input to observe the change of the efficiency of the power system, as shown in Figure 14. The battery SOC is set to 0.95 and the initial terminal voltage of the SC is set to 108 V when simulating.

As shown in Figure 14, the power system efficiency of rule-based strategy in driving and braking process is higher than that of the single power mode. It can be seen from Figure 14b that the power system efficiency is hardly improved in the case of a small power demand in an urban cycle condition; however, the efficiency of the hybrid power system is improved to a certain extent because it can enter the dual driving mode in time in the case of high power demand. Under braking process, the power system enters the regenerative braking state, and since the single-cell prototype does not have the regenerative braking function, so the efficiency of hybrid power system is higher in braking process. From Figure 14c, it can be observed that the system efficiency of dual driving mode is higher at the time because the vehicle speed in suburban cycle condition is higher. If the SC terminal voltage is above the median, the hybrid power system will be switched to dual driving mode in time to improve the efficiency of power system. Therefore, the system has greater efficiency improvement under the condition of higher power demand, such as suburban cycle condition.

As shown in Figure 15a, rule-based strategy can enter dual driving mode timely, therefore, battery current of rule-based strategy is smaller than battery-only mode. However, rule-based strategy cannot enter dual driving mode under suburban driving condition due to low SC voltage, which can be solved through increasing SC usage, but which will bring cost increase. Furthermore, as shown in Figure 15b, rule-based strategy enters into SC single driving mode concerning high SC voltage and small power demand at about 20s.

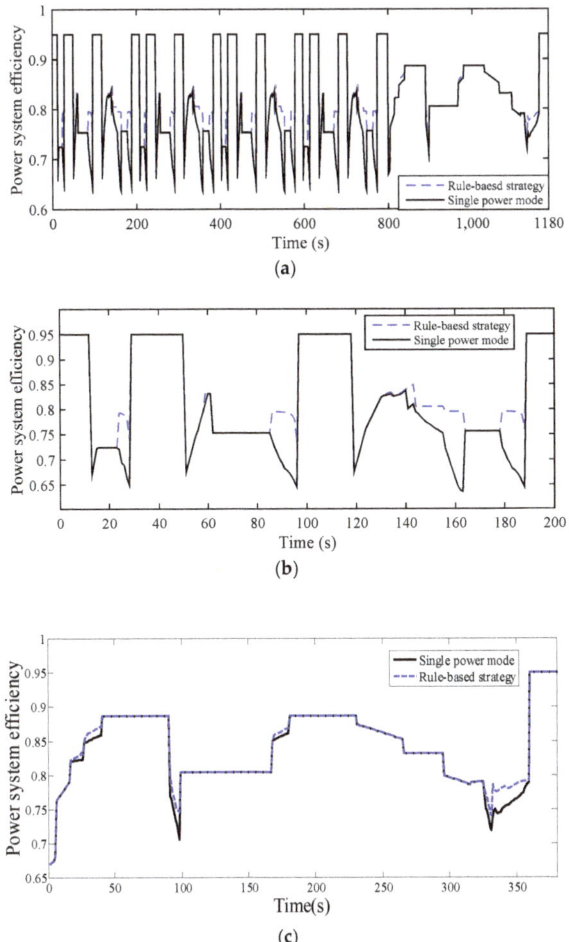

Figure 14. Comparison of power system efficiency along NEDC. (**a**) Power system efficiency along New European Driving Cycle (NEDC); (**b**) Power system efficiency along an urban driving cycle; (**c**) Power system efficiency along a suburban driving cycle.

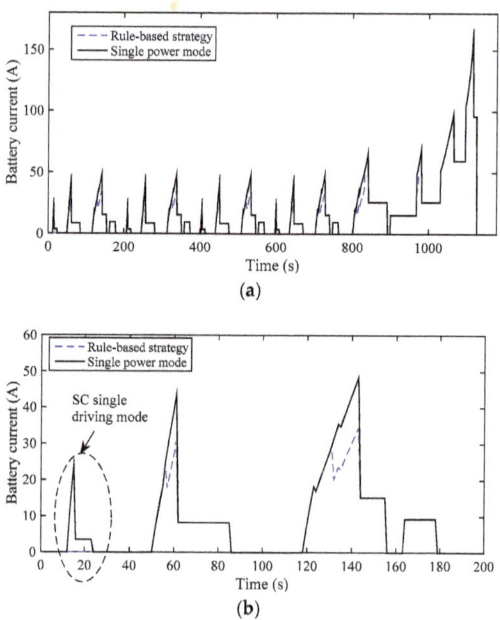

Figure 15. Battery current along NEDC. (**a**) Battery current along NEDC; (**b**) Battery current along an urban driving cycle.

3.2. The Results of DP Strategy

Compared with rule-based energy management control strategy, the effectiveness of the proposed energy optimization strategy based on DP algorithm is verified. The initial SOC was set to 0.95, and the initial SC terminal voltage was 108 V. The main program of the DP algorithm was then run; power system efficiency comparison of DP and rule-based strategy is shown in Figure 16. Compared with rule-based control strategy, the energy optimization control strategy based on DP algorithm has more efficiency optimization space. Figure 16b is a partial enlarged view of the system efficiency of an urban cycle, as shown in Figure 16b, the power system efficiency of DP approach is higher than rule-based strategy as a whole. Besides, Figure 16c is a partial enlarged view of the system efficiency of a suburban cycle, as shown in Figure 16c, concerning the SC is of sufficient electricity, there is a magnificent efficiency increase during the first driving process.

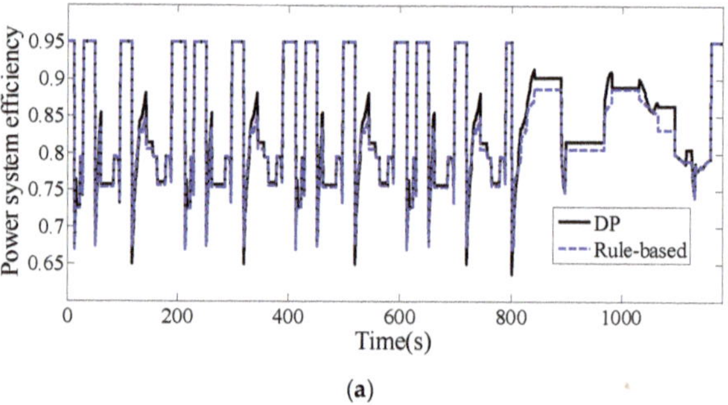

(**a**)

Figure 16. *Cont.*

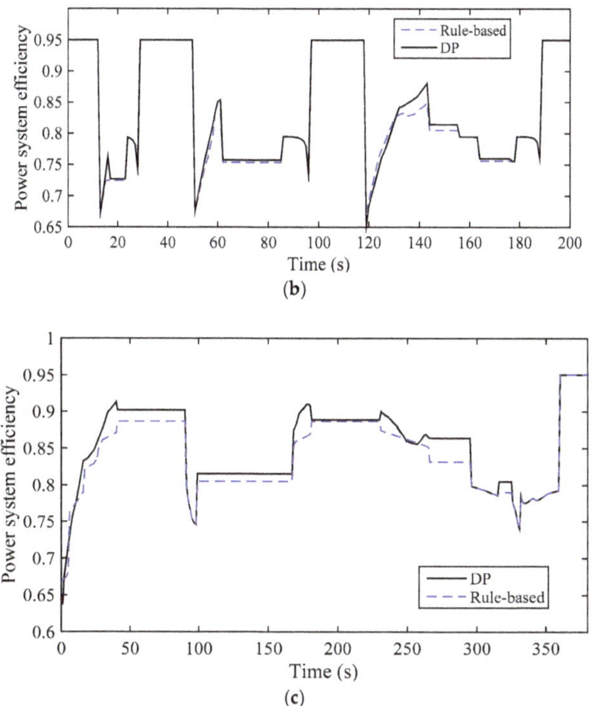

Figure 16. Power system efficiency comparison of DP and rule-based strategy. (**a**) NEDC driving cycle; (**b**) Urban driving cycle; (**c**) Suburban driving cycle.

In order to verify the advantages and disadvantages of the two strategies proposed, the initial SOC of the battery is set to 1 and the voltage of the SC is set to 108 V under the NEDC condition and the actual road driving cycle respectively, and the simulation model is set to circulate until the SOC of the battery is reduced to 0.15. Among them, the whole course of the NEDC condition is 11.022 km, and the whole course of the actual road driving cycle is 11.655 km. The SOC change of battery under single NEDC condition is shown in Figure 17a, the SOC change of battery (1–0.15) under multiple NEDC conditions is shown in Figure 17b; and the SOC change of battery under single actual condition is shown in Figure 18a, and the SOC change of battery (1–0.15) under multiple actual conditions is shown in Figure 18b. The speed in Figure 18a is collected from a real vehicle on the campus road surface, which is called the actual road driving cycle. It can be seen from Figures 17 and 18 that compared with the rule-based energy management strategy, the battery SOC of the energy allocation optimization strategy based on the DP algorithm reduces more slowly. That is to say, the energy allocation optimization strategy based on the dynamic programming algorithm is more economical.

Table 6 shows the different strategies comparison of driving range and energy consumption under NEDC condition. It can be seen from Table 6 that compared with the rule-based energy management strategy, the energy optimization allocation strategy based on the DP algorithm has a longer driving range and a lower energy consumption. Compared to the battery-only EV based on simulation results, the hybrid power system controlled by rule-based strategy can decrease 13.4% of the energy consumption along the NEDC condition, while the power split strategy derived from DP approach can reduce by 17.6%. The results verify the effectiveness of the DP algorithm optimization strategy.

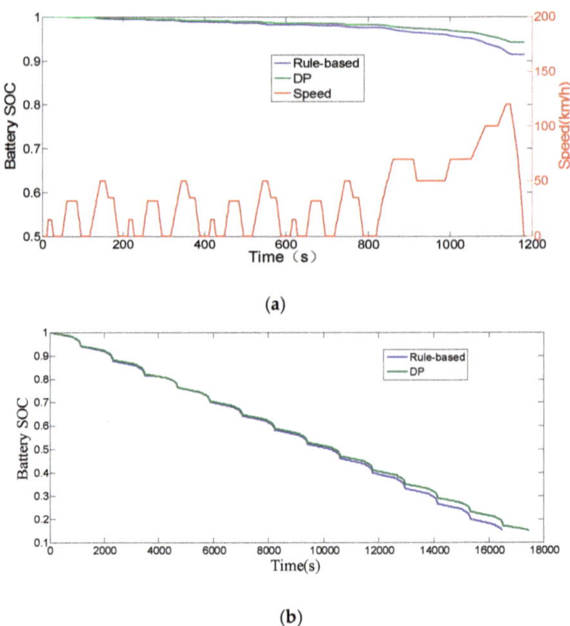

Figure 17. SOC change of battery under NEDC condition. (**a**) SOC change of battery under single NEDC condition; (**b**) SOC change of battery under multiple NEDC condition.

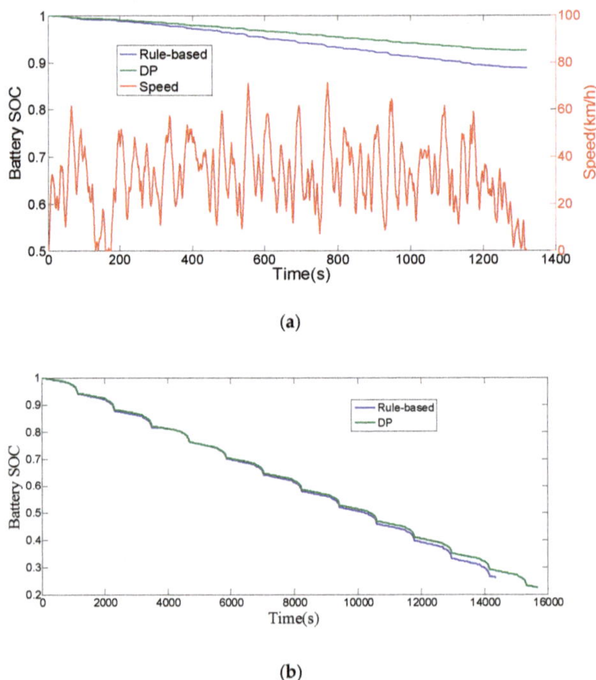

Figure 18. SOC change of battery under actual road driving cycle. (**a**) SOC change of battery under single actual road driving cycle; (**b**) SOC change of battery under multiple actual road driving cycle.

Table 6. Comparison of driving range and energy consumption under NEDC condition.

Strategy	Driving Range (km)	Energy Consumption (Wh/km)
Single battery system	120	104.82
Rule-based	131	90.73
DP approach	138	86.41

4. Conclusions

Based on an optimal efficiency model, a rule-based energy management strategy is proposed, which is based on the projection partition of composite power system efficiency, so it has strong adaptive adjustment ability. In order to explore the maximum energy-saving potential of hybrid power electric vehicles, the DP optimization method is proposed on the basis of the establishment of the whole hybrid power system, which takes into account various energy-consumption factors of the whole system.

System efficiency varies significantly under different conditions due to different battery and SC states. Depending on the working states of battery and SC, the hybrid power system is classified into four conditions in this paper, and the switching rule of single driving mode and dual driving mode under four different conditions are gained. Rule-based strategy is proposed on the basis of the above contents, which can be easily implemented and prominently improves the system efficiency. Compared with the battery-only system, the rule-based energy management strategy has higher power system efficiency, and power system controlled by the proposed rule-based strategy can reduce 13.4% of the energy consumption along NEDC. Meanwhile the effectiveness of the hybrid power system and the availability of the proposed rule-based strategy are validated.

While the rule-based strategy is empirical, it cannot achieve the best efficiency optimization effect. The DP algorithm is utilized to solve the optimal problem. Compared with the rule-based strategy based on simulation results, power system controlled by the strategy deriving from DP approach can reduce 4.8% of the energy consumption along NEDC. Compared to the battery-only EV based on simulation results, the hybrid power system controlled by the power split strategy derived from DP approach can reduce by 17.6%. As a result, the availability of the strategy deriving from DP approach is validated. The results show that compared with rule-based strategy, the optimized DP strategy has higher system efficiency and lower energy consumption.

Author Contributions: Conceptualization, L.C.(Long Chen) and L.C.(Liao Chen); Methodology, C.P.; Software, Y.L.; Validation, C.P., Y.L. and Liao Chen; Formal Analysis, Y.L.; Investigation, L.C.(Long Chen); Resources, L.C.(Long Chen); Data Curation, C.P. and Y.L.; Writing—Original Draft Preparation, C.P. and Y.L.; Writing—Review & Editing, C.P. and Y.L.; Visualization, L.C.(Liao Chen); Supervision, L.C.(Long Chen); Project Administration, C.P.; Funding Acquisition, C.P. and L.C.(Long Chen).

Funding: This research was funded by the National Natural Science Foundation of China, grant number [51105178 and 51707084], and funded by the Natural Science Foundation of Jiangsu Province, grant number [BK20171300 and BK20160529]. This research was also supported by the China Postdoctoral Science Foundation, grant number [2016M591775].

Conflicts of Interest: The authors declare no conflict of interest to the paper. The funders had no role in the design of the study; in the collection, analysis, or interpretation of data; in the writing of the manuscript, and in the decision to publish the results.

References

1. Zhou, M.; Wei, L.; Wen, J. The Parameters Matching and Simulation of Pure Electric Vehicle Composite Power Supply Based on CRUISE. *Appl. Mech. Mater.* **2014**, *602*, 2836–2839. [CrossRef]
2. Griffo, A.; Wang, J. Modeling and Stability Analysis of Hybrid Power Systems for the More Electric Aircraft. *Electr. Power Syst. Res.* **2012**, *82*, 59–67. [CrossRef]
3. Salah, I.B.; Bayoudhi, B.; Diallo, D. EV energy management strategy based on a single converter fed by a hybrid battery/supercapacitor power source. In Proceedings of the IEEE International Conference on Green Energy (ICGE), Sfax, Tunisia, 25 March 2014; pp. 246–250.

4. Mastragostino, M.; Soavi, F. Strategies for high-Performance supercapacitors for HEV. *J. Power Sources.* **2007**, *174*, 89–93. [CrossRef]
5. Yi, H.; Ma, F. Simulation Study of Compound Power Supply System for Electric Vehicles. *Power World* **2013**, *1*, 30–33.
6. Xu, R.; Wang, Y. Simulation of composite electric power for electric vehicles. In Proceedings of the IEEE International Conference on Mechatronics and Automation (ICMA), Takamatsu, Kagawa, Japan, 6–9 August 2017; pp. 967–972.
7. Cezar, B.; Onea, A. A rule-based energy management strategy for parallel hybrid vehicles with supercapacitors. In Proceedings of the International Conference on System Theory, Control and Computing (ICSTCC), Sinaia, Romania, 14–16 October 2011; pp. 1–6.
8. Shi, Q.; Zhang, C.; Cui, N. Optimal Control of Energy Management Problems for a New Dual Energy Source Pure Electric Vehicle. *Trans. China Electrotechnical Soc.* **2008**, *23*, 137–142.
9. Trovão, J.P.; Antunes, C.H. A comparative analysis of meta-heuristic methods for power management of a dual energy storage system for electric vehicles. *Energy Convers. Manage.* **2015**, *95*, 281–296. [CrossRef]
10. Chen, Z.; Mi, C.C.; Xiong, R.; Xu, J.; You, C. Energy management of a power-split plug-in hybrid electric vehicle based on genetic algorithm and quadratic programming. *J. Power Sources* **2014**, *248*, 416–426. [CrossRef]
11. Padmarajan, B.V.; Mcgordon, A.; Jennings, P.A. Blended Rule-Based Energy Management for PHEV: System Structure and Strategy. *IEEE Trans. Veh. Technol.* **2016**, *65*, 8757–8762. [CrossRef]
12. Banvait, H.; Anwar, S.; Chen, Y. A rule-based energy management strategy for plug-in hybrid electric vehicle (PHEV). In Proceedings of the American Control Conference, St. Louis, MO, USA, 10–12 June 2009; pp. 3938–3943.
13. Hemi, H.; Ghouili, J.; Cheriti, A. A real time energy management for electrical vehicle using combination of rule-based and ECMS. In Proceedings of the IEEE Electrical Power and Energy Conference (EPEC), Halifax, NS, Canada, 21–23 August 2013; pp. 1–6.
14. Hung, Y.-H.; Wu, C.-H. An integrated optimization approach for a hybrid energy system in electric vehicles. *Appl. Energy* **2012**, *98*, 479–490. [CrossRef]
15. Hu, X.; Johannesson, L.; Murgovski, N.; Egardt, B. Longevity-conscious dimensioning and power management of the hybrid energy storage system in a fuel cell hybrid electric bus. *Appl. Energy* **2015**, *137*, 913–924. [CrossRef]
16. Laura, V.P.; Guillermo, R.B.; Diego, M.; Guillermo, O.G. Optimization of power management in an hybrid electric vehicle using dynamic programming. *Math. Comput. Simul.* **2006**, *73*, 244–254.
17. Chen, B.-C.; Wu, Y.-Y.; Tsai, H.-C. Design and analysis of power management strategy for range extended electric vehicle using dynamic programming. *Appl. Energy* **2014**, *113*, 1764–1774. [CrossRef]
18. Ansarey, M.; Panahi, M.S.; Ziarati, H.; Mahjoob, M. Optimal energy management in a dual-storage fuel-cell hybrid vehicle using multi-dimensional dynamic programming. *J. Power Sources* **2014**, *250*, 359–371. [CrossRef]
19. Zou, Y.; Chen, R.; Hou, S.; Hu, X. Energy Management Strategy for Hybrid Electric Tracked Vehicle Based on Stochastic Dynamic Programming. *J. Mech. Eng.* **2012**, *48*, 91–96. [CrossRef]
20. Hou, C.; Ouyang, M.; Xu, L.; Wang, H. Approximate Pontryagin's minimum principle applied to the energy management of plug-in hybrid electric vehicles. *Appl. Energy* **2014**, *115*, 174–189. [CrossRef]
21. Kim, N.; Cha, S.; Peng, H. Optimal Control of Hybrid Electric Vehicles Based on Pontryagin's Minimum Principle. *IEEE Trans. Control Syst. Technol.* **2011**, *19*, 1279–1287.
22. Song, Z.; Hofmann, H.; Li, J.; Han, X.; Ouyang, M. Optimization for a hybrid energy storage system in electric vehicles using dynamic programming approach. *Appl. Energy* **2015**, *139*, 151–162. [CrossRef]
23. Xu, L.; Ouyang, M.; Li, J.; Yang, F.; Lu, L.; Hua, J. Optimal sizing of plug-in fuel cell electric vehicles using models of vehicle performance and system cost. *Appl. Energy* **2013**, *103*, 477–487. [CrossRef]
24. Bai, Z.; Cao, L.; Yang, J. Study on the Performance Evaluation Method of Power Batteries for Pure Electric Vehicles. *J. Hunan Univ.: Nat. Sci. Ed.* **2006**, *33*, 48–51.
25. Pan, C.; Chen, L.; Chen, L.; Huang, C.; Xie, M. Research on energy management of dual energy storage system based on the simulation of urban driving schedules. *Int. J. Electr. Power Energy Syst.* **2013**, *44*, 37–42. [CrossRef]

Article

Simulation of Thermal Behaviour of a Lithium Titanate Oxide Battery

Seyed Saeed Madani *, Erik Schaltz and Søren Knudsen Kær

Department of Energy Technology, Aalborg University, DK-9220 Aalborg, Denmark; esc@et.aau.dk (E.S.); Skk@et.aau.dk (S.K.K.)

* Correspondence: ssm@et.aau.dk

Received: 29 January 2019; Accepted: 15 February 2019; Published: 20 February 2019

Abstract: One of the reasonable possibilities to investigate the battery behaviour under various temperature and current conditions is the development of a model of the lithium-ion batteries and then by employing the simulation technique to anticipate their behaviour. This method not only can save time but also they can predict the behaviour of the batteries through simulation. In this investigation, a three-dimensional model is developed to simulate thermal and electrochemical behaviour of a 13Ah lithium-ion battery. In addition, the temperature dependency of the battery cell parameters was considered in the model in order to investigate the influence of temperature on various parameters such as heat generation during battery cell operation. Maccor automated test system and isothermal battery calorimeter were used as experimental setup to validate the thermal model, which was able to predict the heat generation rate and temperature at different positions of the battery. The three-dimensional temperature distributions which were achieved from the modelling and experiment were in well agreement with each other throughout the entire of discharge cycling at different environmental temperatures and discharge rates.

Keywords: thermal modelling; thermal behaviour; lithium titanate oxide batteries

---

1. Introduction

Lithium-ion batteries are one of the most developing categories of batteries on the market these days because of their high energy density and capacity. A large amount of energy is stored inside them and they have great sensitivity to the operating conditions. Therefore, safety is an important issue in lithium-ion batteries. In addition, demands on safety of these batteries is increasing with their utilization in more applications.

With the intention of reaching out to safety requirements of the lithium-ion batteries on electronic device applications, researchers are resuming to do supplementary investigations on the essential issues in relation to the lithium-ion batteries.

System safety, cycle life, and cell performance are influenced by temperature distribution in the cell. Consecutively, it depends on heat dissipation rate at surface of the cell and heat generation rate within the cell.

Although lithium-ion batteries are susceptible to extreme heat load under severe or abnormal functional conditions, thermal management has been one of the considerable issues in developing lithium-ion batteries in hybrid electric vehicle and battery system applications.

A pseudo 2D electrochemical model for modelling electrochemical systems subject to realistic automotive operation situations was proposed [1]. The model was developed for a lithium ion battery. It consists of complicated electrochemical phenomena, which were generally eliminated in online battery performance forecasters such as over potentials owing to mass transport restrictions and the full current-over potential relation and variable double layer capacitance. The model was able to simulate battery cell behaviour under dynamic procedures [1].

Electrochemical characteristics of layered transition metal oxide cathode materials for lithium ion batteries were investigated by considering different parameters such as thermal properties, surface, and bulk behaviour [2].

The electrochemical behaviour of vapour grown carbon nanofibers was optimized for lithium-ion batteries by hydrothermal and thermal treatments and impregnation [3]. It was concluded that the surface of the untreated carbon nanofibers experiences an aging process during the earliest cycles [3].

In recent years, research on lithium-ion batteries heat loss has become very popular. However, most of the previous studies did not quantify the reversible and irreversible heat sources in lithium-ion batteries. A simple transformation of coordinates was proposed which simplifies the efficient simulation of the non-isothermal lithium-ion pseudo 2D battery cell model [4].

Model reformulation and efficient simulation of two-dimensional electrochemical thermal behaviour of lithium-ion batteries were investigated [5]. The two dimensional battery model was presented and developed by using Chebyshev-based orthogonal collocation. It was concluded that great changes in internal variables could appear, even under approximately mild situations [5].

A coupled continuum formulation for the mechanical processes—thermal, chemical, and electrostatic—in battery materials was proposed [6]. The main improvement was to model the evolution of porosity because of strains, which was induced by mechanical stresses, thermal expansion, and intercalation [6].

A mathematical model was developed to anticipate the time dependent behaviour of a cell [7]. It was concluded that the reaction current was concentrated neighbouring the terminals at the start of the discharging process, continuously became more homogeneous over the electrode surface, and developed into a concentrated situation underneath the electrode neighbouring the ending of the discharge process [7].

A 1D model appertaining to electrochemical and physical processes of a lithium ion cell was employed to explain hybrid pulse power characterization and constant current data from a battery cell [8]. It was designed for hybrid electric vehicle utilization. It was concluded that depending on battery cell operating situation and design, the end of discharge pulse might be attributable to positive electrode solid phase Li saturation, electrolyte phase Li discharge, or negative electrode solid phase Li discharge [8].

Electrodes modelling was accomplished for three different battery cell geometries to investigate the influence of the positioning of current collecting tabs and the aspect ratio of the electrodes on the discharge behaviours of the battery [9]. In addition, with the intention of predicting the thermal behaviour of the lithium-polymer battery cell the heat generation rate as a function of the location on the electrodes and discharge time was determined. The modelling outcomes were compared with the experimental discharge curves at different discharge rates [9]. It was concluded that that the parameters, which were adjusted for the electrodes of one geometry, could be used for the electrodes of other geometries. It should be noted that to accomplish this the manufacturing processes, compositions, and materials of the electrodes should be the same [9].

Four distinct battery cell designs were investigated to appraise the effects of cell stack aspect ratio, size, and tab configuration for similar electrode-level designs [10]. The model outcomes demonstrated that the internal battery cell kinetics is considerably affected by the macroscopic battery cell design for heat transport and electrical current [10].

The current density and potential distribution on the electrodes of a lithium-polymer battery were investigated by employing the finite element procedure [11]. The outcomes demonstrated that the placing and size of current collecting tabs and the aspect ratio of the electrodes have a considerable impact on the current density and potential distribution on the electrodes to affect the SOC distribution on the electrodes, hence influencing the homogeneous usage of the active material of electrodes [11].

A procedure was designed for dependency modelling of the discharge behaviour of a lithium-ion battery cell on the environmental temperature [12]. The two-dimensional modelling of the potential was validated by the modelling outcomes. The heat generation rates as a function of the position on the electrodes and the discharge time were determined in order to anticipate the temperature distributions of the lithium-ion battery [12]. This was according to the modelling outcomes of current density and the potential distributions. The temperature distributions, which were achieved from experimental measurements, were in good agreement with the modelling [12].

A battery cell model, which is flexible to investigate the thermal, electrochemical, physical phenomena, and advance over extensive length scales in battery cell systems of different assemblies, is necessary.

Unfortunately, thermal parameter measurement explanations and electrical parameters determination for lithium-ion batteries were not conveniently found in the literature. Many researchers commonly address the thermal parameters without reporting measurement procedures. A detailed description of thermal parameter measurement is reported in this investigation. Notwithstanding, to the author's best knowledge, only very few publications [13–15] are available in the literature that discuss the thermal simulation of lithium-ion batteries by considering all of the influential parameters such as thermal, electrical, and chemical processes on the thermal behaviour of the lithium-ion batteries. In addition, most of the previous studies did not take into account all of the electrochemical phenomena. In this investigation, the Multi-Scale Multi-Dimensional (MSMD) battery module was used for a lithium titanate oxide battery, which to the author´s best knowledge, it has not been done yet. The investigated model is able to determine the surface temperature distribution of the battery cell at various operating conditions with high accuracy.

2. The Battery Modelling

A 13 Ah pouch type commercial lithium-ion battery cell with dimensions of 204 mm width, 129 mm length, and 7.7 mm thickness and a lithium titanate oxide based anode was modelled for all simulations. The picture of the battery cell inside fixture, which was chosen for this investigation, is illustrated in Figure 1.

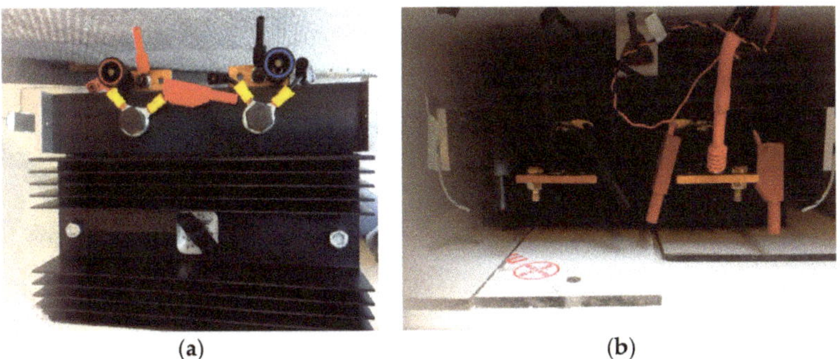

(a) (b)

Figure 1. 13 Ah pouch type commercial lithium-ion battery cell. (**a**) The battery cell inside fixture. (**b**) The battery cell and fixture inside the Maccor chamber.

In accordance with the construction and geometry of the battery cell, a three-dimensional model was constructed in ANSYS (2018) and the battery geometry was generated and analysed in an appropriate manner for additional analysis. Different components of positive and negative current tabs are illustrated in Figure 2. The negative tab and positive tab are used to accumulate the current flow via the battery cell. As shown in the figure five parts comprise the model:

(1) Positive current tab meshing structure;
(2) Interior part of positive current tab;
(3) Positive current tab;
(4) Skin of positive current tab;
(5) Contact region of positive current tab.

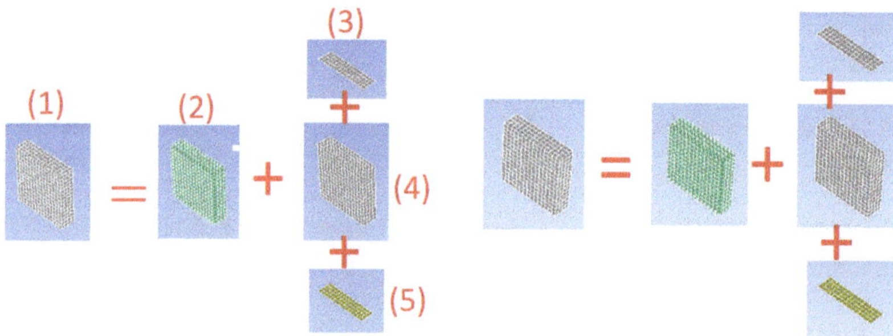

Figure 2. Different components of (**a**) positive current tab (**b**) negative current tabs.

The active volume, contact region, and skin of the battery cell are illustrated in Figure 3. The active volume demonstrates the stacked construction, comprising separator layers, negative and positive active materials, and aluminium foils. The thin skin enclosing the active volume. Geometrical structured meshing of lithium titanate oxide battery cell is shown in Figure 4. As shown in Figure 3 four parts comprise the model:

(A) Meshing of battery cell stacked construction;
(B) Skin of active volume;
(C) Active volume;
(D) Contact region of active volume.

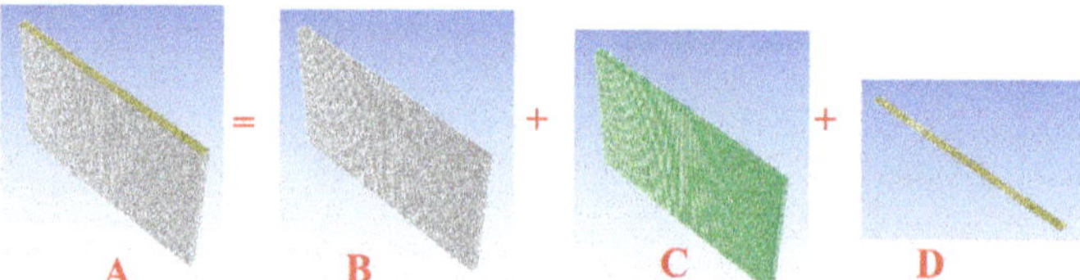

Figure 3. Battery cell geometry, which was analysed in this investigation.

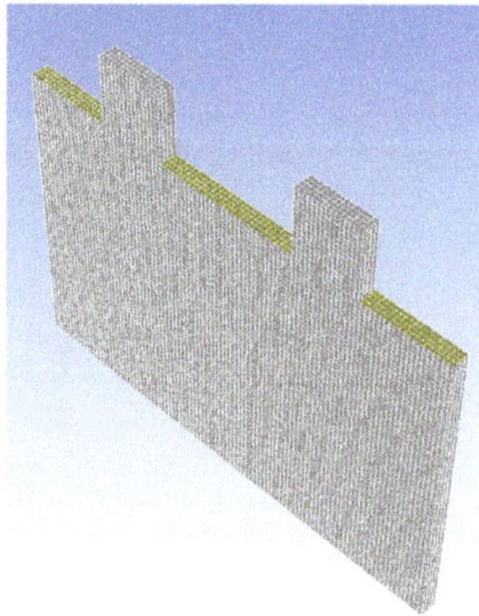

Figure 4. Geometrical structured meshing of lithium titanate oxide battery cell.

3. Identification of Model Parameters

3.1. Determination of Thermal Parameters

The general energy balance differential equation, which explains the distribution of generated local, conducted, and accumulated heats and the variation of temperature within a battery cell or pack, can be written as follows:

$$\frac{\partial}{\partial x}\left(k_x\frac{\partial T}{\partial x}\right)+\frac{\partial}{\partial y}\left(k_y\frac{\partial T}{\partial y}\right)+\frac{\partial}{\partial z}\left(k_z\frac{\partial T}{\partial z}\right)=\rho C_p\frac{\partial T}{\partial t}-\lambda \quad (1)$$

where

C_p: Specific heat capacity
λ: Volumetric heat generation
ρ: Physical mass
k_i: Thermal conduction in direction i
T: Temperature

Free convection is the main heat transfer process from the surfaces of the battery cell. The dissipated heat flux from the battery surface to the surrounding can be considered by both the convection and the radiation heat contributions:

$$Q_s = h(T_s - T_a) + \varepsilon\sigma({T_s}^4 - {T_a}^4) \quad (2)$$

where

σ: Stefan–Boltzmann constant
ε: Emissivity of the battery cell surface
T_a: Ambient temperature
h: Convective heat transfer
T_s: Battery surface temperature

Natural convection on a surface is a function of the orientation besides the geometry of the surface. In addition, it depends on the thermos physical properties of the fluid and the variation of temperature on the surface [16]. The complicatedness of the fluid flow makes it hard to achieve straightforward analytical relations for natural convection. The Rayleigh number, which controls the flow regime in natural convection, is defined as the product of the Prandtl and Grashof numbers [16]:

$$Ra = Pr \times Gr = \frac{g\beta(T_s - T_a)\delta^3}{v^2}Pr \quad (3)$$

where

v: Kinematics viscosity of the fluid
β: Coefficient of volume expansion
g: Gravitational acceleration
δ: Characteristic length of the geometry

The Nusselt number for natural convection could be determined in the following form [16]:

$$\mathrm{Nu} = \frac{h\delta}{k} = CRa^n = 0.54Ra^{1/4} \quad (4)$$

where the constants n and C depend on the flow and the geometry of the surface.

In this investigation natural convection process was considered for heat transfer from the surfaces of the battery cell. A procedure was used for the determination of the thermal parameters of the lithium-ion batteries. The method is able to determine a single thermal parameter such as specific heat capacity or thermal conductivity.

The heat capacity is quantifiable physical parameters of a substance, which characterizes the amount of heat, which is required to alter the temperature of the substance by one degree. In other words, heat capacity or specific heat is the amount of heat, which is needed to heat or cool 1 kg of a material by 1 °C. It determines how quickly a battery heats or cools down from its primary temperature in a given surrounding condition. The outcome of the heat capacity characterization process depends on the amount of absorbed heat, which consecutively depends on the ambient circumstances of the process.

In this investigation, isothermal battery calorimeter was employed for determination of specific heat. In order to measure the heat capacity of the battery cell with known mass (m) a thermal procedure was selected. At first, the battery was placed in the Maccor chamber for 3 h to be equilibrated at initial temperature T_1. The chamber temperature was set to T_1. Then it was placed rapidly in the isothermal battery calorimeter. The isothermal battery calorimeter temperature was set to T_2. After a period, the battery cell temperature reached to the chamber temperature. This procedure was repeated several times and the average amount was considered. The heat (Q) which was transferred between the isothermal battery calorimeter and battery cell was measured by using heat flux sensors inside chamber. Consequently, by having the temperature difference ($T_1 - T_2$) and the mass (m) of the battery cell the heat capacity was calculated by using the following equation:

$$C_P = Q/m(T_1 - T_2) \tag{5}$$

Transient and steady state methods are the main methods, which could be used for the determination of the thermal conductivity of batteries. Guarded hot plate is a steady state method, which can approximately estimate the thermal conductivity. In this method, the battery is placed between a heat sink and heat source. The thermal conductivity could be determined by knowing the battery cell thickness, temperature difference across the battery, and heat flux.

$$K = \frac{q\Delta x}{\Delta T} \tag{6}$$

where

q: Heat flux
ΔT: Temperature gradient
Δx: Thickness

To assure of the accuracy of the results different ways were used for determining the battery cell density. The battery density was calculated based on the battery material. In addition, the density of the lithium-ion battery was calculated by measuring the mass and volume of the battery cell. In order to determine precisely the physical and thermal parameters of the lithium titanate oxide battery cell and to assure of the accuracy of the previous estimations a procedure was selected. In this method, the battery cell was divided to different parts such as negative current collector, negative electrode, separator, positive electrode and positive current collector. The battery cell cross-section along with the thickness of different layers is illustrated in Figure 5. The material properties of the battery cell were estimated by employing the following formulations [17]:

$$K = \frac{0.5(K_{P_c})(T_{P_c}) + (K_{P_e})(T_{P_e}) + (K_S)(T_S) + (K_{N_e})(T_{N_e}) + 0.5(K_{N_c})(T_{N_c})}{0.5(T_{P_c}) + T_{P_e} + T_S + T_{N_e} + 0.5(T_{N_c})} \tag{7}$$

$$\rho = \frac{0.5(\rho_{P_c})(T_{P_c}) + (\rho_{P_e})(T_{P_e}) + (\rho_S)(T_S) + (\rho_{N_e})(T_{N_e}) + 0.5(\rho_{N_c})(T_{N_c})}{0.5(T_{P_c}) + T_{P_e} + T_S + T_{N_e} + 0.5(T_{N_c})} \quad (8)$$

$$C_P = \frac{0.5(C_{PP_c})(T_{P_c}) + (C_{PP_e})(T_{P_e}) + (C_{PS})(T_S) + (C_{PN_e})(T_{N_e}) + 0.5(C_{PN_c})(T_{N_c})}{0.5(T_{P_c}) + T_{P_e} + T_S + T_{N_e} + 0.5(T_{N_c})} \quad (9)$$

$$\sigma_p = \frac{0.5(\sigma_{P_c})(T_{P_c}) + (\sigma_{P_e})(T_{P_e})}{0.5(T_{P_c}) + T_{P_e} + T_S + T_{N_e} + 0.5(T_{N_c})} \quad (10)$$

$$\sigma_n = \frac{0.5(\sigma_{N_c})(T_{N_c}) + (_{N_e})(T_{N_e})}{0.5(T_{P_c}) + T_{P_e} + T_S + T_{N_e} + 0.5(T_{N_c})} \quad (11)$$

$$\alpha = \frac{k}{\rho C_p} \quad (12)$$

where

α: Thermal diffusivity
T_{N_c}: Thickness of negative current collector
T_{N_e}: Thickness of negative electrode
T_S: Thickness of separator
T_{P_e}: Thickness of positive electrode
T_{P_c}: Thickness of positive current collector
K_{P_c}: Thermal conductivity of positive current collector
K_{P_e}: Thermal conductivity of positive electrode
K_S: Thermal conductivity of separator
K_{N_e}: Thermal conductivity of negative electrode
K_{N_c}: Thermal conductivity of negative current collector
ρ_{P_c}: Density of positive current collector
ρ_{P_e}: Density of positive electrode
ρ_S: Density of separator
ρ_{N_e}: Density of negative electrode
ρ_{N_c}: Density of negative current collector
C_{PP_c}: Heat capacity of positive current collector
C_{PP_e}: Heat capacity of positive electrode
C_{PS}: Heat capacity of separator
C_{PN_e}: Heat capacity of negative electrode
C_{PN_c}: Heat capacity of negative current collector
σ_{P_c}: Electric conductivity of positive current collector
σ_{P_e}: Electric conductivity of positive electrode
σ_{N_c}: Electric conductivity of negative current collector
σ_{N_e}: Electric conductivity of negative electrode

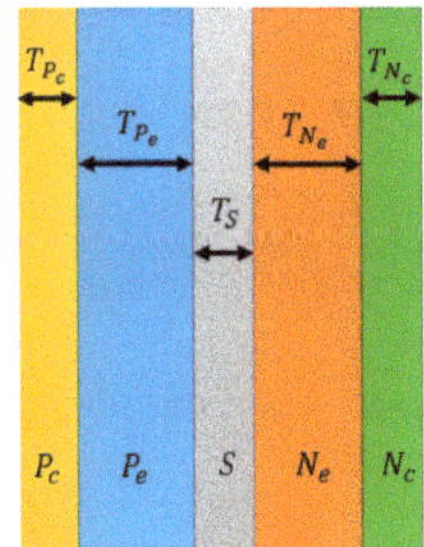

Figure 5. The lithium-ion battery cell cross-section.

3.2. Determination of Electrical Parameters

A 2RC equivalent circuit model was used in this investigation. The model is presented in Figure 6. To achieve the parameters of the equivalent circuit model different loading profile were applied to the battery cell. The loads consist of charge and discharge cycles with different C-rates. The voltage variation of load profiles at different temperature and C-rates is illustrated in Figure 7. An example of the voltage response when a 4 C-rate (52 Ah) current is applied to the battery cell at 27 °C is illustrated in Figure 8.

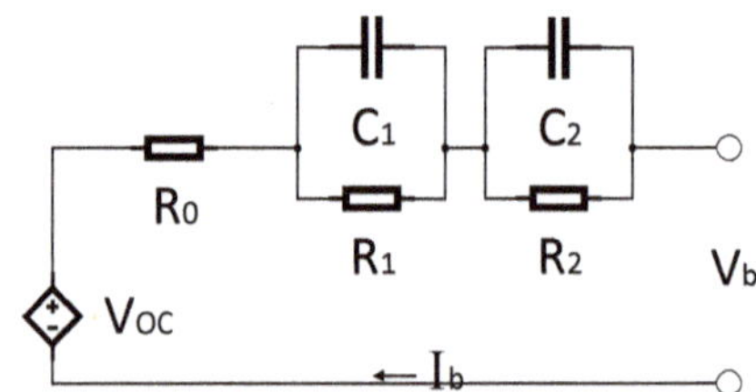

Figure 6. Equivalent circuit model of the battery.

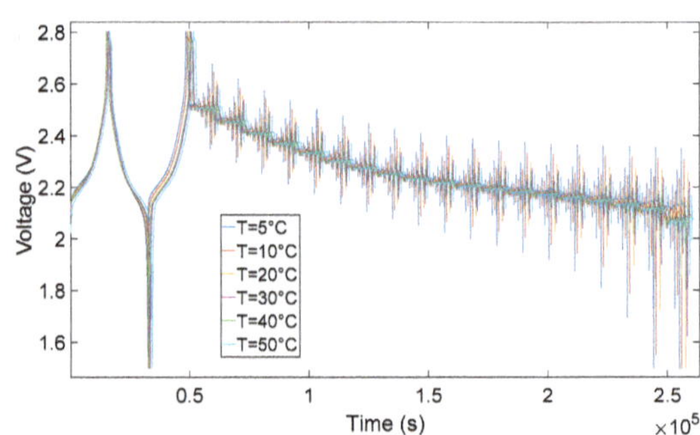

Figure 7. The voltage response of the battery cell to charging and discharging current pulses.

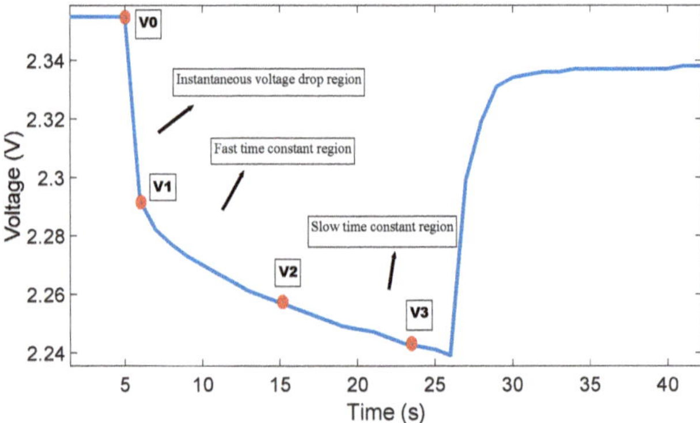

Figure 8. The voltage response of the battery cell for 4 C discharge at 30 °C.

The values of open circuit voltage, resistances, and capacitances of the 2RC equivalent circuit model are illustrated in Figure 9. The parameters of the 2RC equivalent circuit model were determined by using the following equations [18,19]:

$$R_0 = \frac{V_0 - V_1}{I},\ R_1 = \frac{V_1 - V_2}{I},\ R_2 = \frac{V_2 - V_3}{I} \tag{13}$$

$$C_1 = \frac{t_1 - t_2}{\ln(\frac{V_1(t_2)}{V_1(t_1)}) \times R_1}, \; C_2 = \frac{t_2 - t_3}{\ln(\frac{V_2(t_3)}{V_2(t_2)}) \times R_2} \tag{14}$$

where

V_0: Battery voltage before the discharge current pulse is applied.
R_0: Ohmic resistance
I: Amplitude of the current pulse.
V_1: Battery voltage one seconds (t_1) after the discharge current is applied.
R_1, C_1: Resistance and capacitance of the first RC network
V_2: Battery voltage ten seconds (t_2) after the discharge current is applied.
V_3: Battery voltage eighteen seconds (t_3) after the discharge current is applied.
R_2, C_2: Resistance and capacitance of the second RC network

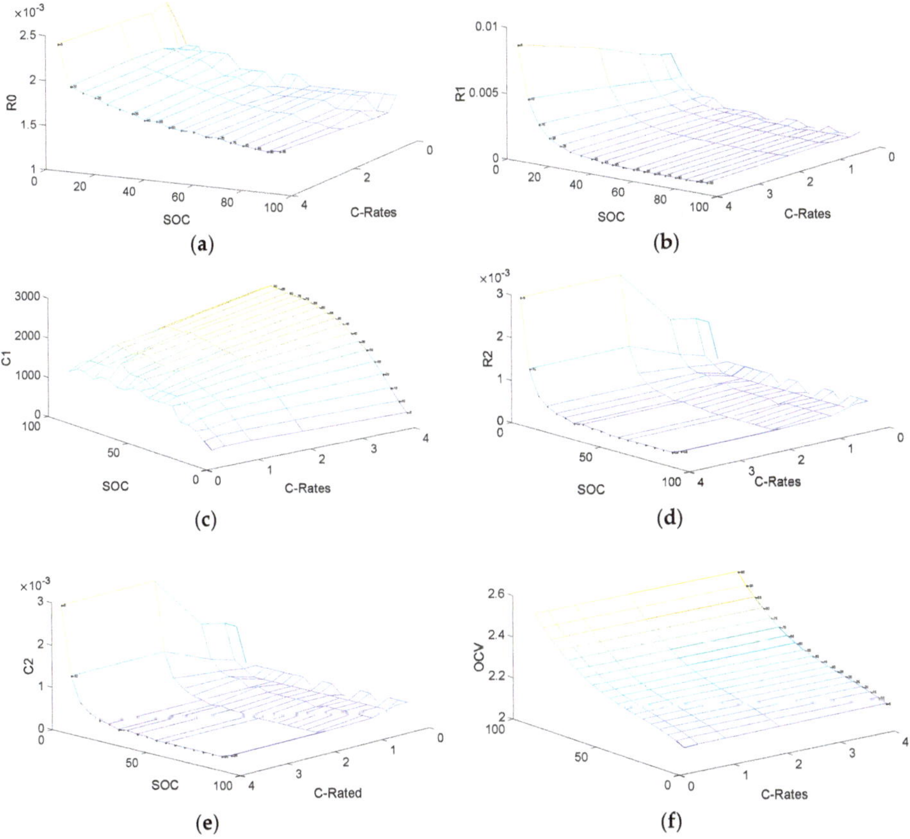

Figure 9. Amount of (**a**) R_1, (**b**) R_2, (**c**) C_1, (**d**) C_2, (**e**) V_0 and (**f**) OCV.

For the investigated battery cell, the open circuit voltage, resistances, and capacitances of the 2RC equivalent circuit model were considered as functions of the battery state of charge (SOC).

The coefficients of fifth order polynomial are illustrated in Table 1. These functions were expressed in fifth order polynomial form:

$$\begin{bmatrix} R_0 \\ R_1 \\ R_2 \\ C_1 \\ C_2 \\ V_0 \end{bmatrix} = \begin{bmatrix} A1 & A2 & A3 & A4 & A5 & A6 \\ B1 & B2 & B3 & B4 & B5 & B6 \\ C1 & C2 & C3 & C4 & C5 & C6 \\ D1 & D2 & D3 & D4 & D5 & D6 \\ E1 & E2 & E3 & E4 & E5 & E6 \\ F1 & F2 & F3 & F4 & F5 & F6 \end{bmatrix} \times \begin{bmatrix} 1 \\ SOC \\ (SOC)^2 \\ (SOC)^3 \\ (SOC)^4 \\ (SOC)^5 \end{bmatrix} \tag{15}$$

Table 1. Coefficients of fifth order Polynomial.

Parameters	A1	A2	A3	A4	A5	A6
R_0-0.25C	-3.5809×10^{-9}	2.5521×10^{-7}	-6.7798×10^{-6}	8.3191×10^{-5}	−0.00046957	0.0025463
R_0-0.5	-1.866×10^{-8}	9.4976×10^{-7}	-1.7805×10^{-5}	0.00015173	−0.00061228	0.0027413
R_0-1	-9.7461×10^{-9}	5.0726×10^{-7}	-9.8659×10^{-6}	8.9925×10^{-5}	−0.00041996	0.0025851
R_0-2	-6.1113×10^{-8}	3.5975×10^{-6}	-8.0772×10^{-5}	0.00086392	−0.0044514	0.01021
R_0-4	-9.1265×10^{-8}	5.3384×10^{-6}	−0.00011908	0.0012635	−0.0064278	0.013877
R_1-0.25C	-2.114×10^{-8}	1.3532×10^{-6}	-3.3195×10^{-5}	0.00039194	−0.0022685	0.0064137
R_1-0.5	-3.2789×10^{-8}	1.9467×10^{-6}	-4.4325×10^{-5}	0.00048576	−0.0026136	0.0068707
R_1-1	-3.9476×10^{-8}	2.3809×10^{-6}	-5.4759×10^{-5}	0.00060016	−0.0031773	0.0078591
R_1-2	-6.1113×10^{-8}	3.5975×10^{-6}	-8.0772×10^{-5}	0.00086392	−0.0044514	0.01021
R_1-4	-9.1265×10^{-8}	5.3384×10^{-6}	−0.00011908	0.0012635	−0.0064278	0.013877
R_2-0.25C	-2.6294×10^{-8}	1.433×10^{-6}	-2.8897×10^{-5}	0.00026512	−0.0010917	0.0021711
R_2-0.5	-1.7408×10^{-8}	1.0479×10^{-6}	-2.3627×10^{-5}	0.00024636	−0.0011691	0.002445
R_2-1	-2.1097×10^{-8}	1.1844×10^{-6}	-2.509×10^{-5}	0.00024926	−0.0011648	0.002562
R_2-2	-4.1644×10^{-8}	2.4069×10^{-6}	-5.2546×10^{-5}	0.00053695	−0.0025564	0.0049448
R_2-4	-4.1644×10^{-8}	2.4069×10^{-6}	-5.2546×10^{-5}	0.00053695	−0.0025564	0.0049448
C_1-0.25C	−0.0092475	0.45755	−7.8046	47.083	34.95	311.78
C_1-0.5	−0.0019811	0.067608	−0.38574	−12.919	240.28	158.33
C_1-1	−0.0059824	0.25455	−3.4037	6.1343	215.73	184.13
C_1-2	−0.0054655	0.25365	−3.9189	13.296	219.94	136.56
C_1-4	−0.0079124	0.40473	−7.3001	44.494	157.33	113.13
C_2-0.25C	0.032132	−1.7143	33.427	−293.41	1165.2	239.18
C_2-0.5	−0.011611	0.35172	−0.50313	−72.798	747.78	550.81
C_2-1	0.012608	−0.72394	15.828	−172.34	1019.5	111.45
C_2-2	−0.01571	0.6205	−6.7757	−20.924	827.98	−182.01
C_2-4	−0.029829	1.5623	−30.689	258.86	−513.18	1515
OCV-0.25C	6.5341×10^{-8}	-5.1732×10^{-6}	0.00022863	−0.0032124	0.028214	2.0797
OCV-0.5	-1.963×10^{-8}	-8.8693×10^{-7}	0.00014772	−0.0024952	0.025067	2.0873
OCV-1	-8.6373×10^{-8}	2.5889×10^{-6}	7.9331×10^{-5}	−0.001866	0.022391	2.0925
OCV-2	-4.6552×10^{-8}	6.1797×10^{-7}	0.00011227	−0.0020581	0.022357	2.0958
OCV-4	-7.1791×10^{-8}	2.0946×10^{-6}	7.9307×10^{-5}	−0.0017074	0.020533	2.101

4. Modelling Method

The model was solved in ANSYS by employing the Multi-Scale Multi-Dimensional (MSMD) battery module. The model combines the principal design parameters of the battery cell such as corresponding physical parameters, materials, and dimensions to computational fluid dynamics and heat transfer. In addition, the battery model is able to simulate a single battery cell or a battery pack to investigate their electrochemical and thermal behaviour. At the solution phase of the model, unsteady state problem and the thermal time interdependent were solved numerically assuming the heat generation in the battery cell as a dynamic source. The amount of heat generation inside a lithium-ion battery cell, which was proportionate to the temperature and current rate, was measured by an isothermal battery calorimeter and was considered as an input to the thermal model. The thermal and electrical fields were solved by using the following equations [17]:

$$\begin{gathered} \frac{\partial \rho C_p T}{\partial t} - \nabla(k \nabla T) = G + \dot{Q} \\ G = \sigma_+ |\nabla \phi_+|^2 + \sigma_- |\nabla \phi_-|^2 \\ \nabla(\sigma_+ \nabla \varphi_+) = \nabla(\sigma_- \nabla \varphi_-) = J \end{gathered} \tag{16}$$

where

J: Volumetric transfer
C_p: Heat capacity
$\dot{Q}$: Heat generation
k: Thermal conductivity
σ: Effective electric conductivities
ϕ: Phase potential
T: Temperature

5. Measurement of Heat Generation Rates

Heat generation inside batteries is a complicated process, which could be divided into reversible and irreversible parts. Experimental setup, which was used in this investigation, is illustrated in Figure 10. Isothermal battery calorimeter was employed for heat loss measurement. The detailed experimental procedures and setup, as well as corresponding equipment and materials, could be found in [20]. Maccor automated system was used as a battery cycler for the whole experiments. The battery cycler charged and discharged the battery with different current rates. Heat flux determination of lithium titanate oxide battery cell by using isothermal calorimeter is shown in Figure 11.

Figure 10. Experimental setup.

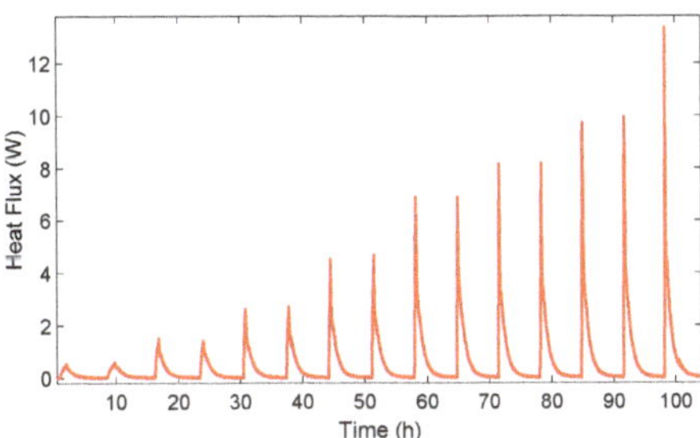

Figure 11. A heat generation analysis of lithium titanate oxide battery cell for 300 K.

6. Experimental Validation

Several sets of experiments were accomplished to validate model performance at different temperature by using a 13 Ah battery, which was fabricated by Altairnano and is shown in Figure 4. A 45A discharge process was used to validate the temperature, which was anticipated by the model. FLIR thermal camera and contact thermocouples were used to monitor the surface temperature of the battery cell at different positions. Experimental temperature results and temperature simulation for

45 A discharge process is illustrated in Figure 12. It could be seen from the figure that the temperature increase of the simulation is in good agreement with the experimental data. This demonstrate that the model is capable of simulating the real battery cell. The experimental temperature increase data, which was used in this investigation, was the average amount of several thermocouples and FLIR thermal camera measurements. In addition, the simulation temperature increase was the average value of entire battery cell temperature.

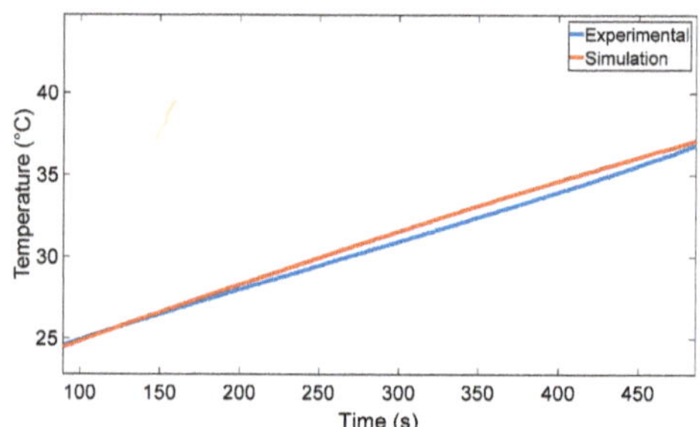

Figure 12. Experimental temperature results and temperature simulation for 45 A discharge process.

During the charge and discharge cycling, the battery cell was made to be isothermal by situating it in an isothermal battery calorimeter. The calorimetric measurements were used as a heat generation source in the battery cell. A Maccor automated test system was employed as the discharge apparatus with the intention of monitoring the current and voltage. The simulated and experimental outcomes for 52 A discharge process and 300 K environmental temperature are illustrated in Figure 13. The experimental temperature data, which were used for model validation, are the value of surface temperature of the battery at four different locations, which were measured by four contact thermocouples. Temperature value of the experimental data compare good to the simulation, demonstrating that the model could simulate the real battery cell.

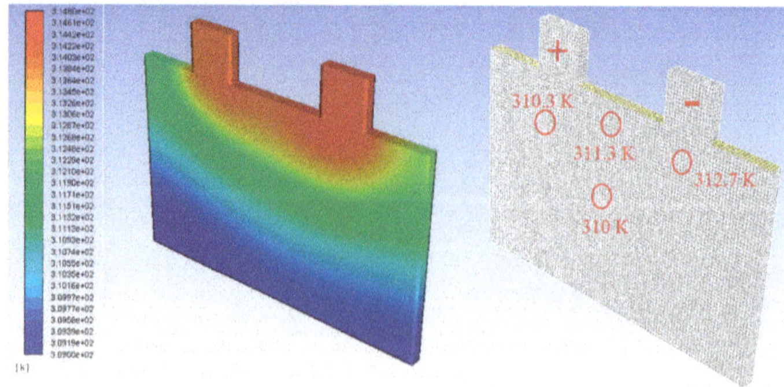

Figure 13. Temperature distribution and experimental results at the end of 52 A discharge.

7. Simulation Results and Discussion

In this investigation, several quantities for the volumetric heat generation were considered. The values were measured by an isothermal battery calorimeter for various load profiles. The model is simulated in both time dependent and steady state environment to determine the temperature spatial distribution over the battery surface. In addition, the modelling is able to show the maximum value of surface temperature of the battery as a function of time for different environment temperature and discharge current rates.

The temperature distributions of the battery cell were determined as a function of time at different discharge rates. As could be seen in Figures 12 and 13, the overall temperature distributions, which were achieved from the model and experiment, are in good agreement with each other.

The heat, which is dissipated from the battery, and the heat generated inside it are approximately equal during low current rates. Therefore, fast equilibrium could be reached. In another word, most of generated heat will be transferred to the surrounding through free or forced convection. The evolution of the uttermost temperature of the battery cell is confined at low current rates. The phenomenon demonstrates the minor rises in the surface temperature. Notwithstanding, the difference among the maximum temperatures, which were attained, from the modelling and experiment was lower than that between the corresponding minimum ones.

The simulation was accomplished at different discharge current rates ranging from 0.25 C to 9 C with 0.25 C interval. The modelling discharge profiles agree good with those, which were gained from experimental. The corresponding heat loss from the battery cell was shown in Figure 11. In accordance with the findings, the temperature increase sharply to a specific point. As anticipated, the position of the hottest area is seen near the negative tab of the battery cell throughout the discharge process. In addition, non-uniform temperature propagation was observed.

Temperature distribution of the battery cell at different discharge rates ranging from 0.25 C to 9 C is illustrated in Figure 14. As could be seen from the distributions of temperature over the volume of the battery cell, the temperature contours showed moderate slopes at low current rates. On the contrary, the temperature contours demonstrate sharper slopes at higher current rates. It demonstrates quicker temperature increase during the discharge of the battery cell, which is due to the higher heat generation inside the battery cell. At high current rates, achieving equilibrium occurs in a longer time. This phenomenon demonstrates that the modelling discharge curves agree well with those which were achieved from the experiments.

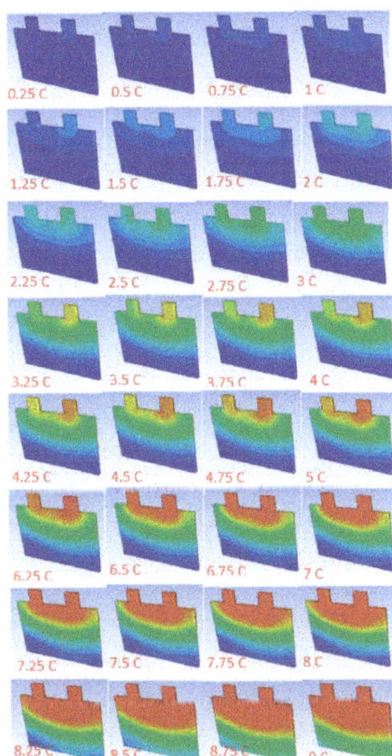

Figure 14. Temperature distribution of the battery cell at different discharge rates from 0.25 C to 9 C.

Although the current flows in the neighbourhood of the tabs of both the negative and positive electrodes are correspondingly great, the electrical conductivity of the active material of the negative electrode is much higher than that of the positive electrode. This phenomenon leads to lower temperatures in the neighbourhood of the current collecting tab of the positive electrode compared to the negative electrode [21–23].

Electrochemical reaction rate was increased due to higher temperature gradient. This phenomenon could be described by higher current rates in some parts of the cell owing to high temperature gradient. The minimum and maximum temperatures, which, were collected from the modelling and experiment, are in good agreement with each other over the entire scope of battery cell surface at the different discharge rates. Notwithstanding, some difference was seen among the discharge curves, which were achieved from the experiment and model in close proximity to the end period of discharge. The highest discrepancy was seen for high discharge rates.

8. Conclusions

The principal objective of this investigation was to develop a precise, computationally efficient and simplified Dual Potential Multi-Scale Multi-Dimensional (MSMD) Battery model. A procedure was used to simulate the thermal behaviour of a lithium-ion battery at different current rates and environmental temperature. The three-dimensional temperature distribution of the battery was anticipated as a function of the discharge time by using the model. By comparing the modelling discharge curves with the experimental outcomes at different environmental temperatures and discharge rates the modelling was validated. The parameters of equivalent circuit model were determined from multi-pulse charge and discharge data. An average specific heat capacity was considered for the battery cell in the time dependent and unsteady state simulation. To assist the suggested modelling method, calorimetric experiments were accomplished. By using the heat generation, which was measured by the isothermal calorimeter, the model was simulated to demonstrate the temperature distribution. A great temperature discrepancy was seen in battery surface at high current rates. This phenomenon could be described by high amount of heat generation due to higher temperature gradient. In addition, the value of surface temperature of the battery was determined by the model, was compared to the experimental data, and was in good agreement with the model data. Greater temperature gradients were seen at the battery cell surfaces owing to higher current rate. Subsequently, design of an appropriate thermal management system specifically during high current rates charging and discharging could play a fundamental role in preventing great temperature growth of the battery cell. The simulation methodology, which was demonstrated in this investigation, might contribute to the development of a battery cell thermal management system, which enables the temperature evolution of lithium-ion batteries as a subordinate of time to be more precisely anticipated. In addition, it can assist to anticipate the evolution of the thermal, electrical, and chemical processes.

Author Contributions: S.S.M. proposed the idea of the paper; S.S.M. wrote the paper; E.S. provided suggestions on the content and structure of the paper; S.K.K. and E.S. has been reviewing the draft manuscripts.

Funding: This research received no external funding.

Conflicts of Interest: The authors declare no conflict of interest.

References

1. von Srbik, M.; Marinescu, M.; Martinez-botas, R.F.; Offer, G.J. A physically meaningful equivalent circuit network model of a lithium-ion battery accounting for local electrochemical and thermal behaviour, variable double layer capacitance and degradation. *J. Power Sources* **2016**, *325*, 171–184. [CrossRef]
2. Tian, C.; Lin, F. Electrochemical Characteristics of Layered Transition Metal Oxide Cathode Materials for Lithium Ion Batteries: Surface, Bulk Behavior, and Thermal Properties. *Acc. Chem. Res.* **2018**, *2*, 89–96. [CrossRef]
3. Ortiz, G.F.; Alcántara, R.; Lavela, P.; Tirado, J.L. Optimization of the Electrochemical Behavior of Vapor Grown Carbon Nanofibers for Lithium-Ion Batteries by Impregnation, and Thermal and Hydrothermal Treatments. *J. Electrochem. Soc.* **2005**, *152*, A1797–A1803. [CrossRef]

4. Northrop, P.W.C.; Ramadesigan, V.; De, S.; Subramanian, V.R. Coordinate Transformation, Orthogonal Collocation, Model Reformulation and Simulation of Electrochemical-Thermal Behavior of Lithium-Ion Battery Stacks. *J. Electrochem. Soc.* **2011**, *158*, A1461–A1477. [CrossRef]
5. Northrop, P.W.C.; Pathak, M.; Rife, D.; De, S.; Santhanagopalan, S.; Subramanian, V.R. Efficient Simulation and Model Reformulation of Two-Dimensional Electrochemical Thermal Behavior of Lithium-Ion Batteries. *J. Electrochem. Soc.* **2015**, *162*, A940–A951. [CrossRef]
6. Wang, Z.; Siegel, J.; Garikipati, K. Intercalation Driven Porosity Effects in Coupled Continuum Models for the Electrical, Chemical, Thermal and Mechanical. *J. Electrochem. Soc.* **2017**, *164*, A2199–A2212. [CrossRef]
7. Gu, H. Mathematical Analysis of a Zn/NiOOH Cell R. *J. Electrochem. Soc.* **1983**, *130*, A1459–A1464. [CrossRef]
8. Smith, K.; Wang, C.Y. Solid-state diffusion limitations on pulse operation of a lithium ion cell for hybrid electric vehicles. *J. Power Sources* **2006**, *161*, 628–639. [CrossRef]
9. Kim, U.S.; Shin, C.B.; Kim, C.S. Effect of electrode configuration on the thermal behavior of a lithium-polymer battery. *J. Power Sources* **2008**, *180*, 909–916. [CrossRef]
10. Kim, G.-H.; Smith, K.; Lee, K.-J.; Santhanagopalan, S.; Pesaran, A. Multi-Domain Modeling of Lithium-Ion Batteries Encompassing Multi-Physics in Varied Length Scales. *J. Electrochem. Soc.* **2011**, *158*, A955–A969. [CrossRef]
11. Kwon, K.H.; Shin, C.B.; Kang, T.H.; Kim, C.S. A two-dimensional modeling of a lithium-polymer battery. *J. Power Sources* **2006**, *163*, 151–157. [CrossRef]
12. Yi, J.; Kim, U.S.; Shin, C.B.; Han, T.; Park, S. Modeling the temperature dependence of the discharge behavior of a lithium-ion battery in low environmental temperature. *J. Power Sources* **2013**, *244*, 143–148. [CrossRef]
13. Tang, Y.; Wu, L.; Wei, W.; Wen, D.; Guo, Q.; Liang, W.; Xiao, L. Study of the thermal properties during the cyclic process of lithium ion power batteries using the electrochemical-thermal coupling model. *Appl. Therm. Eng.* **2018**, *137*, 11–22. [CrossRef]
14. Panchal, S.; Mathew, M.; Fraser, R.; Fowler, M. Electrochemical thermal modeling and experimental measurements of 18650 cylindrical lithium-ion battery during discharge cycle for an EV. *Appl. Therm. Eng.* **2018**, *135*, 123–132. [CrossRef]
15. Mastali, M.; Foreman, E.; Modjtahedi, A.; Samadani, E.; Amirfazli, A.; Farhad, S.; Fraser, R.A.; Fowler, M. Electrochemical-thermal modeling and experimental validation of commercial graphite/LiFePO4pouch lithium-ion batteries. *Int. J. Therm. Sci.* **2018**, *129*, 218–230. [CrossRef]
16. Bahrami, M.; Consider, N.C.; Water, S. Natural Convection. *Eng. Thermodyn. Heat Transf.* **2011**, *388*, 1–7.
17. ANSYS, Inc. *ANSYS FLUENT Manual*; ANSYS, Inc.: Cannonsburg, PA, USA, 2018.
18. Hentunen, A.; Lehmuspelto, T.; Suomela, J. Time-domain parameter extraction method for thévenin-equivalent circuit battery models. *IEEE Trans. Energy Convers.* **2014**, *29*, 558–566. [CrossRef]
19. Stroe, A.I.; Stroe, D.I.; Swierczynski, M.; Teodorescu, R.; Kær, S.K. Lithium-Ion battery dynamic model for wide range of operating conditions. In Proceedings of the 2017 International Conference on Optimization of Electrical and Electronic Equipment (OPTIM) & 2017 International Aegean Conference on Electrical Machines and Power Electronics (ACEMP), Brasov, Romania, 25–27 May 2017; pp. 660–666.
20. Madani, S.; Schaltz, E.; Kær, S.K.; Madani, S.S.; Schaltz, E.; Kær, S.K. Heat Loss Measurement of Lithium Titanate Oxide Batteries under Fast Charging Conditions by Employing Isothermal Calorimeter. *Batteries* **2018**, *4*, 59. [CrossRef]
21. Wu, B.; Li, Z.; Zhang, J. Thermal Design for the Pouch-Type Large-Format Lithium-Ion Batteries: I. Thermo-Electrical Modeling and Origins of Temperature Non-Uniformity. *J. Electrochem. Soc.* **2014**, *162*, A181–A191. [CrossRef]
22. Guo, M.; White, R.E. A distributed thermal model for a Li-ion electrode plate pair. *J. Power Sources* **2013**, *221*, 334–344. [CrossRef]
23. Yi, J.; Lee, J.; Shin, C.B.; Han, T.; Park, S. Modeling of the transient behaviors of a lithium-ion battery during dynamic cycling. *J. Power Sources* **2015**, *277*, 379–386. [CrossRef]

Article

Optimal Sizing of Storage Elements for a Vehicle Based on Fuel Cells, Supercapacitors, and Batteries

José Luis Sampietro [1], Vicenç Puig [2,*] and Ramon Costa-Castelló [2]

[1] Instituto de Robótica e Informática Industrial (IRI), CSIC-UPC, C/Llorens i Artigues, 4-6, 08028 Barcelona, Spain; jlsampietro@iri.upc.edu

[2] Departament d'Enginyeria de Sistemes, Automàtica i Informàtica Industrial, Universitat Politècnica de Catalunya, C/Pau Gargallo, 5, 08028 Barcelona, Spain; ramon.costa@upc.edu

* Correspondence: vicenc.puig@upc.edu; Tel.: +34-937-398-562

Received: 3 December 2018; Accepted: 4 March 2019; Published: 10 March 2019

Abstract: To achieve a vehicle-efficient energy management system, an architecture composed of a PEM fuel cell as the main energy source and a hybrid storage system based on battery banks and supercapacitors is proposed. This paper introduces a methodology for the optimal component sizing aiming at minimizing the total cost, achieving a cheaper system that can achieve the requirements of the speed profiles. The chosen vehicle is an urban transport bus, which must meet the Buenos Aires Driving Cycle, and the Manhattan Driving Cycle. The combination of batteries and supercapacitors allows a better response to the vehicle's power demand, since it combines the high energy density of the batteries with the high power density of the supercapacitors, allowing the best absorption of energy coming from braking. In this way, we address the rapid changes in power without reducing the global efficiency of the system. Optimum use of storage systems and fuel cell is analyzed through dynamic programming.

Keywords: optimal control; supercapacitors; batteries; fuel cell; hybrid vehicle

---

1. Introduction

Today, one of the topics of interest in scientific research is the depletion of the planet's natural resources. The energy that comes from fossil fuels such as coal and oil, among others, will be exhausted in the next future. Moreover, this type of energy produces environmental pollution and greenhouse gases, which are responsible for the biggest damage to the ozone layer. Energy consumption in the transport sector is known to be very large, around 29.5% of the total energy consumed [1]. In particular, vehicles are responsible for most of the energy consumed [2]. For this reason, environmental deterioration is one of the main causes of the development of energy management research in vehicles. Hybrid vehicles have been a step forward in this direction, and the advantages of hybridizing a system [3] can be summarized as:

- The vehicle can recover a fraction of the kinetic energy while braking (regenerative breaking)
- The main power source might be shut down during idle periods and low-load phases without compromising vehicle drivability
- The main power source can operate at high efficiency points independently of the vehicle trajectory.
- The main power source can be designed with a slightly lower capacity.

Articles like [4–8], use a battery as an auxiliary energy recovery system, while others as [9–11] use a supercapacitor for that purpose. Currently, there are combinations of both. Combining the energy density of the batteries with the power density of the supercapacitors increases fuel economy [12–14].

To replace combustion engines, other devices as fuel cells have been introduced [15–17]. According to [18], many previous studies have shown the effectiveness of fuel-cell-based vehicles. In addition, zero emissions and low noise generation make fuel cells a tempting energy converter for automotive powertrains. As an example of this, fuel cell-powered bus projects report that since 2011 there are approximately 100 of such buses distributed around the world.

Fuel cells have relatively high efficiency compared to internal combustion engines [19]. Ref. [20] contains a comparison between fuel cells and internal combustion engines in the transportation sector.

As shown in [21], the most common types of fuel cells on the market are proton exchange membrane fuel cells (PEMFC), direct methanol fuel cells, alkaline fuel cells, phosphoric acid fuel cells, molten carbonate fuel cells, solid oxide fuel cells, and microbial fuel cells. In this article, we will focus on the use of PEMFC. The sizing of the fuel cell systems and associated storage elements is a problem that must be treated with care, because its cost in the market is still high.

Usually, Fuel Cell Electrical Vehicles (FCEV) are composed of a fuel cell acting as main power source and an energy storage system (ESS). The ESS can contribute to improving the performance of an FCEV [22,23], reduce the FC size [24], improve the operating efficiency of the system [25,26], and extend the service life of the elements [27]. In FCEV, ESS is usually composed of a battery and/or supercapacitors. However, this hybridization involves a greater complexity of the system, which highlights the importance of energy management [28,29].

This paper proposes a methodology to obtain an optimal sizing of the ESS, composed of a battery and supercapacitors, in an urban transport FCEV. The combination of batteries and supercapacitors allows a better response to the vehicle's power demand, since it combines the high energy density of the batteries with the high power density of the supercapacitors, allowing the best absorption of the energy coming from the braking. In this way, we address the rapid changes in power without reducing the global efficiency of the system.

Component optimal sizing aims to minimize the total cost while achieving the required performance. It is well-known that the vehicle performance depends a lot on the speed profile. For this reason, in this work two different urban driving profiles will be used as reference. In particular, the Buenos Aires Driving Cycle and the Manhattan Driving Cycle will be considered.

The proposed methodology will proceed as follows: firstly, the optimal energy evolution will be obtained using dynamic programming when following the considered speed profiles. This procedure will be repeated for different battery and supercapacitor sizes. Then, from the obtained results, optimal sizing will be determined.

The remainder of the paper is organized as follows: In Section 2, the vehicle architecture and the models of the components are described. In Section 3, the driving profiles are introduced and the theoretical amount of energy that can recover from regenerative braking is presented. Section 4 describes the ESS optimal sizing methodology based on dynamic programming. Section 5 presents the results of the sizing of the components in the considered vehicle with the proposed methodology. Finally, in Section 6, the main conclusions are drawn, introducing further research paths.

2. Vehicle Architecture

Vehicle architecture of HEV refers to the topological relationship and energy flow between its components [30,31]. The main configurations are the series, parallel, and series-parallel. Designing and selecting the architecture of an HEV's is a critical procedure, as it influences future design, control, and optimization. As a first step, we will define the total power that the components of the vehicle's propulsion system must deliver. The dynamics of the vehicle are based on the energetic balance of the forces that contribute to the movement of the vehicle, and those that oppose to it [32]. Then, we can express the mechanical power as a product of the forces and the speed of the vehicle. The inherent power of motion is deducted from the kinetic energy stored in it. The forces opposing the movement, are called dissipative forces, which are aerodynamic drag, frictional resistance to the ground and the resistance force due to the inclination of the road. Then, the mechanical power required to move the vehicle will be:

$$p_v = \frac{1}{2}pv(sc_x)v^3 + mgvc_{rr} + mgv\sin(\alpha) + mv\frac{dv}{dt} \quad (1)$$

where m is the mass of the vehicle, α is the slope of the road, v is its speed, p is the air density, s is the front area of the vehicle, c_x is the aerodynamic drag coefficient, g is gravity and c_{rr} is the coefficient of rolling resistance. The parameters are based on a service bus, and are those shown in Table 1, being obtained from [33]. The total mass of the vehicle includes the mass of the chassis, the propulsion system, the components and the weight of the passengers. It should be noted that as the weight of the vehicle increases with the increase in the weight of its components and the number of occupants, more power is needed to reach the speed profile, because there are higher power peaks. Figure 1 shows the components of the propulsion system that will be part of the vehicle's energy management. The main unidirectional source of energy is the fuel cell, which is connected to a DC converter. The storage elements (batteries and supercapacitor) are considered bidirectional, as they can deliver power to the movement, and at the same time, they can store the energy recovered from breaking. These elements are also associated with a DC converter. The speed profile can be placed as a power profile, and must be fulfilled by the sources. Then, the power balance can be expressed as:

$$p_v = p_{sup} + p_{bat} + p_{fc} + p_{break} \quad (2)$$

where p_{sup} is the supercapacitor power, p_{bat} is the battery power, p_{fc} is the fuel cell power and p_{break} is the power dissipated in the mechanical brake. As expressed in Equation (2), the sum of the powers of the elements must be equal to the mechanical power.

Table 1. Parameters of the vehicle.

Name	Symbol	Value	Unit
Air density	p	1.2	kg/m^3
Coefficient of resistance to movement	c_{rro}	0.008	s/u
Coefficient of resistance to movement	c_{rrl}	0.00012	s^2/m^2
Aerodynamic coefficient	c_x	0.65	s/u
Front area	s	8.06	m^2
Total mass	m	14,000	kg
Gravity	g	9.8	m/s^2

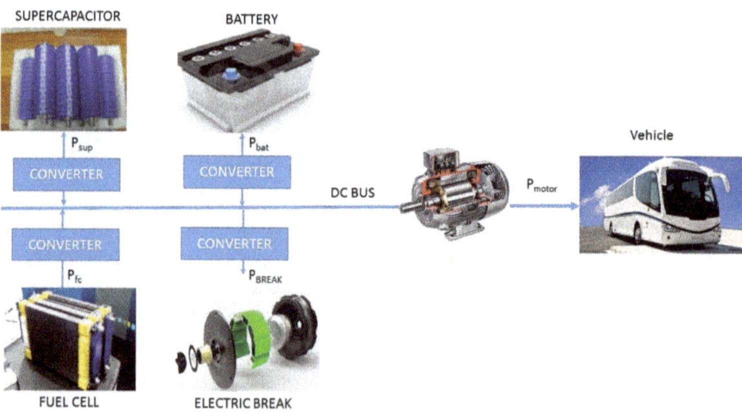

Figure 1. Vehicle architecture.

2.1. Battery Modelling

Electrochemical batteries are one of the key components in hybrid electric vehicles. Batteries, for specific energy management, will be characterized mainly in terms of power and energy. They are characterized by their nominal capacities, and by the state of charge (SOC), which describes the remaining energy stored in the battery, expressed as a percentage of its maximum capacity. Some desirable attributes of batteries for EV and HEV applications are high energy density and cycle life [34]. The energy density is a measure of the total amount of energy that a battery can store for a given mass. These elements can store considerable amounts of energy. Other features include long service life, low initial and replacement costs, high reliability, wide operating temperature range, and robustness. Battery operation is typically defined by a certain SOC window, whose limits are the minimum SOC during discharge and the maximum SOC during charging. Internal resistance is the factor that limits the battery's charge and discharge efficiency. Resistance has different values under load and discharge conditions. Resistance and open-circuit voltage are non-linear functions of the battery SOC. A battery model can be derived from an equivalent circuit, where the battery is regarded as an open-circuit voltage source, in series with an internal resistor.

Depending on the amount of voltage/current, we connect a set of batteries, in series, in parallel, or a mixed connection series-parallel. For a series connection, the voltage supplied by the assembly is equal to the sum of the voltages. In parallel, the current increases as the sum of the number of batteries inserted. In both cases, the capacity always increases. According to [33], the equations for battery power charging and discharging should be taken into account as a function of SOC, where p_{cb} is the charge power and p_{db} is the discharge power

$$p_{cb}(k) = -\frac{n_{bats}u_{cmax}^2 - u_{oc}(k)u_{cmax}}{r_i}n_{batp} \tag{3}$$

$$p_{db}(k) = \frac{-n_{bats}u_b(k)^2 + u_{oc}(k)u_{cmin}}{r_i}n_{batp} \tag{4}$$

where u_b is the battery voltage, u_{oc} is the battery open-circuit voltage, r_i is the battery internal resistance, where n_{batp} is the number of parallel cells and n_{bats} is the number of serial cells and k is the discrete-time. The supercaps open-circuit voltage is a function of the battery charge. The total power of the battery will be the sum of p_{db} and p_{cb}, and is called p_{bat}^*. The battery is also associated with a converter efficiency δ_{bat}, which represents the losses in the converters and takes a value of 0.98. Then, we can define the total battery power p_{bat} as shown.

$$p_{bat}(k) = \delta_{bat}p_{bat}^*(k). \tag{5}$$

The considered battery is a prismatic Ni-MH one in a resin case. Battery parameters are shown in Table 2 and taken from [35]:

Table 2. Battery parameters.

Parameter	Data
Manufacturer	PEVE
Shape	Prismatic
Case	Plastic
Cell capacity (Ah)	6.5
Cell voltage (V)	7.2
Specific energy (Wh/kg)	46
Specific power (W/kg)	1300
Mass (kg)	1.04
Operation temperature (°C)	−20 to 50
Cost (€/kg)	33.88

2.2. Supercapacitor Model

Supercapacitor are energy accumulators. The specific power, or instantaneous power, that can deliver is greater than that of batteries, but their specific energy, or the amount of energy that can store is substantially less. In some cases, supercapacitors are used as primary ESSs, while in other cases, such as in this paper, they can be placed as a secondary storage system. This allows improvement of the performance of the main power system and the ESS. The equivalent circuit of supercapacitor consists of a capacitor that represents the capacitance and a series resistor that represents the ohmic losses in the electrodes and electrolyte [36].

In the model, we will redefine equations based on the capacitor state of energy (SOE). A detailed study of the process can be found in [34]. Then, the SOE is defined by:

$$SoE(k) = \frac{e_{sc}(k)}{e_{sc,t}} \tag{6}$$

where $e_{sc,t}$ is the total storable energy and e_{sc} is the instantaneous energy. Then, e_{sc} is defined by:

$$e_{sc}(k) = \frac{1}{2} c_{sc} q_{sc}^2(k) \tag{7}$$

where q_{sc} is the capacitor voltage expressed in (V), and c_{sc} is the capacitance expressed in (F).

According to [34], the charging p_{cs} and discharging power p_{ds} is given by

$$p_{cs}(k) = \frac{n_{sc} u_{sc,max} (u_{sc}(k) - u_{sc,max})}{r_{sc}} \tag{8}$$

$$p_{ds}(k) = \frac{n_{sc} u_{sc,min} (u_{sc}(k) - u_{sc,min})}{r_{sc}} \tag{9}$$

where n_{sc} is the number of elements, $u_{sc,max}$, and $u_{sc,min}$ are the supercapacitor voltage limits, u_{sc} is the open-circuit voltage and r_{sc} is the circuit resistance. A more detailed analysis and parameters can be found at [34]. The parameters used are from Maxwell 125 V Heavy transportation module, and are shown in Table 3.

Table 3. Supercapacitor parameters.

Parameter	Data
Manufacturer	Maxwell Technologies
Packaging	Bulk
Cell capacitance (F)	3000
Rated Voltage (V)	125
Temperature (°C)	−40 to 65
Mass (kg)	1.3
Specific power (W/Kg)	1700
Specific energy (Wh/Kg)	2.3
SOE_{max}	1
SOE_{min}	0
Cost (€/Kg)	88.34

The power of the supercapacitor p_{sup}^* is the sum of p_{cs} and p_{ds}. The supercapacitor system is associated with an efficiency of the converter shown in Figure 1, δ_{sup}, which represents the losses in the converters. In the current work, this parameter will take a value of 0.95. Then, the total output power of the supercapacitor system p_{sup} is given by

$$p_{sup}(k) = \delta_{sup} p_{sup}(k)^*. \tag{10}$$

2.3. Fuel Cell Model

The PEMFC has two electrodes: The anode where the fuel is oxidized and the cathode where the oxidant is reduced [37]. The electrolyte simultaneously acts as an electrical insulator and a proton conductor. It also separates the cathode and anode reactions. Electrons go from the anode to the cathode through an external circuit generating electrical current, while protons do so through the electrolyte. In the cathode, electrons, protons, and oxidant are reduced, generating sub-products. Hydrogen is often used as an oxidizing agent and oxygen as reducing agent in this type of fuel cells.

The potential difference generated by a single unit or mono cell is less than one volt, so several mono-cells must be connected in series to obtain the appropriate voltage for the required application. However, although the fuel cell is the main part of a fuel cell system, the entire system typically involves the following subsystems:

- Supply of oxidant.
- Fuel supply.
- Heat management.
- Water management.
- Power conditioning, instrumentation, and controls.

The fuel and oxygen inlet lines to each cell are connected in parallel to achieve similar pressure in the anode and cathode. Impedance is a function of fuel pressure, membrane moisture, and catalyst status. To characterize the model to be used, we know that power is the product of current and potential. The power density is the product of the potential and current density, so it can be represented by:

$$p_{fc} = v_{fc} i_{fc}. \tag{11}$$

The power density is usually drawn with current density using the so-called polarization curves and indicate that there is a maximum power density that a fuel cell can reach. It is not always possible to operate the fuel cells at their maximum power levels. The polarization curve and power-current curve used in this work for the fuel cell are shown in Figure 2. In this work, we have taken the curve of the fuel cell for a BALLARD XD6 FCvelocity module fuel cell system, which is dimensionalized according to the maximum power of the driving profiles to be used, which are explained in the next section. Therefore, there is a maximum power that the cell can reach, because the efficiency of the fuel cell is directly proportional to the potential of the cell. Fuel cell efficiency is defined by:

$$n_{fc} = \frac{p_{fc}}{p_{H_2}} \tag{12}$$

where p_{fc} is the electrical power produced and p_{H_2} is the theoretical power associated with the hydrogen consumed, which is defined as

$$p_{H_2} = \frac{p_{fc} + p_{com}}{\eta_{therm} \cdot \eta_{util} \cdot \eta_{fci}} \tag{13}$$

where p_{com} is the power that the compressor demands, η_{therm} is thermodynamic efficiency (0.98 at 298 K), η_{util} is the efficiency of cell use, defined as a relationship between the mass of fuel that reacted and the mass that entered in the fuel cell; and η_{fci} is the efficiency of each cell, calculated as the relationship between the cell voltage v_{fc}, and the open-circuit voltage E_{oc}. This relationship can also be expressed as a function of cell voltage and current

$$n_{fc} = \frac{v}{1.482} \frac{i}{(i + i_{loss})}. \tag{14}$$

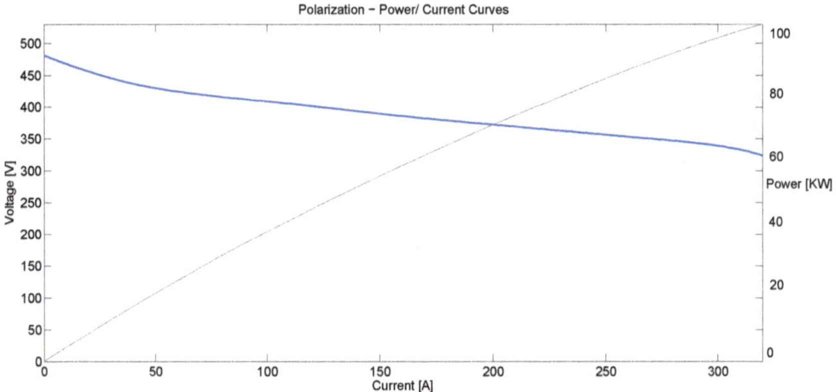

Figure 2. Polarization and power-current curve of fuel cell.

The losses of i, called i_{loss} are usually small. Greater efficiency can be achieved with the same fuel cell, with significantly lower power density level. This means that for a required power, a fuel cell can be expanded (with a larger active area) and be more efficient [38].

An electric model characterizing the fuel cell can be obtained using voltage and current equations

$$u_{fc} = E_{oc} - u_{act} - u_{ohmic} \tag{15}$$

where u_{fc} is the system voltage output, u_{ohmic} is the voltage of ohmic losses, and u_{act} is the activation voltage drop. E_{oc}, the open-circuit voltage, is defined by

$$E_{oc} = K_c[E_o + (T_{fc} - T_{ref})\frac{-\alpha_{Tref}}{zF} + \frac{RT}{zF}ln(P_{H_2}P_{O_2}^{1/2})] \tag{16}$$

where α_{Tref} is a temperature constant, E_o is the electromotive force under standard pressure conditions, T_{ref} is the temperature of reference, K_c is the rated voltage constant, T_{fc} is the operating temperature, z is the electron transfer number, which can be obtained as shown in [39], P_{H_2}, P_{O_2} are the gas pressure, F is the Faraday constant and R is the gas constant. The activation drop, u_{act}, is given by:

$$u_{act} = \frac{1}{\tau s + 1} N A_{nom} ln(\frac{i_{fc}}{i_o}) \tag{17}$$

where τ is the voltage time constant, and N is the number of cells. The ohmic voltage drop, u_{ohmic} is expressed by:

$$u_{ohmic} = r_{internal} i_{fc} \tag{18}$$

where i_{fc} is the cell output current and $r_{internal}$ is the inner resistance of fuel cell system. The parameters of the fuel cell stack are shown in Table 4. Finally, the hydrogen consumption is defined by:

$$m_{H_2} = \frac{N M_{H_2} i \lambda}{nF} \tag{19}$$

where m_{H_2} is the mass of hydrogen consumed, M_{H_2} is the molar mass of hydrogen, λ is the ratio of excess hydrogen and n is the number of electrons acting on the reaction.

Table 4. HD 100 FCvelocity Ballard fuel cell parameters.

Parameter	Data
Maximum voltage	580 V
Maximum current	288 A
Number of cells	560
Operating temperature	330 °K
Nominal air pressure	2.24 bar
Maximum power	100 kW
Mass	285 kg
Temperature of reference	298 °K
Temperature constant	44.43
Cost	100 k€

3. Driving Profiles

A driving cycle consists of a speed profile which defines the route that must follow the vehicle. Some types of vehicles track specific cycles, such as urban transport, which follow and predefined urban routes. Different driving cycles have been created that represent the driving conditions of vehicles with greater accuracy [40]. For example, the *ECE*15, which is the European cycle, whose main problem is the smooth accelerations; the USFTP 72 cycle, which represents the conditions of circulation in the Angeles; the USFTP 75, used for emissions certification in the USA.

However, in this paper we will present two specific driving cycles, the transport Driving Cycle in Buenos Aires (BADC), and the Manhattan Driving Cycle (Manhattan DC), because they are driving cycles designed for city buses, such as those indicated in Table 1, in which the driving conditions of these buses are considered. They have several stops and decelerations, which allows recovery of a significant amount of energy.

3.1. *Buenos Aires City Driving Cycle*

For the construction of the Buenos Aires Driving Cycle (BADC), 30 h of GPS data have been acquired, which are related to 51 bus trips covering a total of 313.6 km. The BADC was validated on a reference diesel bus widely used in Buenos Aires, and comparing the results obtained from fuel consumption to those reported by the bus line operator. The speed profile is shown in Figure 3, and its main characteristics are presented in Table 5.

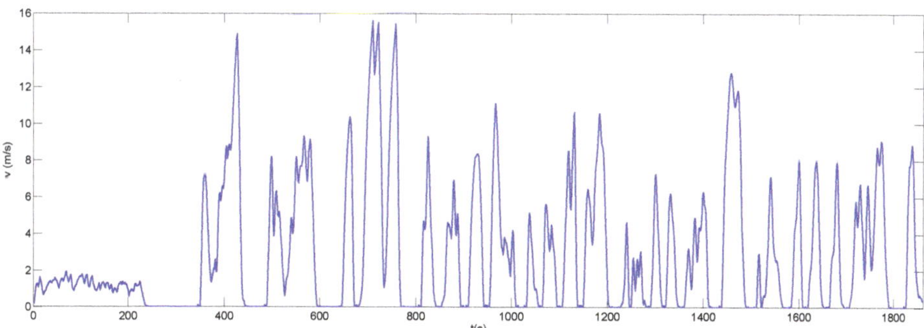

Figure 3. BADC driving cycle.

Table 5. BADC driving cycle parameters.

Parameter	Value
Total cycle time	1864 s
Average Speed	3.92 m/s
Maximum speed	15.6 m/s
Maximum acceleration	9.2155×10^{-5} m/s^2
e_v^+	22,678.62 kJ
e_v^-	11,870.63 kJ

Using Equation (1), we can obtain the instantaneous power needed to follow this profile. The equation allows us to obtain the power values, p_v^+, which are the instantaneous values that need to be delivered to produce the movement. The sum of these power values, for the complete profile, becomes the energy needed to produce movement, e_v^+. We can also obtain the power values that we can recover by means of regenerative braking, p_v^-. Analogously, the total sum of these power values, for the complete profile, will be the energy recovered by braking e_v^-. In the same way, the equation allows us to obtain the maximum instantaneous power that must be contributed p_{maxv}^+, and the maximum instantaneous power that can be recovered from braking p_{minv}^-, which is useful for dimensionalizing the storage systems.

Using the ratio indicated in Equation (19), we can obtain the maximum amount of energy that can be recovered when there are no losses. For the BADC, this amount is 52.34%.

$$\%recovery = \frac{e_v^-}{e_v^+}. \tag{20}$$

3.2. Manhattan Driving Cycle

This driving cycle used for bus testing in New York has a profile travel distance of 3.30 km, with a maximum acceleration of 2.04 $\frac{m}{s^2}$ and a driving time of 1089 s. Figure 4 shows the Manhattan profile velocity cycle. Table 6 shows the most relevant parameters of the profile.

In the same way as for the BADC profile, and using ratio Equation (20), the maximum amount of energy that can be recovered from braking for this profile is 58.84%.

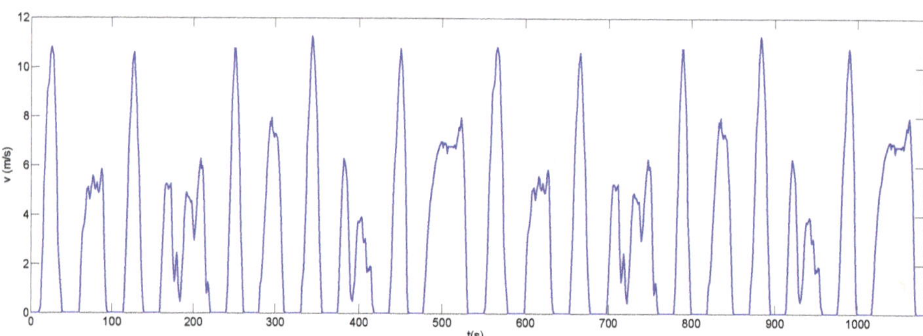

Figure 4. Manhattan Driving Cycle.

Table 6. Manhattan Driving Cycle parameters.

Parameter	Value
Total cycle time	1089 s
Average Speed	3.033 m/s
Maximum speed	11.24 m/s
Maximum acceleration	2.044 m/s^2
e_v^+	13,747.04 kJ
e_v^-	8090.08 kJ

4. Dynamic Programming

Dynamic programming is a very powerful numerical tool for solving optimal control problems, as indicated in [41,42]. One of the advantages over other methods is that the solution of the optimal control can be found in the complete time horizon. However, in some cases the computational effort grows exponentially with the number of state variables and inputs of the dynamic system. When the problem includes state constraints, any control input trajectory is limited to keep the system operating in the space delimited by them.

The optimal problem for the energy management in vehicles is posed in a constant time interval, with fixed initial conditions and a speed profile known to priori. The proposed optimal control problem can be generically formulated considering the cost function

$$J = h_N(x(N)) + \sum_{k=0}^{N-1} h_k(x(k), u(k)) \tag{21}$$

where the first term $h_N(x(N))$, refers to the final cost. The second term $h_k(x(k), u(k))$, refers to the cost of reaching a proposed state $x(k)$, applying a control signal $u(k)$, in an instant k, considering that system dynamics is represented in discrete-time state space as

$$x(k+1) = f_k(x(k), u(k)). \tag{22}$$

Please note that second term of Equations (21) and (22) depend on k, therefore their value varies with each iteration.

In case of the vehicle energy system, the states, $x(k)$, are the battery SOC, the supercapacitor SOE, and the fuel cell energy e_{fc}, while $u(k)$ are the power of the elements p_{sup}, p_{bat}, p_{fc}, and p_{break}.

Then, the discrete-time model of the system is defined by

$$x_1(k+1) = SOC_{bat}(k) + \frac{p_{bat}(k)}{e_{bat}} \tag{23}$$

$$x_2(k+1) = SoE_{sup}(k) + \frac{p_{sup}(k)}{e_{sc}} \tag{24}$$

$$x_3(k+1) = e_{fc}(k) + p_{fc}(k). \tag{25}$$

Excessive computational cost can be avoided by expressing the final system as follows. To make this reduction, the power ratio shown in Equation (2) is taken into account.

$$x_1(k+1) = SOC_{bat}(k) + \frac{p_v(k) - p_{fc}(k) - p_{sup}(k) - p_{break}(k)}{e_{bat}}. \tag{26}$$

$$x_2(k+1) = SoE_{sup}(k) + \frac{p_{sup}(k)}{e_{sc}}. \tag{27}$$

Constraints will be imposed on the battery SOC and supercapacitor SOE, as state restrictions, as follows

$$0.4 \leq SOC_{bat}(k) \leq 0.8 \tag{28}$$

$$0 \leq SoE_{sup}(k) \leq 1 \tag{29}$$

$$SOC_{bat,0} = SOC_{bat,N} \tag{30}$$

$$SoE_{sup,0} = SoE_{sup,N} \tag{31}$$

where $SOC_{bat,0}$, is the SOC of the battery at the initial instant and $SOC_{bat,N}$ is the SOC of the battery at the end of the driving cycle. In the same way $SoE_{sup,0}$ is the SOE of supercapacitor at the initial instant and $SoE_{sup,N}$ is the SOE of the supercapacitor at the end of the driving cycle.

The safety threshold [0.4, 0.8] applied to the battery SOC, which theoretically could vary in the ranges [0, 1] as the supercapacitor, is included to extend its useful life, avoiding deep discharges. Constraints on control signals, $u(k)$, are also included as follows

$$P_{lowerbat} \leq P_{bat}(k) \leq P_{maxbat} \tag{32}$$

$$P_{lowersup} \leq P_{sup}(k) \leq P_{maxsup} \tag{33}$$

$$P_{lowerfc} \leq P_{fc}(k) \leq P_{maxfc}. \tag{34}$$

The maximum and lower power and energy values will be taken from the tables indicated in the models of the elements.

4.1. Cost Function

When defining the particular expression of the cost function, Equation (21), for the energy management system, we will take into account the following considerations:

- The operational life of the elements.
- The amount of hydrogen consumed.

In the case of the operational life of elements, such as batteries, the parameters that are evaluated to characterize the main causes for degradation are: (a) temperature, (b) depth of discharge, and (c) rate of discharge [43–49]. Degradation can be avoided by limiting rapid power changes and preventing the instantaneous value from reaching the maximum value, which would result in deep discharges. In the case of the fuel cell, high current peaks and rapid variation in current should be avoided [38,50–52].

Finally, the cost function according to the control objectives is defined with the following terms.

1. To preserve the operational life of the elements (state of health of the elements) abrupt variations
   (a) in the power delivered by the fuel cell $p_{fc}(k) - p_{fc}(k-1)$ and batteries $p_{bat}(k) - p_{bat}(k-1)$ and
   (b) in the SOC of the battery $SOC_{bat}(k) - SOC_{bat}(k-1)$, [53], should be avoided [54].
2. The amount of hydrogen consumed by the fuel cell, expressed as a function of the power delivered, $p_{fc}(k)$, which determines the economic cost should be minimized.

Thus, the cost function is finally defined as

$$\begin{aligned} J = w_{u1}[p_{bat}(k) - p_{bat}(k-1)]^2 + w_{u2}[p_{fc}(k) - p_{fc}(k-1)]^2 + w_{SOH}p_{bat}(k)^2 + \\ w_{soc}[SOC_{bat}(k) - SOC_{bat}(k-1)]^2 + w_{\alpha}p_{fc}(k)^2 \end{aligned} \tag{35}$$

where the weights w_{α}, w_{u1}, w_{u2}, w_{SOH}, w_{soc}, have been determined based on of a sweep of these coefficients as explained in the following section.

4.1.1. Coefficient Sweep for BADC

Once the structure of the cost function was defined, we should proceed to make a sweeping of the weights to determine the Pareto front that allows choosing those that allow reduction of the power delivered by the fuel cell in order to reduce hydrogen consumption, as proposed in [55]. In addition, the one that allows a smoother variation of the SOC of the battery to preserve its useful life should be selected. To adjust the cost function coefficients based on sweeping of the weights, an initial sizing of the system is required to solve the control problem proposed in the previous section. This sizing is done with a storage element size shown in Table 7. The reason for choosing this initial size of the storage system is that the literature recommends that the size of the storage system be about 30% of the size of the main source. The fuel cell used is the one detailed in Table 4.

Table 7. Initial sizing for the calculation of pareto coefficients.

Component	Mass	Power	Energy
Battery	8 kg	10.4 kW	368 Wh
Supercapacitor	12 kg	20.4 kW	27.6 Wh

It is considered that the sum of the weights $w_{\alpha}, w_{u1}, w_{u2}, w_{SOH}, w_{soc}$, will always satisfy

$$w_{\alpha} + w_{u1} + w_{u2} + w_{SOH} + w_{soc} = 1 \tag{36}$$

The coefficients w_{u2} and w_{α}, affect the behavior of the fuel cell and the w_{u1}, w_{SOH}, w_{soc} coefficients affect the behavior of the battery. Then, it starts with a value of the coefficients $w_{u2} = 0$ and $w_{\alpha} = 0$, while the coefficients of the terms referring to the battery are maximum with a value of $w_{u1} = 0.33$, $w_{SOH} = 0.33$, and $w_{soc} = 0.33$. In this first case, the condition of Equation (36) is fulfilled. In a second iteration, the coefficients related to the fuel cell take the values of $w_{u2} = 0.05$ and $w_{\alpha} = 0.05$, with an increase of 0.05 with respect to the first iteration. The coefficients related to the battery take the value of $w_{u1} = 0.3$, $w_{SOH} = 0.3$, and $w_{soc} = 0.3$. All the coefficients related to the battery have the same value that is calculated by:

$$w_{u1} = w_{SOH} = w_{soc} = \frac{1}{3} - \frac{(w_{\alpha} + w_{u2})}{3} \tag{37}$$

In this case, the second iteration also complies with Equation (35). In the last iteration with the increase of 0.05 to the fuel cell related coefficients in each new iteration, the coefficients have the values of $w_{u2} = 0.4$, $w_{\alpha=0.4}$, $w_{u1} = 0.067$, $w_{SOH} = 0.067$, and $w_{soc} = 0.067$. All combinations (iterations) of the coefficients can be seen in Table 8. For each of these combinations of coefficients, there is an amount of energy contributed by each element of the propulsion system. The power generated by the fuel cell, battery and supercapacitor must be equal to the power needed for motion fulfilling Equation (2). For a better understanding, the energy of each element will be expressed as a percentage. In the case of the battery and supercapacitor, this percentage will be the amount of energy they give to the system with respect to the maximum possible that they can recover in braking, calculated in Equation (20) for the BADC profile. As mentioned, the SOC and energy of these elements are equal at the beginning and at the end, being the energy recovered from the braking, equal to the energy delivered. In the case of the fuel cell, the percentage of energy saved by hybridization is shown compared to a pure fuel cell system without storage elements. These results, for each iteration of weights can also be observed in Table 8. In the last configuration of coefficients shown in Table 8, it is observed that there is the lowest fuel consumption with a fuel cell energy consumption reduction of 27.22%. In the same configuration, the power delivered by the battery is the highest in the table with 21.71%. Being one of the control objectives that the variation of the SOC is not abrupt, it is necessary to choose a configuration of weights in which the variation of the SOC is not the highest. For this reason, the configuration of coefficients chosen will be (a) $w_{\alpha} = 0.3$, (b) $w_{u1} = 0.13$, (c) $w_{u2} = 0.3$, (d) $w_{SOH} = 0.13$, (e) $w_{soc} = 0.13$, where the power delivered by the battery is 19.96%, being the same lower than 21.71% which is the maximum value. In this configuration, the energy savings delivered by the fuel cell is 26.22%. As it can be seen, the fuel saving is still significant in this configuration, being only 1% less than the maximum saving case. This configuration of coefficients achieves a better balance of the proposed control objectives. Please note that the criterion for choosing the coefficients is based on the fact that the energy delivered by the battery is not the maximum and that the variation between the maximum consumption in the use of the fuel cell and the chosen value should be similar to 1%.

Table 8. Variation in the weight of the cost function for BADC.

Weights					Energy		
w_{u2}	w_{u1}	w_{SOH}	w_{soc}	w_{α}	Battery (%)	Supercapacitor (%)	Fuel cell (%)
0	0.33	0.33	0.33	0	13.24	23.84	19.41
0.05	0.3	0.3	0.3	0.05	16.79	27.74	23.31
0.1	0.267	0.267	0.267	0.1	18.00	29.35	24.78
0.15	0.23	0.23	0.23	0.15	18.76	29.48	25.25
0.2	0.2	0.2	0.2	0.2	18.99	29.57	25.32
0.25	0.167	0.167	0.167	0.25	19.68	29.80	25.73
0.3	0.13	0.13	0.13	0.3	19.96	30.15	26.22
0.35	0.1	0.1	0.1	0.35	20.96	30.19	26.72
0.4	0.067	0.067	0.067	0.4	21.71	30.84	27.22

Figures 5 and 6 illustrate the increase in fuel savings when the amount of power recovered by the storage elements increases for each combination of cost function coefficients. Figure 5 presents this relationship for the fuel cell and battery, while Figure 6 shows this relationship for the fuel cell and supercapacitors. Figure 5 shows in the lower left-hand corner, the case in which the coefficients take the values of $w_{\alpha} = 0$, $w_{u1} = 0.33$, $w_{u2} = 0$, $w_{SOH} = 0.33$, and $w_{soc} = 0.33$. For this case, the energy reduction delivered by the fuel cell is 19.41%, while the energy recovered by the battery is 13.24% and for the supercapacitor is 23.84%. This is the case where the fuel cell delivers the most energy to the system. In the upper right corner, there is the case where the coefficients take a value of $w_{\alpha} = 0.4$, $w_{u1} = 0.067$, $w_{u2} = 0.4$, $w_{SOH} = 0.067$, and $w_{soc} = 0.067$. For this case, the energy reduction delivered by the fuel cell is 27.22%, and the energy recovered by the battery is 21.71% and for the supercapacitor is 30.84%. This is the case where the fuel cell delivers the smallest amount of energy. The intermediate cases are taken from Table 8. In Figure 6, the same cases as for Figure 5 are shown, with the difference that the power delivered by the supercapacitor and not that of the battery is shown.

When the fuel cell delivers less energy to the movement, and the battery also delivers less energy (of the regenerative brake's recovered power), due to the weights of the cost function, the supercapacitor delivers an increasing amount of energy when other sources are restricted. The indicated behavior between the supercapacitor and the fuel cell is shown in Figure 6.

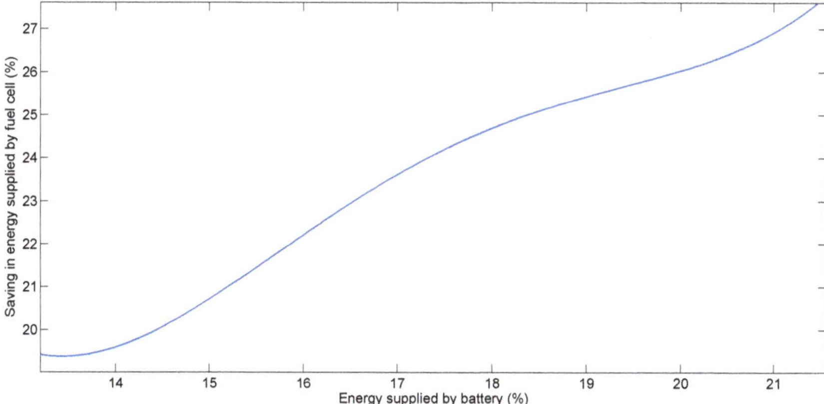

Figure 5. Saving in energy supplied by fuel cell and energy supplied by batteries for the different combination of coefficients of the cost function for BADC.

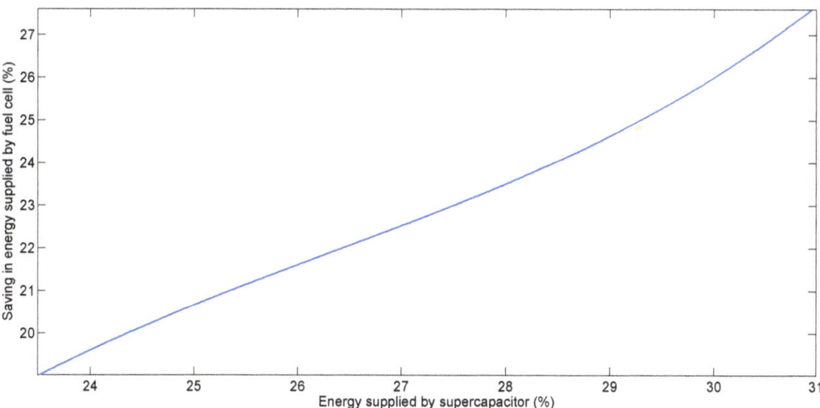

Figure 6. Saving in energy supplied by fuel cell and energy supplied by supercapacitor for the different combination of coefficients of the cost function for BADC.

4.1.2. Coefficient Sweep for Manhattan Driving Cycle

For the Manhattan Driving Cycle, the procedure is similar to BADC. The design of the propulsion system is the same as for the previous case. Once the cost function is known, we vary the weight tuning, to find the best combination between the use of its elements, focusing on hydrogen saving, and preserving the useful life of the elements.

The variation in the weights of the cost function is similar to that of the BADC profile, and a summary of the region of interest is shown in Table 9.

Table 9. Variation in the weight of the cost function for Manhattan DC.

Weights					Energy		
w_{u2}	w_{u1}	w_{SOH}	w_{soc}	w_{α}	Battery (%)	Supercapacitor (%)	Fuel cell (%)
0	0.33	0.33	0.33	0	11.02	25.66	19.57
0.05	0.3	0.3	0.3	0.05	13.20	25.89	20.33
0.1	0.267	0.267	0.267	0.1	14.52	26.36	21.16
0.15	0.23	0.23	0.23	0.15	15.51	27.89	22.73
0.2	0.2	0.2	0.2	0.2	15.84	28.08	23.56
0.25	0.167	0.167	0.167	0.25	16.43	28.47	23.76
0.3	0.13	0.13	0.13	0.3	17.21	28.83	24.38
0.35	0.1	0.1	0.1	0.35	18.15	29.23	24.58
0.4	0.067	0.067	0.067	0.4	21.29	30.72	25.19

Figure 7 shows the behavior of the battery with respect to the fuel cell saving. In this profile, with the combination of coefficients $w_{\alpha} = 0$, $w_{u1} = 0.33$, $w_{u2} = 0$, $w_{SOH} = 0.33$, and $w_{soc} = 0.33$, the energy delivered by the battery is 11.02%, the energy delivered by the supercapacitor is 25.66%, while the reduction in fuel cell use is 19.57%. This is the lower left-hand corner of Figure 7, which corresponds to the case where the fuel cell delivers the most energy to the movement. With the combination of coefficients $w_{\alpha} = 0.4$, $w_{u1} = 0.067$, $w_{u2} = 0.4$, $w_{SOH} = 0.067$, and $w_{soc} = 0.067$, the energy delivered by the battery is 21.29%, the energy delivered by the supercapacitor is 30.72%, while the reduction in fuel cell use is 25.19%, which corresponds to the upper right-hand corner of Figure 7 and the case where the fuel cell delivers the least amount of energy to movement.

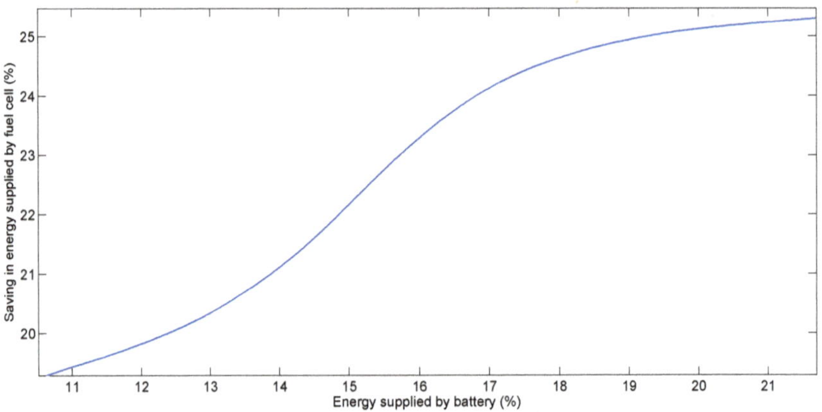

Figure 7. Saving in energy supplied by fuel cell and energy supplied batteries for the different combination of coefficients of the cost function for Manhattan Driving Cycle.

Figure 8 shows the behavior of the fuel cell saving and supercapacitors for the same cases in Figure 7, summarized in Table 9.

As for the BADC profile, with the latest configuration of coefficients from Table 9, the energy delivered by the battery is 21.19%, while the reduction in fuel cell usage is 25.19%. In this case, the fuel economy is maximum and the variation of the battery SOC is also highest. For this reason, to have a smaller variation in the SOC of the battery, we use the configuration of coefficients (a) $w_{\alpha} = 0.2$, (b) $w_{u1} = 0.2$, (c) $w_{u2} = 0.2$, (d) $w_{SOH} = 0.2$, (e) $W_{soc} = 0.2$, where the energy delivered by the battery is 15.84% and the reduction in the use of the fuel cell is 23.56%. In this way, we have a smaller variation of the SOC, and the fuel saving is about 1% of the maximum possible.

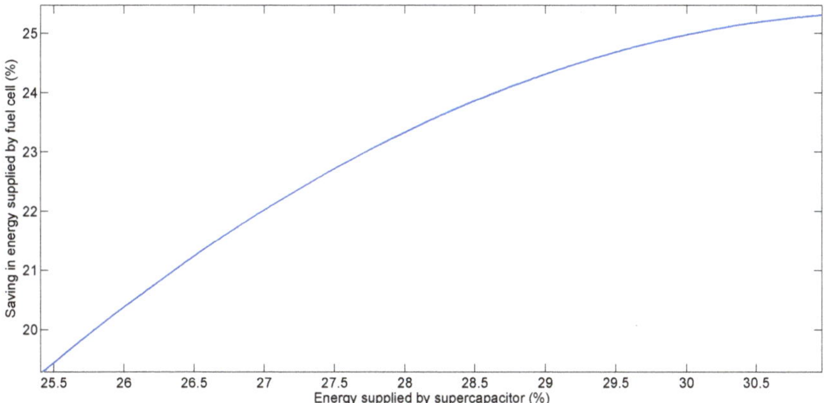

Figure 8. Saving in energy supplied by fuel cell and energy supplied batteries for the different combination of coefficients of the cost function for Manhattan Driving Cycle.

5. Results

Considering BADC and Manhattan profiles as case studies, the power profile for each driving cycle will be simulated. The combination of power generated by the fuel cell, and the energy recovered by the ESS, should be sufficient to reach the power profile, derived from the velocity profile. Additionally, to the parameters shown in Tables 2–4, we will use for each profile the coefficients resulting from the sweeping of parameters of the cost function corresponding to each one. The control problem is the same as described in Section 4. When the mass of ESS is equal to zero, the power will be generated with the fuel cell only, and that will be a first case of analysis. In this first case, we must properly dimensionalize the fuel cell to reach the required velocity at all times. In a second case, we will use the ESS, as mentioned above. To properly dimensionalizing of the system, in Tables 3 and 4, power and energy are expressed as a function of mass. Then, we will increase their mass to analyze the optimal configuration and price.

5.1. Fuel Cell Operation Only

For the proposed analysis, no batteries or supercapacitors is assumed. The fuel cell provides all the power needed to reach both profiles. The cost of power production for high volume cells is 1 €/W.

5.1.1. Buenos Aires Driving Cycle

For this profile, we will use a fuel cell system with a power of 200 kW, with a cost of 200 k€. In this case study, the fuel cell must be able to fulfill the highest power peaks. This is the reason for using a system of 200 kW of total power. Figure 9 shows the power profile derived from the BADC speed cycle and the power delivered by the fuel cell.

In red, we can see the power required to fulfill the speed profile. In blue, the instantaneous power delivered by the cell. The graphs in blue and red are the same, because they have the same power values at each instant. As it can be observed, with this fuel cell sizing, we can also fulfill the profile. Then, the total cost of the propulsion system to achieve the BADC profile with fuel cell operation will be 2000 k€.

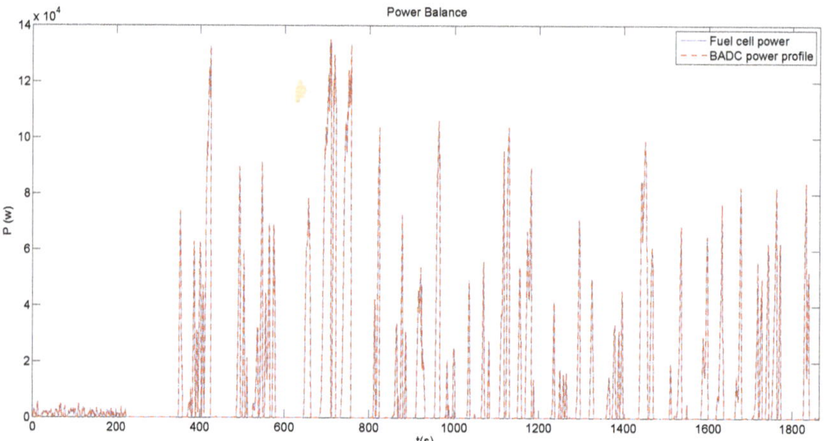

Figure 9. Power profile BADC with only fuel cell operation.

5.1.2. Manhattan Driving Cycle

For this profile, we will use the same configuration of fuel cell than in BADC profile. The maximum power of the system will be 200 kW, whose cost is 200 k€. Figure 10 shows in red, the power derived from the Manhattan velocity profile, and in blue, the power delivered by the fuel cell system. The values of instantaneous power as for the previous case are the same, so the blue and red graphs are the same.

As in the previous case, with this dimensionalizing, we were able to reach the required speed. The cost of the propulsion system is 2000 k€.

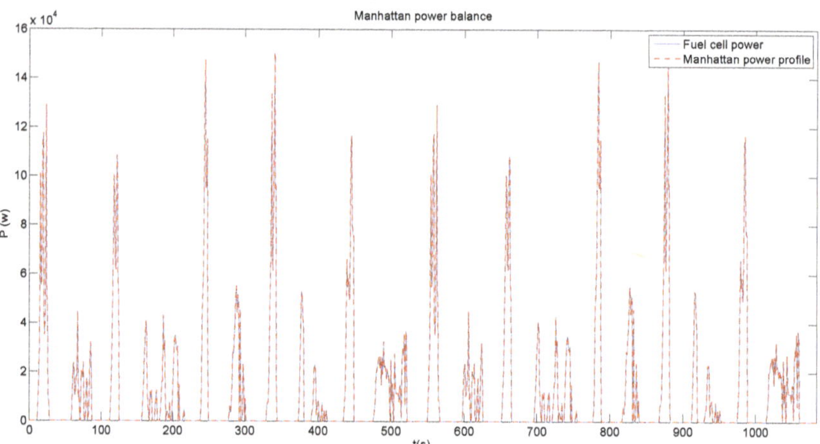

Figure 10. Power profile Manhattan with only fuel cell operation.

5.2. Hybrid Operation

When the mass of ESS increases, we can increase their capacity in power and energy and provide a significant reduction in fuel cell use. This will allow the fuel cell not to give the maximum power peaks of each profile, but to be able to give the average power of each one. The analysis of this variation, expressed in percentages of energy supplied by the storage elements and in the reduction of fuel cell use, will be presented below. Furthermore, the power profile of battery, supercapacitor, and fuel cell will also be plotted for a particular ESS mass value. Finally, in each profile, the monetary cost involved in increasing the power of the ESS will be reported in a graph.

BADC Driving Profile

During the sizing process, the total mass of the storage elements should be constant

$$m_{bat} + m_{sup} = m_{ess} \tag{38}$$

where m_{ess} is the total mass and is constant and m_{bat} and m_{sup} are the ones that are going to vary. As indicated, supercapacitors allow recovery of a greater amount of power from braking, but they are more expensive than the battery. If the storage system is composed only of supercapacitors, the power of the fuel cell used in the system decreases, but the momentary cost of the storage system increases. Then, the objective is to find the mass of batteries and supercapacitors to reduce the cost of the storage system, but without forgetting the objectives of fuel economy control and SOC variation. For this reason, the case where the storage system has the lowest cost will not be optimal. This optimal case will depend on the compendium of the cost of the storage system and the other control objectives. A system with only supercapacitors ($m_{bat} = 0$) is initially dimensionalized and mass is added to the batteries in each iteration. This is done to decrease the cost associated with the storage system in each iteration and to know how the fuel saving varies. Then, the initial configuration will be $m_{bat} = 0$ and $m_{sup} = m_{ess}$. In order to fulfill with the power profile using the fuel cell described in Table 4, the minimum mass of supercapacitors should be 30 kg. Otherwise, if it is lower, the power profile is not fulfilled.

In the total mass, the mass of each element varies with respect to the other as follows. For example, in case 1: (a) When the mass of the supercapacitor is 30 kg, the battery mass should be 0 kg; (b) when the mass of the supercapacitor is 29 kg, the battery mass should be 1 kg. For each mass variation in batteries or supercapacitors, there is a new cost involved, and a new power and energy capacity. For example, for the same examples, in case 1, the cost of the battery is 0 €, while that of the supercapacitor is 2650 €. For case 2, the cost of the battery is 33.87 €, while the cost of the supercapacitor is 2561.67 €. As we can see, the total mass remains constant, but the economic value varies for each case. The final case will be when we have 28 kg of battery and 2 kg of supercapacitors, with a cost of 948.39 €, and 176.67 €, respectively. The configuration of 29 kg of batteries and 1 kg of supercapacitors is not considered, because with this configuration the power profile derived from the speed profile is not fulfilled. The weight, power and cost of the fuel cell remains constant for each configuration of batteries and supercapacitors in this scenario. The weight of the battery, supercapacitor and fuel cell, is added to the total mass for calculating the power profile, shown in Equation (1), to achieve a more realistic scenario. The cost of fuel cell FCveloCity-HD is 100 k€.

Contrarily to the case without hybridization, if the mass of the storage elements is different from 0, with a certain minimum value, we can reduce the size of the fuel cell. For the first case, where the mass of the supercapacitor is 30 kg ($m_{sup} = 30$) and the mass of the battery is 0 kg ($m_{bat} = 0$) the reduction in fuel cell usage is the highest with 46.98%. The cost of the storage system for this same case is also the highest with a value of 2650 €. For the last possible case, in which the mass of the supercapacitors is 2 kg ($m_{sup} = 2$) and the mass of the battery is 28 kg ($m_{bat} = 28$) the reduction in the consumption of the fuel cell is 30.4% and the storage system has the lowest cost, with 1125.05 €. Although 1125.05 €, is the cheapest cost of the storage system, the reduction in fuel cell usage is only 30.4%, while the battery delivers 55.98% of energy, being the same the highest of all configurations. This causes the variation of the SOC to be increased.

Figure 11 shows graphically the reduction in fuel cell consumption as a percentage of energy, compared to the percentage of energy recovered by the battery for each configuration. Even though the percentage of energy recovered by the battery increases, the reduction in fuel cell consumption decreases because the mass of the supercapacitors decreases. This shows that although the mass of the battery increases, the system does not absorb large peaks of power, so the fuel cell must provide more power. Figure 12 shows the same behavior of the fuel cell with the supercapacitor. Since supercapacitors have a high power density, they allow the system to recover the highest power peaks of the profile and the reduction in fuel cell consumption increases.

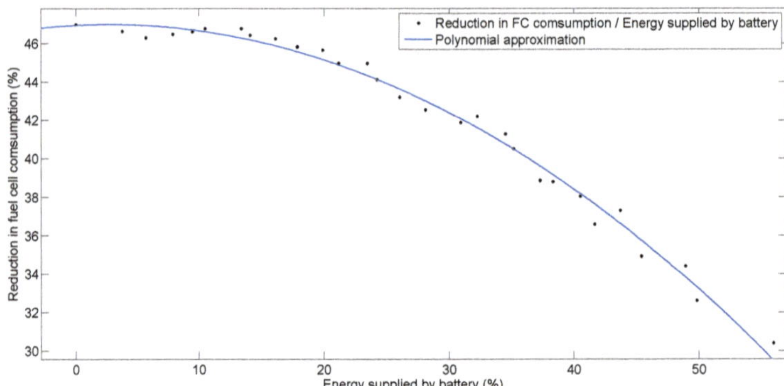

Figure 11. Reduction in fuel cell consumption versus energy supplied by battery for BADC profile.

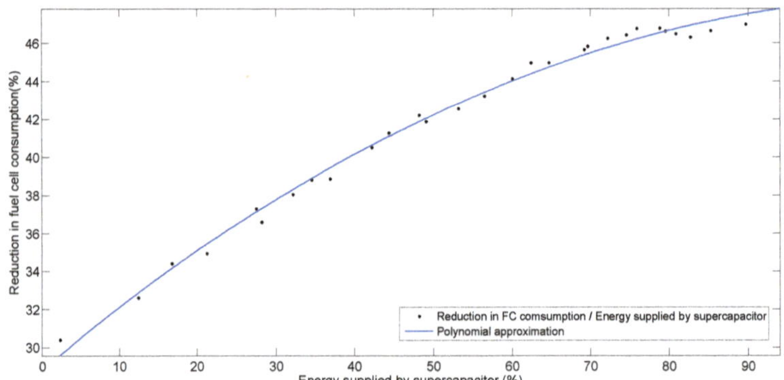

Figure 12. Reduction in fuel cell consumption versus energy supplied by supercapacitor for BADC profile.

Figure 13 shows the variation of hydrogen consumption Equation (19) in relation to the cost of the storage system. The BADC profile has 1864 seconds of operation (31.06 min). A bus normally rolls 15 h per day. In one day, it would roll 29 times the profile, in one month it would roll 870 times and in a year 10.585 times.

In the Y axis of the Figure 13, the variation of the cost of hydrogen is indicated for a year of operation of the bus, and in the X axis the cost of the storage system is indicated. From the figure, it can be observed that with the lowest cost of the storage system (1125.05 €), a greater amount of hydrogen is consumed. This corresponds to the point of 28 kg of batteries and 2 kg of supercapacitors. Increasing the cost of the storage system reduces the consumption of hydrogen. In the maximum point the cost is 2650 € with 30 Kg of supercapacitors and 0 Kg of batteries.

However, it can be observed that from 2200 €, with the increase in the mass of the storage system, the decrease in hydrogen consumption is almost linear. This point corresponds to 9 Kg of batteries and 21 Kg of supercapacitors. According to this analysis, this will be the optimum point. In this configuration, the fuel cell consumption reduction is 45.82% (average reduction in fuel cell consumption for BADC mass variation), and 87.54% of the energy from the regenerative brake is recovered.

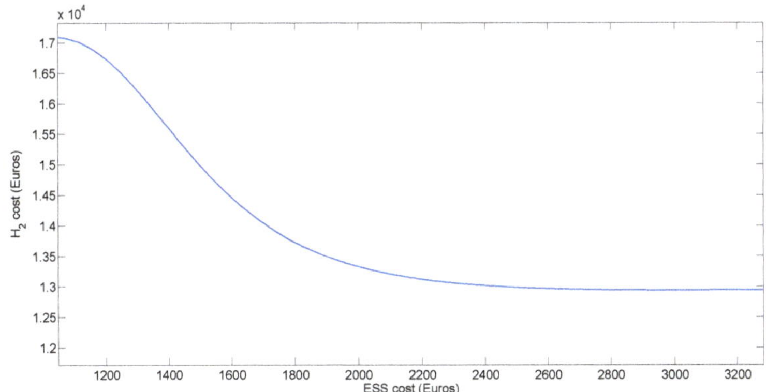

Figure 13. Cost of the storage and power delivered by fuel cell system in BADC profile.

Therefore, for case 1 with a storage element cost of 2650 €, the energy delivered by the fuel cell is 53.02%. For the case 2 with a storage element cost of 2595.54 €, the energy delivered by the fuel cell is 53.37%. While for the last case, with a storage system cost of 1125.05 €, the fuel cell delivers 69.55% of energy to the movement. It can be observed how the fuel cell delivers a greater amount of energy, given the price decrease of the total storage system. In this sense, when we decrease the size of the supercapacitor system, the power can be recovered from regenerative braking decreases, and therefore, the fuel cell must provide more power to achieve the profile.

Figure 14 shows the supplied power by each element, while Figure 15, shows the battery SOC and supercapacitor SOE variation. The SOC has a slower variation than the SOE, due to the penalty of the cost function.

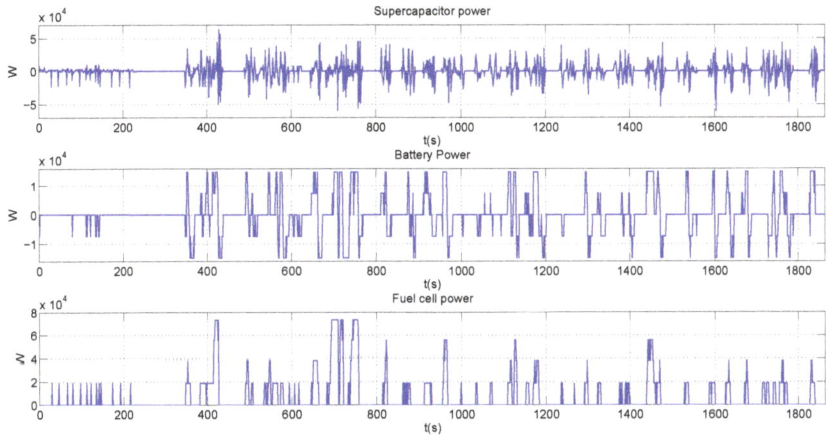

Figure 14. Power of the propulsion system for the BADC profile.

The sum of the battery, supercapacitor, and fuel cell powers in Figure 15 are equal to the power required to reach the BADC speed profile.

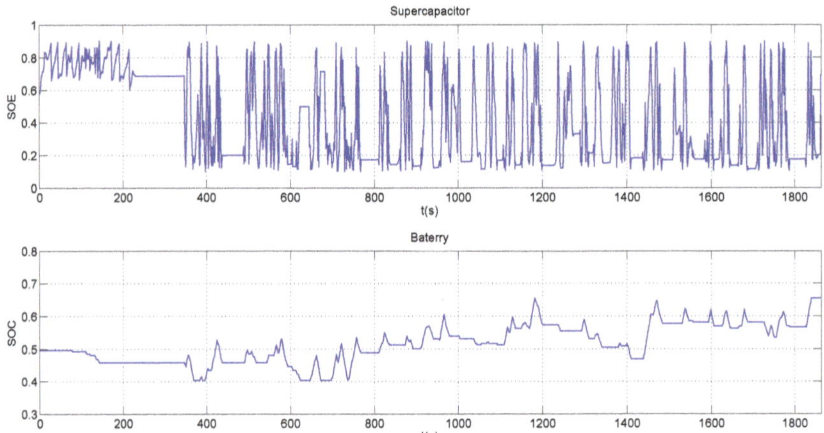

Figure 15. SOC and SOE for the BADC profile.

5.3. Manhattan Driving Profile

In the case of the Manhattan profile, the maximum power is higher than BADC. To satisfy Equation (38) and the initial condition of starting only with mass of supercapacitors ($m_{bat} = 0$ and $m_{sup} = m_{ess}$), the minimum mass of supercapacitors must be 32 kg. With this initial mass, the control problem is feasible. In the first case must start with a mass of supercapacitors of 32 kg, and 0 kg of batteries, with a total cost of 2826.67 €. The fuel cell is the same as for the BADC profile, in cost and maximum power. The second case, to keep the total mass of 32 kg constant, we use 31 kg of supercapacitors and 1 kg of batteries, with a total cost of 2772.20 €, and so on in the other cases. The last case is with 29 kg of batteries and 3 kg of supercapacitors, with a cost of 1247.26 €. The configuration of 30 kg batteries and 2 kg supercapacitors is not possible, because the control problem is not feasible.

Figure 16 shows that the increase in the mass of the batteries and the decrease in the mass of the supercapacitors produces a decrease in the value of the reduction in the consumption of the fuel cell. Therefore, even though the energy recovered from battery for the braking increases, the fuel consumption increases.

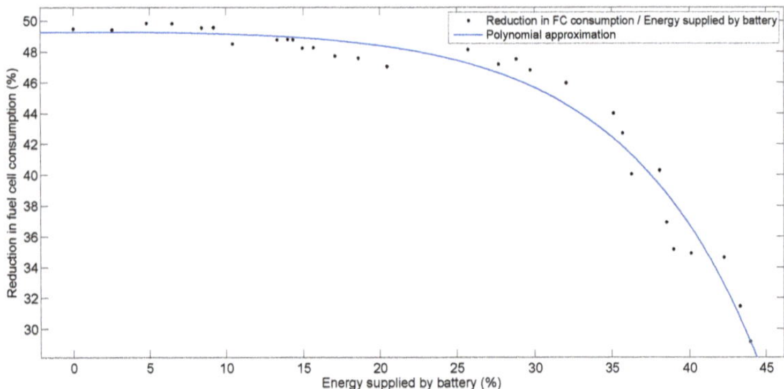

Figure 16. Reduction in fuel cell consumption versus energy supplied by battery for Manhattan profile.

Figure 17 shows the energy contributed to the movement by the supercapacitors and the reduction in fuel cell consumption. With a higher mass of supercapacitors, fuel consumption decreases, even if the battery mass is reduced.

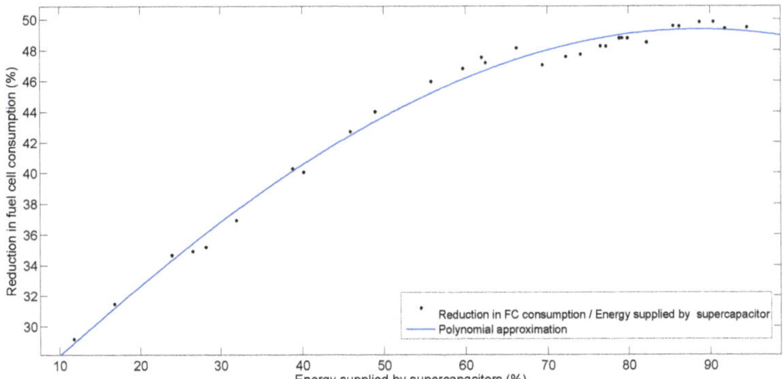

Figure 17. Reduction in fuel cell consumption versus energy supplied by supercapacitor for Manhattan profile.

Figure 18 shows in the X axis the cost of the storage system, while in the Y axis, the cost of the hydrogen consumption. In the same way as for the BADC, the duration of the Manhattan profile is 1089 s (18.15 min). In a year, a bus rolls 18,250 times the profile.

On the left side of the figure, can be observed that the cost of the storage system is lower, and hydrogen consumption is higher. The figure begins with a cost of 1247.26 € for the storage system and 18 K€ for hydrogen consumption. This is given with 29 Kg of batteries and 3 Kg of supercapacitors. Moving to the right of the figure, it is observed that as the cost of the storage system increases, the consumption of hydrogen decreases. With 32 kg of supercapacitors and 0 Kg of batteries, hydrogen consumption will be the lowest with almost 12.500 K€, while the cost of the storage system will be maximum with 2826.67 €. However, can be seen that from a value of 2200 €, the reduction in hydrogen consumption is almost linear. Then, the configuration of 12 Kg of batteries and 20 Kg of supercapacitors, is the optimal point of dimensionalizing. In this configuration, the fuel cell delivers 52.28% of the total energy of the movement.

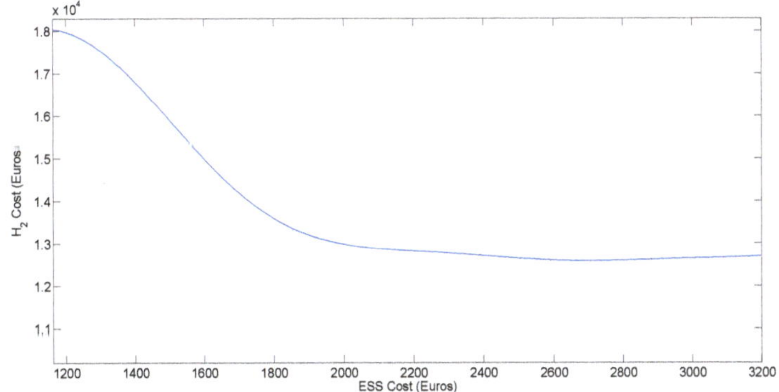

Figure 18. Cost of the storage system in Manhattan profile.

In the optimal point, can be recovered 91.17% of the braking energy. The reduction of the fuel cell consumption is 47.72% (average reduction in fuel cell consumption for Manhattan Driving Cycle mass variation) and the cost is 2.173.12 €. As for the BADC, the case of the minimum cost was not taken into account as the optimal case because the energy delivered by the battery is the maximum and increases the variation of the SOC.

Figure 19 shows the power of each propulsion system element for the desired configuration, in addition Figure 20 presents the SOC and SOE variation.

In the same way, as for the previous case, the SOC variation is softer than SOE, due to the penalization conditions imposed on the cost function for the battery. Also, the sum of the three instantaneous powers reach the power required for the Manhattan speed profile.

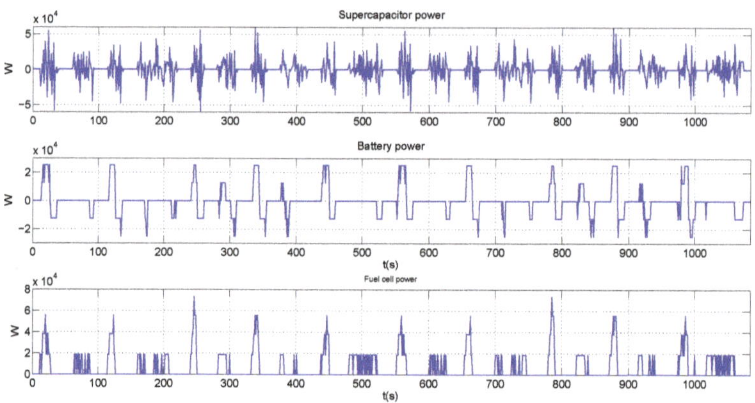

Figure 19. Power of the propulsion system for the Manhattan profile.

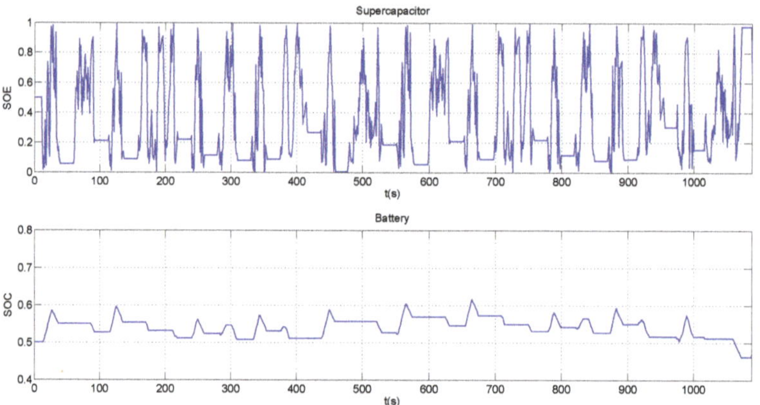

Figure 20. SOC and SOE for the Manhattan profile.

6. Conclusions

In the considered HEV, the propulsion system without the use of energy storage elements requires a fuel cell system capable of generating all the power required for the BADC profile and Manhattan in each case. This means that the cost is high, as it is 200 k€, in each case (only fuel cell system). The inclusion of energy storage elements such as batteries and supercapacitors allows us to reduce fuel cell usage and reduce fuel cell size. For both profiles, the fuel cell can be dimensionalized with 100 kW of power instead of 200 kW, with a cost of 100 k€. Then, the increase in mass on ESS allows reduction of the consumption of hydrogen from the fuel cell. With higher mass of supercapacitors, greater savings in fuel consumption and reduction in power generated by the fuel cell are achieved. In the same way, the cost of the storage system is higher. For this reason, it is concluded that for both profiles the pure use of supercapacitors allows greater savings of hydrogen, but makes the storage system more expensive. The inclusion of batteries allows this cost to be reduced, but at the same time increases the use of the fuel cell.

A larger battery size allows the SOC to vary less abruptly, which helps to improve the operational life of the batteries. Several mass configurations of the storage elements were simulated to find the best cost of the storage system compared with a maximum quantity (supercapacitor system only), and at the same time, reduce the hydrogen consumption. In this way, we can take advantage of the battery's energy density and supercapacitor power density characteristics. For the BADC speed profile, the optimal configuration was 9 Kg of batteries, and 21 Kg of supercapacitors. In this case, the energy recovered by the ESS is 87.54%, while the energy delivered by the fuel cell is 54.18%. The reduction of the cost, with the configuration (most expensive case), where only supercapacitors are presented as an ESS system, is 18.52%. For the Manhattan speed profile, the optimal configuration was 12 Kg of batteries, and 20 Kg of supercapacitors. In this case, the energy recovered by the ESS is 91.17 %, while the energy delivered by the fuel cell is 52.28%. The reduction of the cost, with the configuration (most expensive case), where only supercapacitors are presented as an ESS system, is 23.12%.

The behavior of the system with increasing mass of the storage elements is similar in both profiles. Its inclusion can be considered a significant contribution to hydrogen savings, and improves fuel cell efficiency.

As future work, additional objectives will be considered as the battery and fuel life degradation following the ideas proposed in [56] as well as the effect of the uncertainty to include robustness in the sizing.

Author Contributions: All the authors have equally contributed in the research presented in this manuscript as well as in its preparation.

Funding: This work has been partially funded by the Spanish national projects MICAPEM (ref. DPI2015-69286-C3-2-R, MINECO/FEDER) and SCAV (ref. DPI2017-88403-R MINECO/FEDER). This work is supported by the Spanish State Research Agencythrough the María de Maeztu Seal of Excellence to IRI (MDM-2016-0656). This work is partially funded by AGAUR of Generalitat de Catalunya through the Advanced Control Systems (SAC) group grant (2017 SGR 482).

Conflicts of Interest: The authors declare no conflict of interest.

Abbreviations

The following abbreviations are used in this manuscript:

DP	Dynamic Programing
ESS	Energy storage system
SC	Supercapacitor
FC	Fuel cell
P_{bat}	Battery power
P_{sup}	Supercapacitor power
P_{fc}	Fuel cell power
P_{break}	Break power
SOC	Battery state of charge
SOH	Battery state of health
SOE	Supercapacitor state of energy
EV	Electric vehicle
HEV	Hybrid electric vehicle
BADC	Buenos Aires Driving Cycle

References

1. Jia, S.; Peng, H.; Liu, S.; Zhang, X. Review of Transportation and Energy Consumption Related Research. *J. Transp. Syst. Eng. Inf. Technol.* **2009**, *9*, 6–16. [CrossRef]
2. Mahlia, T.; Saidur, R.; Memon, L.; Zulkifli, N.; Masjuki, H. A review on fuel economy standard for motor vehicles with the implementation possibilities in Malaysia. *Renew. Sustain. Energy Rev.* **2010**, *14*, 3092–3099. [CrossRef]
3. Sciarretta, A.; Guzzella, L. Control of hybrid electric vehicles. *IEEE Control Syst.* **2007**, *27*, 60–70. [CrossRef]
4. Mesbahi, T.; Khenfri, F.; Rizoug, N.; Chaaban, K.; Bartholomeüs, P.; Moigne, P.L. Dynamical modeling of Li-ion batteries for electric vehicle applications based on hybrid Particle Swarm–Nelder–Mead (PSO–NM) optimization algorithm. *Electr. Power Syst. Res.* **2016**, *131*, 195–204. [CrossRef]
5. Hu, X.; Moura, S.J.; Murgovski, N.; Egardt, B.; Cao, D. Integrated Optimization of Battery Sizing, Charging, and Power Management in Plug-In Hybrid Electric Vehicles. *IEEE Trans. Control Syst. Technol.* **2016**, *24*, 1036–1043. [CrossRef]
6. Hoke, A.; Brissette, A.; Smith, K.; Pratt, A.; Maksimovic, D. Accounting for Lithium-Ion Battery Degradation in Electric Vehicle Charging Optimization. *IEEE J. Emerg. Sel. Top. Power Electron.* **2014**, *2*, 691–700. [CrossRef]
7. Redelbach, M.; Özdemir, E.D.; Friedrich, H.E. Optimizing battery sizes of plug-in hybrid and extended range electric vehicles for different user types. *Energy Policy* **2014**, *73*, 158–168. [CrossRef]
8. Sakti, A.; Michalek, J.J.; Fuchs, E.R.; Whitacre, J.F. A techno-economic analysis and optimization of Li-ion batteries for light-duty passenger vehicle electrification. *J. Power Sources* **2015**, *273*, 966–980. [CrossRef]
9. Hemi, H.; Ghouili, J.; Cheriti, A. Combination of Markov chain and optimal control solved by Pontryagin's Minimum Principle for a fuel cell/supercapacitor vehicle. *Energy Convers. Manag.* **2015**, *91*, 387–393. [CrossRef]
10. Zou, Z.; Cao, J.; Cao, B.; Chen, W. Evaluation strategy of regenerative braking energy for supercapacitor vehicle. *ISA Trans.* **2015**, *55*, 234–240. [CrossRef] [PubMed]
11. Rodatz, P.; Paganelli, G.; Sciarretta, A.; Guzzella, L. Optimal power management of an experimental fuel cell/supercapacitor-powered hybrid vehicle. *Control Eng. Pract.* **2005**, *13*, 41–53. [CrossRef]
12. Ayad, M.; Becherif, M.; Henni, A. Vehicle hybridization with fuel cell, supercapacitors and batteries by sliding mode control. *Renew. Energy* **2011**, *36*, 2627–2634. [CrossRef]
13. Shen, J.; Dusmez, S.; Khaligh, A. Optimization of Sizing and Battery Cycle Life in Battery/Ultracapacitor Hybrid Energy Storage Systems for Electric Vehicle Applications. *IEEE Trans. Ind. Inform.* **2014**, *10*, 2112–2121. [CrossRef]
14. Choi, M.E.; Lee, J.S.; Seo, S.W. Real-Time Optimization for Power Management Systems of a Battery/Supercapacitor Hybrid Energy Storage System in Electric Vehicles. *IEEE Trans. Veh. Technol.* **2014**, *63*, 3600–3611. [CrossRef]
15. Thounthong, P.; Chunkag, V.; Sethakul, P.; Davat, B.; Hinaje, M. Comparative Study of Fuel-Cell Vehicle Hybridization with Battery or Supercapacitor Storage Device. *IEEE Trans. Veh. Technol.* **2009**, *58*, 3892–3904. [CrossRef]
16. Hannan, M.; Azidin, F.; Mohamed, A. Hybrid electric vehicles and their challenges: A review. *Renew. Sustain. Energy Rev.* **2014**, *29*, 135–150. [CrossRef]
17. Hemi, H.; Ghouili, J.; Cheriti, A. A real time fuzzy logic power management strategy for a fuel cell vehicle. *Energy Convers. Manag.* **2014**, *80*, 63–70. [CrossRef]
18. Fotouhi, A.; Yusof, R.; Rahmani, R.; Mekhilef, S.; Shateri, N. A review on the applications of driving data and traffic information for vehicles energy conservation. *Renew. Sustain. Energy Rev.* **2014**, *37*, 822–833. [CrossRef]
19. Hu, X.; Murgovski, N.; Johannesson, L.M.; Egardt, B. Optimal Dimensioning and Power Management of a Fuel Cell ;Battery Hybrid Bus via Convex Programming. *IEEE/ASME Trans. Mechatron.* **2015**, *20*, 457–468. [CrossRef]
20. Sharaf, O.Z.; Orhan, M.F. An overview of fuel cell technology: Fundamentals and applications. *Renew. Sustain. Energy Rev.* **2014**, *32*, 810–853. [CrossRef]
21. Fatás, E.; Pérez-Flores, J.C.; Ocón, P. Pilas de combustible: una alternativa limpia de producción de energía. *Revista Española de Física* **2013**, *27*, 26–34.

22. Tie, S.F.; Tan, C.W. A review of energy sources and energy management system in electric vehicles. *Renew. Sustain. Energy Rev.* **2013**, *20*, 82–102. [CrossRef]
23. Ren, G.; Ma, G.; Cong, N. Review of electrical energy storage system for vehicular applications. *Renew. Sustain. Energy Rev.* **2015**, *41*, 225–236. [CrossRef]
24. Gao, L.; Dougal, R.A.; Liu, S. Power enhancement of an actively controlled battery/ultracapacitor hybrid. *IEEE Trans. Power Electron.* **2005**, *20*, 236–243. [CrossRef]
25. Schupbach, R.M.; Balda, J.C.; Zolot, M.; Kramer, B. Design methodology of a combined battery-ultracapacitor energy storage unit for vehicle power management. In Proceedings of the 34th Annual Power Electronics Specialist Conference, Acapulco, Mexico, 15–19 June 2003; Volume 1, pp. 88–93. [CrossRef]
26. Nielson, G.; Emadi, A. Hybrid energy storage systems for high-performance hybrid electric vehicles. In Proceedings of the 2011 IEEE Vehicle Power and Propulsion Conference, Chicago, IL, USA, 6–9 September 2011; pp. 1–6. [CrossRef]
27. Qu, X.; Wang, Q.; Yu Y. Power Demand Analysis and Performance Estimation for Active-Combination Energy Storage System Used in Hybrid Electric Vehicles. *IEEE Trans. Veh. Technol.* **2014**, *63*, 3128–3136. [CrossRef]
28. Chen, Z.; Mi, C.C.; Xu, J.; Gong, X.; You, C. Energy Management for a Power-Split Plug-in Hybrid Electric Vehicle Based on Dynamic Programming and Neural Networks. *IEEE Trans. Veh. Technol.* **2014**, *63*, 1567–1580. [CrossRef]
29. Jeong, J.; Kim, N.; Lim, W.; Park, Y.I.; Cha, S.W.; Jang, M.E. Optimization of power management among an engine, battery and ultra-capacitor for a series HEV: A dynamic programming application. *Int. J. Automot. Technol.* **2017**, *18*, 891–900. [CrossRef]
30. Sabri, M.M.; Danapalasingam, K.; Rahmat, M. A review on hybrid electric vehicles architecture and energy management strategies. *Renew. Sustain. Energy Rev.* **2016**, *53*, 1433–1442. [CrossRef]
31. Wu, G.; Zhang, X.; Dong, Z. Powertrain architectures of electrified vehicles: Review, classification and comparison. *J. Frankl. Inst.* **2015**, *352*, 425–448. [CrossRef]
32. Feroldi, D.; Serra, M.; Riera, J. Energy management strategies based on efficiency map for fuel cell hybrid vehicles. *J. Power Sources* **2009**, *190*, 387–401. [CrossRef]
33. Carignano, M.G.; Adorno, R.; van Dijk, N.; Nieberding, N.; Nigro, N.; Orbaiz, P. Assessment of Energy Management Strategies for a Hybrid Electric Bus. In Proceedings of the 5th International Conference on Engineering Optimization, Iguassu Falls, Brazil, 19–23 June 2016.
34. Feroldi, D.; Carignano, M. Sizing for fuel cell/supercapacitor hybrid vehicles based on stochastic driving cycles. *Appl. Energy* **2016**, *183*, 645–658. [CrossRef]
35. Aditya, J.P.; Ferdowsi, M. Comparison of NiMH and Li-ion batteries in automotive applications. In Proceedings of the Vehicle Power and Propulsion Conference, Harbin, China, 3–5 September 2008; pp. 1–6.
36. Parvini, Y.; Siegel, J.B.; Stefanopoulou, A.G.; Vahidi, A. Supercapacitor electrical and thermal modeling, identification, and validation for a wide range of temperature and power applications. *IEEE Trans. Ind. Electron.* **2016**, *63*, 1574–1585. [CrossRef]
37. Hoogers, G. *Fuel Cell Technology Handbook*; CRC Press: Boca Raton, FL, USA, 2002.
38. Barbir, F. *PEM Fuel Cells: Theory and Practice*; Academic Press: Cambridge, MA, USA, 2013.
39. Zhou, R.; Zheng, Y.; Jaroniec, M.; Qiao, S.Z. Determination of the electron transfer number for the oxygen reduction reaction: from theory to experiment. *ACS Catal.* **2016**, *6*, 4720–4728. [CrossRef]
40. Tzirakis, E.; Pitsas, K.; Zannikos, F.; Stournas, S. Vehicle emissions and driving cycles: comparison of the Athens driving cycle (ADC) with ECE-15 and European driving cycle (EDC). *Glob. NEST J.* **2006**, *8*, 282–290.
41. Bellman, R. *Dynamic programming*; Dover Publications: Mineola, NY, USA, 2003.
42. Bertsekas, D.P.; Bertsekas, D.P.; Bertsekas, D.P.; Bertsekas, D.P. *Dynamic Programming and Optimal Control*; Athena Scientific: Belmont, MA, USA, 1995; Volume 1.
43. Haifeng, D.; Xuezhe, W.; Zechang, S. A new SOH prediction concept for the power lithium-ion battery used on HEVs. In Proceedings of the Vehicle Power and Propulsion Conference, Dearborn, MI, USA, 7–11 September 2009; pp. 1649–1653.
44. Zou, C.; Manzie, C.; Nešić, D.; Kallapur, A.G. Multi-time-scale observer design for state-of-charge and state-of-health of a lithium-ion battery. *J. Power Sources* **2016**, *335*, 121–130. [CrossRef]

45. Ouyang, M.; Feng, X.; Han, X.; Lu, L.; Li, Z.; He, X. A dynamic capacity degradation model and its applications considering varying load for a large format Li-ion battery. *Appl. Energy* **2016**, *165*, 48–59. [CrossRef]
46. Wang, J.; Liu, P.; Hicks-Garner, J.; Sherman, E.; Soukiazian, S.; Verbrugge, M.; Tataria, H.; Musser, J.; Finamore, P. Cycle-life model for graphite-LiFePO4 cells. *J. Power Sources* **2011**, *196*, 3942–3948. [CrossRef]
47. Hu, X.; Johannesson, L.; Murgovski, N.; Egardt, B. Longevity-conscious dimensioning and power management of the hybrid energy storage system in a fuel cell hybrid electric bus. *Appl. Energy* **2015**, *137*, 913–924. [CrossRef]
48. Johannesson, L.; Murgovski, N.; Ebbesen, S.; Egardt, B.; Gelso, E.; Hellgren, J. Including a battery state of health model in the HEV component sizing and optimal control problem. *IFAC Proc. Vol.* **2013**, *46*, 398–403. [CrossRef]
49. Ebbesen, S.; Elbert, P.; Guzzella, L. Battery State-of-Health Perceptive Energy Management for Hybrid Electric Vehicles. *IEEE Trans. Veh. Technol.* **2012**, *61*, 2893–2900. [CrossRef]
50. Das, V.; Padmanaban, S.; Venkitusamy, K.; Selvamuthukumaran, R.; Blaabjerg, F.; Siano, P. Recent advances and challenges of fuel cell based power system architectures and control—A review. *Renew. Sustain. Energy Rev.* **2017**, *73*, 10–18. [CrossRef]
51. Dicks, A.; Rand, D.A.J. *Fuel Cell Systems Explained*; Wiley Online Library: Hoboken, NJ, USA, 2018.
52. Kongkanand, A.; Mathias, M.F. The priority and challenge of high-power performance of low-platinum proton-exchange membrane fuel cells. *J. Phys. Chem. Lett.* **2016**, *7*, 1127–1137. [CrossRef] [PubMed]
53. Li, L.; You, S.; Yang, C.; Yan, B.; Song, J.; Chen, Z. Driving-behavior-aware stochastic model predictive control for plug-in hybrid electric buses. *Appl. Energy* **2016**, *162*, 868–879. [CrossRef]
54. Song, Z.; Hofmann, H.; Li, J.; Hou, J.; Han, X.; Ouyang, M. Energy management strategies comparison for electric vehicles with hybrid energy storage system. *Appl. Energy* **2014**, *134*, 321–331. [CrossRef]
55. Sockeel, N.; Shi, J.; Shahverdi, M.; Mazzola, M. Pareto Front Analysis of the Objective Function in Model Predictive Control Based Power Management System of a Plug-in Hybrid Electric Vehicle. In Proceedings of the 2018 IEEE Transportation Electrification Conference and Expo (ITEC), Long Beach, CA, USA, 13–15 June 2018; pp. 1–6.
56. Sockeel, N.; Shi, J.; Shahverdi, M.; Mazzola, M. Sensitivity Analysis of the Battery Model for Model Predictive Control: Implementable to a Plug-In Hybrid Electric Vehicle. *World Electr. Veh. J.* **2018**, *9*, 45. [CrossRef]

Article

A Techno-Economic Analysis of Vehicle-to-Building: Battery Degradation and Efficiency Analysis in the Context of Coordinated Electric Vehicle Charging

Stefan Englberger *, Holger Hesse, Daniel Kucevic and Andreas Jossen

Institute for Electrical Energy Storage Technology, Technical University of Munich (TUM), Arcisstr. 21, 80333 Munich, Germany; holger.hesse@tum.de (H.H.); daniel.kucevic@tum.de (D.K.); andreas.jossen@tum.de (A.J.)

* Correspondence: stefan.englberger@tum.de; Tel.: +49-89-289-26969

Received: 13 February 2019; Accepted: 5 March 2019; Published: 12 March 2019

Abstract: In the context of the increased acceptance and usage of electric vehicles (EVs), vehicle-to-building (V2B) has proven to be a new and promising use case. Although this topic is already being discussed in literature, there is still a lack of experience on how such a system, of allowing bidirectional power flows between an EV and building, will work in a residential environment. The challenge is to optimize the interplay of electrical load, photovoltaic (PV) generation, EV, and optionally a home energy storage system (HES). In total, fourteen different scenarios are explored for a German household. A two-step approach is used, which combines a computationally efficient linear optimizer with a detailed modelling of the non-linear effects on the battery. The change in battery degradation, storage system efficiency, and operating expenses (OPEX) as a result of different, unidirectional and bidirectional, EV charging schemes is examined for both an EV battery and a HES. The simulations show that optimizing unidirectional charging can improve the OPEX by 15%. The addition of V2B leads to a further 11% cost reduction, however, this corresponds with a 12% decrease in EV battery lifetime. Techno-economic analysis reveals that the V2B charging solution with no HES leads to strong self-consumption improvements (EUR 1381 savings over ten years), whereas, this charging scheme would not be justified for a residential prosumer with a HES (only EUR 160 savings).

Keywords: battery degradation; battery energy storage system; charging scheme; efficiency; electric vehicle; linear programming; lithium ion battery; operating expenses; residential battery storage; vehicle-to-building

---

1. Introduction

Increasing environmental awareness, technical improvements, and favorable regulatory conditions have all allowed the market for electric vehicles (EVs) in Germany and worldwide to experience an upturn [1,2]. Simultaneously, an increasing number of electricity consumers are investing in renewable energy sources. Photovoltaic (PV) power generators especially benefit from a growing popularity in residential homes, allowing these customers to reduce electricity costs and rendering them as prosumers [3]. A home energy storage system (HES) can be added to further increase self-consumption and self-sufficiency rates [4].

In literature, HESs and EVs are well-researched topics [4–6], however, combined approaches of both storage systems are still a very young research field [7]. While recent literature presents a novel energy management system (EMS) for residential buildings with HES and EV, the contribution comes short on analyzing the technical characteristics of the battery energy storage systems (BESSs) at varying charging schemes [7]. In this work, we analyze how the aforementioned trends may interact, conduct a full techno-economic system analysis and reveal how prosumers with an EV may be able to optimize their electricity expenses. In particular, the degradation and efficiency of the HES and the EV's BESS are discussed. In addition, operating expenses (OPEX) are analyzed in the context of electricity costs for both the building and the vehicle. To increase the comparability of the results, a vehicle with an internal combustion engine (ICE) serves as a reference case.

As illustrated in Figure 1, three different charging strategies for the EV are analyzed and compared: Simple charging (SC) and optimized charging (OC) schemes, which both allow unidirectional power flows from the building to the vehicle, and the vehicle-to-building (V2B) strategy, which is an extension of the OC scheme allowing bidirectional power flows [6,8]. It is known that vehicle usage patterns may vary strongly [9]. For this reason, to make more valid statements about the degradation behavior, efficiency, and OPEX, the vehicle utilization patterns of a commuter and a supplementary vehicle are investigated. These vehicles are characterized by varying plug-in times at the power outlet of the prosumer's residence. As an additional degree of freedom, interaction between the EV battery and an optional stationary HES is examined. Particularly, the influence on the degradation and the efficiency of such a scenario considering two BESSs (EV and HES) is discussed. For the sake of simplicity, throughout this work, a typical German household with corresponding load and PV generator profiles is utilized and price signals of the German energy market are incorporated. However, the methodology can be applied to other profile data and the conclusive results drawn in this contribution are valid for other regions worldwide. An overview of the discussed simulation structure is visualized in Figure 1.

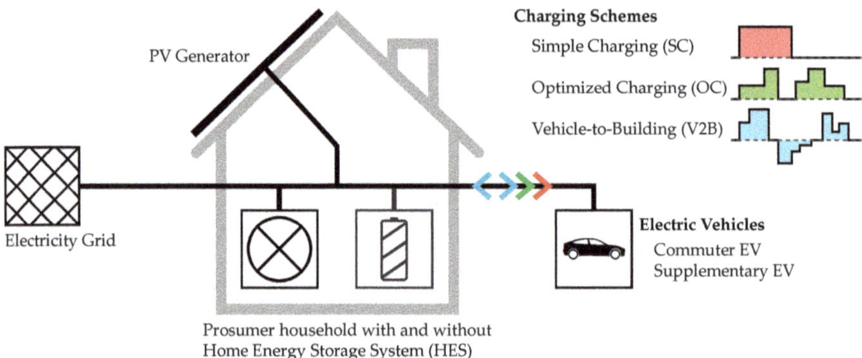

Figure 1. Schematic structure of the simulation environment of a prosumer household with three varying simulation dimensions: Consideration of home energy storage system (HES), two electric vehicle (EV) utilization patterns (commuter and supplementary car), and three different charging schemes (SC, OC, V2B).

The investigated scenarios in this work are simulated using a two-step approach. First, the residential power flow (RPF) model with an underlying linear programming (LP) algorithm optimizes the power flows within the residential multi-node system. Next, the optimized power flows are transferred to the open source simulation tool *SimSES* in order to model the resulting battery degradation and system efficiency [10].

This paper is structured as follows: Section 2 explains the optimization and simulation models as well as the system's topology, Section 3 presents the simulation results, and Section 4 concludes with a summary and discussion.

2. Methods

In order to optimize the electricity exchange between components and analyze the storage systems in a detailed fashion, two solution methods are combined, as is illustrated in Figure 2. First, the power flows between the individual technical units are optimized using the RPF model.

The underlying algorithm is based on LP, derived from the MATLAB optimization toolbox and the Gurobi optimizer [11]. Then, the simulation tool *SimSES* is used, which is capable of simulating the technical parameters of an energy storage system [10]. The results of the linear optimization are transferred to *SimSES* and represent the inputted alternating current (AC) power values of the energy storage system's inverter. By using *SimSES*' integrated operation strategy *PowerFollow*, the predefined time-discrete power values are implemented, and a detailed simulation is carried out. Both tools, the RPF model and the *SimSES* simulations are conducted in MathWorks MATLAB R2018b, operating at a sampling rate of 15 min [5].

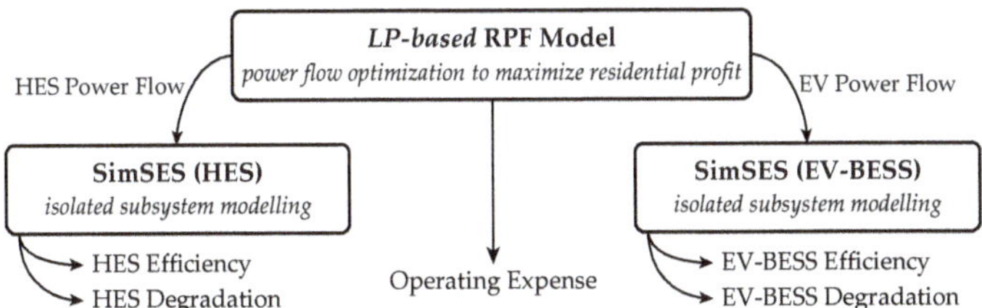

Figure 2. Schematic diagram of the two-step model structure, consisting of a linear programming (LP) based residential power flow (RPF) model, which optimizes the power flows so that the operating expenses (OPEX) are minimized, and the simulation tool *SimSES*, which validates the technical characteristics, round-trip efficiency, and battery degradation of the battery energy storage systems (BESSs).

The profit of a residential electricity prosumer in Germany is computed by simulating several different system configurations: Optional HES, optional EV, three different EV charging schemes, and two vehicle usage patterns.

Depending on the scenario, the RPF model of the investigated household consists of up to six main components, which are illustrated in Figure 3. The household is equipped with a PV generator with 8 kWp peak power, which is a common size for an average German household [12]. The PV generator system is composed of the PV panels, maximum power point tracker (MPPT), and inverter that converts the generator's direct current (DC) power into AC power. The one-year data measured from a PV system installed in Munich, Germany is used as the PV generating profile. To implement the degradation of the PV system, a degradation factor of 0.5% of the PV's peak power per year is assumed [4,13].

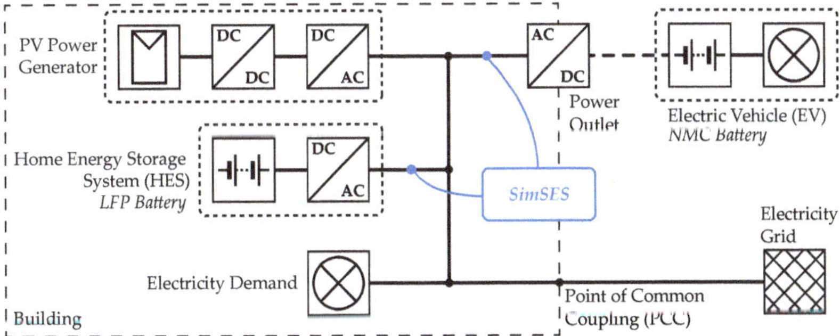

Figure 3. Residential power flow (RPF) model, consisting of the AC-coupled home energy storage system (HES), a photovoltaic (PV) power generator, electricity demand, the power outlet with the connected electric vehicle (EV), and the superordinate electricity grid. The simulation tool *SimSES* is used to validate the technical characteristics of the considered battery energy storage system (BESS).

In order to consider the electricity demand of a typical household, a representative one-year load profile (*profile* 31) out of a freely available set of smart-meter derived household load profiles is used in this study [14]. The annual electricity demand (only of the building, excluding that of the EV) of the considered household is set to 6000 kWh, a value taken from literature and well-suited to an average German household [12].

Further parameters and technical specifications of the household and its stationary HES can be taken from Table 1. The eligibility requirements, according to the German Federal Ministry of Economics and Technology, stipulate a feed-in limitation of 50% for PV generators that are operated in combination with a stationary or decentralized BESS [15]. Furthermore, a fixed feed-in remuneration price of 0.123 EUR/kWh is utilized, which is fixed and guaranteed for a period of twenty years [13]. Due to the projected electricity price of 0.437 EUR/kWh in 2030 and the electricity price of 0.294 EUR/kWh in 2018, a compound annual growth rate (CAGR) of 3.35% is assumed for the electricity purchase price in the simulation [16].

Table 1. Main parameters for the prosumer building and the home energy storage system (HES).

Parameter	Value
Annual electricity demand	6000 kWh [12]
PV peak power	8 kWp [12]
Feed-in limitation	50% [15]
Feed-in remuneration	0.123 EUR/kWh [13]
Initial electricity price	0.294 EUR/kWh [16]
Electricity price CAGR	3.35% [16]
Battery chemistry	lithium iron phosphate (LFP)
Nominal energy content	9 kWh [12,17]
SOC limitation	5%, 95% [12,17]

Lithium ion batteries (LIBs) are assumed for both the EV and HES. The cell chemistry chosen for the stationary HES within the building is based on a lithium iron phosphate (LFP) cathode and graphite anode. This chemistry allows a high cyclic stability [18], which makes it a suitable candidate for stationary applications [17].

The average German household with a HES has a usable energy content of 8.1 kWh [12]. From this the nominal energy content of 9 kWh is derived with the state of charge (SOC) limitations of 5% and 95% [17]. Furthermore, a self-discharge rate of 0.6% of the nominal energy content per month is assumed for the LFP cell [17]. Efficiency losses during charge and discharge processes of the battery are calculated via *SimSES'* equivalent circuit model, which depends on charging and discharging current, battery temperature, and SOC [10].

The semi-empirical degradation model of the LFP cell is also incorporated in *SimSES*. Degradation analysis is based on a superposition of calendar and cycling-related capacity fade [19]. During idle periods only calendar degradation, whereas during load periods also cyclic degradation is occurring [20]. This cyclic degradation is a function of multiple factors, including the depth of cycle (DOC), current, SOC range, and temperature [10]. A constant ambient temperature of 25 °C is assumed throughout the simulation period as the HES is installed within the building.

Since the AC coupling topology is the dominant topology for HESs in Germany [12], this setup is also used in this work. One of the major advantages of this topology over a DC coupling to the PV generator is an easy integration into a building with an existing PV generator, thus ensuring a high level of flexibility [21].

For the power-electronics efficiency, a simplified constant value of 95% is assumed in the RPF model. In order to make more accurate statements about the efficiency of the BESSs, the *SimSES* simulation tool takes into account a concave efficiency curve, which is derived from previous literature [4,22]. This curve considers the dependence on the inverter's output power and the fact that values below 10% of the rated inverter power result in a significantly lower efficiency.

Analogous to the procedure for the stationary HES, the power flows to and from the EV are optimized using the RPF model and then validated in *SimSES*. For all simulations of the EV and the ICE vehicle, a *B-segment* small car is considered [23–25]. An overview of the technical characteristics for the considered vehicles can be found in Table 2.

A nickel manganese cobalt (NMC) based cathode cell chemistry is chosen for the EV's BESS. Compared to other LIB cell chemistries, the NMC cell offers a higher energy density. The nominal and usable energy contents of the chosen EV battery, 21.6 kWh and 18.8 kWh, are closely linked to numbers often stated for EVs widely used in Germany. Derived from the nominal and usable energy contents, SOC boundaries of 8% and 95% are defined [17]. Similar to the LFP cell of the HES, the self-discharge rate of the NMC cell is set to 0.6% of the nominal energy content per month. Both the RPF model and detailed simulations using *SimSES* assume a round-trip efficiency of 95% for the EV battery [26].

In comparison to the highly sophisticated battery model of the LFP cell, the EV's battery is modelled using a more generic approach within *SimSES* [10]. Similar to previous work, a Wöhler curve (i.e., stress-number (S-N) curve) based fatigue model is used as the underlying method to estimate cycling-induced stress in the battery [4]. This method leads to an exponential weighting of DOC, i.e., an increased DOC leads to an overproportional increase in battery stress level, which again results in a reduced amount of equivalent full cycles (EFC) compared to low DOC values; thus, resulting in a shortened battery lifetime [27].

Table 2. Parameters for the electric vehicle (EV) and the internal combustion engine (ICE) vehicle.

Parameter	Value
Vehicle class	B-segment small car [23–25]
Battery chemistry	nickel manganese cobalt (NMC)
Nominal energy content	21.6 kWh [28]
Useable energy content	18.8 kWh [28]
Battery round-trip efficiency	95% [17]
Annually driven distance	13,922 km [29]
Electricity consumption	12.9 kWh/100 km [28]
Fuel consumption	5.3 L/100 km [30]
Initial fuel price	1.45 EUR/L [16]
Fuel price CAGR	2.25% [16]

The annual mileage of a passenger car is based on the German average, which is 13,922 km [29]. Therefore, a comparable EV, which consumes 12.9 kWh/100 km, requires approximately 1800 kWh annually [28]. In this paper, a gasoline-powered vehicle with an average fuel consumption of 5.3 L/100 km is used [30]. Analogous to the electricity costs, a temporally dynamic behavior is also assumed for the fuel price: An initial price of 1.45 EUR/L fuel is assumed for the start of the simulation. Due to the projected gasoline price of 1.89 EUR/L in 2030 and the gasoline price of 1.45 EUR/L in 2018, a CAGR of 2.25% is assumed for fuel prices in the simulation [16].

As part of this work, two EV profiles are created synthetically. The profiles for the two considered EVs (commuter and supplementary vehicle) are based on the US06 driving cycle and 83 charging profiles provided by the Forschungsstelle für Energiewirtschaft e. V., which are used in the federal study *Mobility in Germany* [9,31,32]. Both vehicle utilization patterns consist of a driving profile and a binary time series, which indicates whether the vehicle is connected to the power outlet of the building. It is assumed that the EV is only charged at the residential building and this additional electricity demand is directly allocated to the total electricity consumption of the household.

In Figure 4 an exemplary week (Monday to Sunday) in early summer is illustrated. The dashed areas in the two lower subplots show the plug-in times of the two utilization patterns, where the respective EV is connected to the building. As is immediately apparent, both profiles differ strongly in terms of their total plug-in time and respective daytime behavior: The commuter profile is only rarely connected to the building's power outlet during times of high solar irradiation on weekdays, which makes it more difficult for this vehicle user to directly utilize surplus PV power. Instead, the cumulative plug-in time of the supplementary car is much higher, so the potential of optimizing the power flows between building and vehicle is assumed to be higher.

In order to bring the difference of the vehicle utilization types into a quantifiable context, the quotient between plug-in time and the residual power is formed. Residual power is defined as the difference between PV power and demanded power. For the two types of examined profiles, the resulting correlation coefficients are 7% for the commuter vehicle and 28% for the supplementary car. With the increased plug-in time, the BESS availability of the EV is increased, which increases the degree of freedom for power flow optimization. This increased utilization coefficient leads to a reduction in electricity purchases, which in turn lowers the OPEX of the prosumer. Based on this theory, this metric is introduced and discussed further in the following sections.

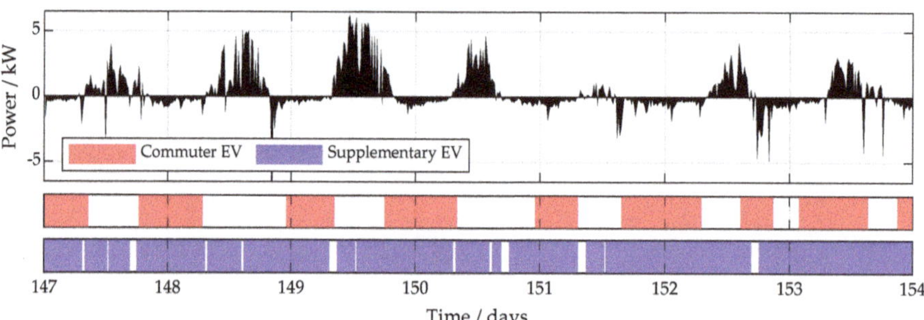

Figure 4. Residual power of exemplary week (Monday to Sunday) where photovoltaic (PV) excess power is characterized by positive values and the associated plug-in times (colored area) of the electric vehicles (commuter EV = red, supplementary EV = blue).

In addition to the two aforementioned vehicle utilization patterns, three different EV charging schemes are introduced. All three strategies are discussed in the context of storage system efficiency, degradation, and economic impact:

- Simple charging (SC): A simple rule-based charging of the EV is applied, where power is delivered unidirectionally from the power outlet of the building to the vehicle. As long as the vehicle is connected to the building and the EV's battery SOC has not reached the maximum SOC limit, the EV gets charged at the maximum allowed charge rate. The RPF model, as well as the simulation tool *SimSES*, are considering constraints for the respective SOC and C-Rate boundaries.
- Optimized charging (OC): Similar to SC the power outlet is used for unidirectional vehicle charging only. An advanced strategy is used that optimizes and controls the amount of energy and the timing of the EV's charging. The controller is fed by input values such as power flows within the building and the plug-in times of the EV.
- Vehicle-to-building (V2B): As an extension of the OC strategy, V2B enables a bidirectional power flow between the EV and building.

The RPF model's objective is to maximize the profit from the electricity sold and purchased throughout the simulation period. This comes down to a minimization of the OPEX of the prosumer. All scenarios use the following base objective function:

$$Max \sum_{i} \left(E_i^{\mathrm{r}} \cdot p_i^{\mathrm{r}} - E_i^{\mathrm{p}} \cdot p_i^{\mathrm{p}} \right) \quad (1)$$

whereby E_i^{r} denotes the amount of electricity that is sold to the superordinate electricity grid at time step i. The purchased electricity per time step is defined by the variable E_i^{p}. The price signals p_i^{r} and p_i^{p} describe the remuneration and purchasing price at time i. Considering changing electricity prices over time, price signals are time-dependent. Besides the objective function, inequality constraints for the BESSs' SOC and C-Rate, as well as equality constraints for the power flows at each node are considered and derived from a previous contribution [33].

Literature shows that the total cost of ownership (TCO) for an EV in Germany depends on many factors [25]. Due to the perennial lifetime of modern BESSs and the complex estimation of future BESS investment costs, capital expenditures (CAPEX) are neglected. In order to make the results as comprehensible as possible, only electricity costs and fuel costs are taken into account.

3. Results

The simulation results are presented and discussed in the following section. In total, fourteen different scenarios are conducted. As shown in Table 3, three different charging schemes, two vehicle usage patterns, and either one or two BESSs within the system are considered. The results are discussed in the context of battery degradation, storage system efficiency, and overall economic assessment, from the perspective of operating expenses for the prosumer.

Table 3. Overview of the fourteen simulated scenarios with three different charging schemes, two vehicle usage patterns, and either one or two BESSs within the prosumer household.

Vehicle Usage Pattern	ICE	ICE w/HES	SC	OC	V2B	SC w/HES	OC w/HES	V2B w/HES
Commuter Supplementary	yes (*ICE*)	yes (*ICE*)	yes yes	yes yes	yes yes	yes yes	yes yes	yes yes

3.1. Economic Assessment of OPEX

As a first metric, the scenarios are evaluated and discussed from an economic perspective. Here, the OPEX for a short-term period of one year and a longer-term ten-year period are considered.

During the first year, even the EV scenario with the highest OPEX, the SC scheme, showed a cost reduction of 31% without HES compared to the ICE vehicle without HES. With the addition of a home energy storage system to the scenarios, the OPEX reduction when using the SC scheme is 39% (EUR 571) in comparison to the ICE vehicle with the same HES.

As illustrated in Figure 5, strong differences between EV charging strategies can be detected. Both without and with HES, the implementation of an optimized charging (OC) scheme leads to a reduction in OPEX. Further cost improvements can be gained by allowing bidirectional power flows (V2B) between the building and the EV. This impact of optimized charging schemes (unidirectional and bidirectional) is particularly strong if there is no additional HES, leading to cost reductions of 14% and 23% in comparison to the SC strategy. The same ratios, with the addition of a HES, are reduced to 12% and 13% respectively.

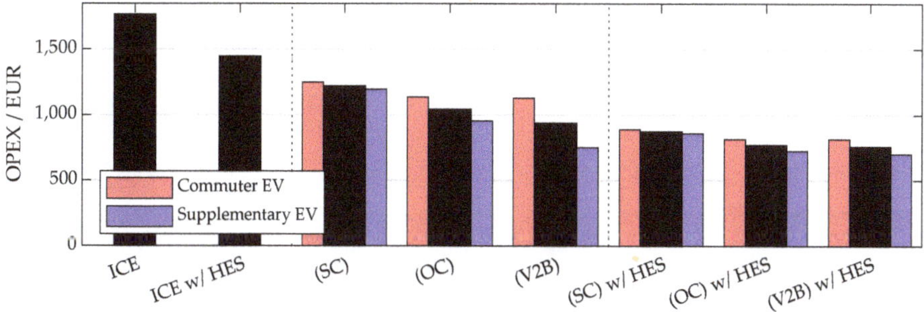

Figure 5. Operating expenses (OPEX) for one year. The dark-grey column represents the average value.

On average, OPEX decrease by 25% if, in addition to an EV, a stationary HES is available, resulting in EUR 115 cost reduction for the observed setting and year. Furthermore, the results for the commuter and supplementary car in the V2B scenario without HES showed a strong difference. Due to the relatively higher plug-in time of the supplementary car (especially during periods of high PV power), more self-generated energy can be stored in the vehicle, which results in higher self-consumption and self-sufficiency rates that are illustrated in Figure 6. Additionally, the scenarios of the supplementary car, with or without an additional HES, result in almost the same costs. Again, the supplementary car's high amount of plug-in time increases the utilization of the vehicle battery, thus making the stationary HES almost obsolete.

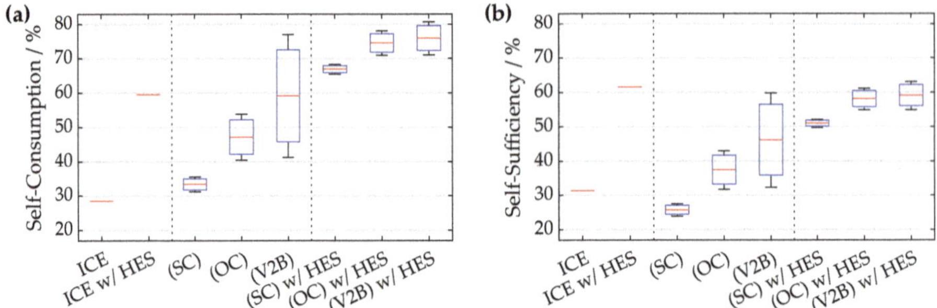

Figure 6. (**a**) Self-consumption and (**b**) self-sufficiency rate for the investigated scenarios. For both metrics, the top edge of each boxplot represents the supplementary car. The lower values of the boxplots are defined by the commuter car, which has a shorter plug-in time compared to the supplementary car.

As shown in Figure 7, the relative differences between the six EV scenarios remain almost the same as in the one-year view. The slight differences are due to the CAGR effect of rising electricity prices. However, the OPEX relationships between the ICE vehicle and EV changed because the expected fuel price increase is lower than that of electricity. A more detailed picture of the OPEX and their seasonal development over ten years can be seen in Figure A1 in the Appendix A.

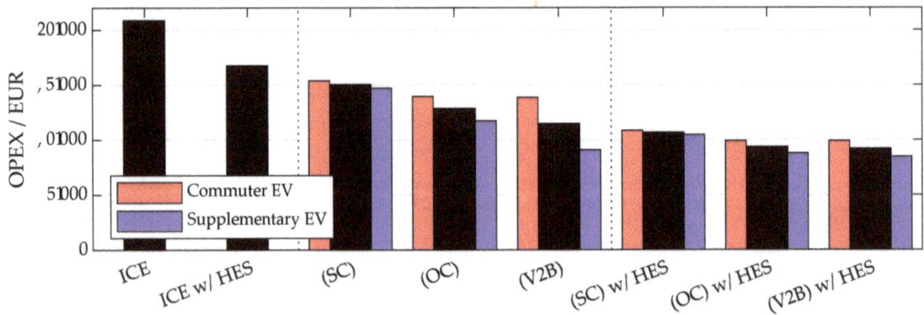

Figure 7. Operating expenses (OPEX) for ten years. Compound annual growth rate (CAGR) of energy costs are considered, so that costs for ten years are more than ten times the one-year costs. The dark-grey column represents the average value.

3.2. Battery Lifetime and Degradation

A common procedure when determining the end of life (EOL) of BESSs is reaching a certain capacity value. Specifically, values between 70% and 80% of the nominal battery capacity are often used to describe the EOL of the BESS [34,35]. In this work, the threshold of 80% is defined as EOL criteria, for both the HES and the EV battery. Figure 8 shows the battery degradation for both BESSs and the simulated scenarios. A more detailed evaluation of the degradation of the two battery types is discussed in the following paragraphs.

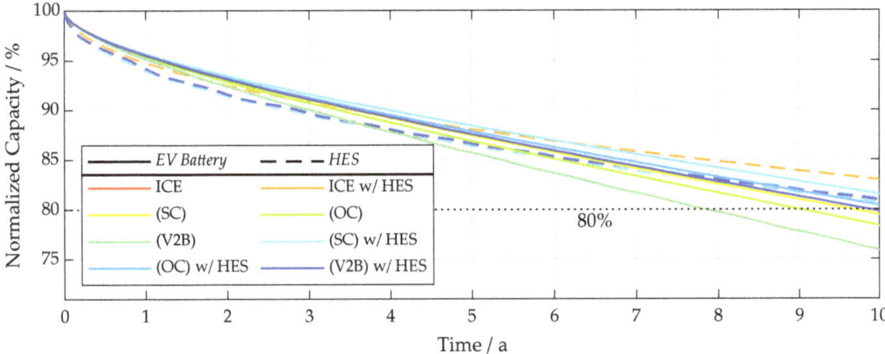

Figure 8. Remaining capacity of electric vehicle's battery (nickel manganese cobalt (NMC) cell chemistry, solid line) and home energy storage system (HES) (lithium iron phosphate (LFP) cell chemistry, dashed line) over ten years, with the highlighted end of life (EOL) threshold at 80% nominal battery capacity.

3.2.1. Home Energy Storage System

As visualized in Figure 9a the results of the observed scenarios show a lifetime between 10.7 years and 13.6 years for the battery of the HES. It is noticeable that the highest lifetime is achieved in the scenario of the ICE vehicle combined with a HES. For the EV scenarios, the lifetime is reduced by about 20%, whereby the simple charging scheme shows the shortest lifespan of 10.7 years. A further trend that can be seen in all three EV scenarios is that the battery lifetime in the scenarios with the supplementary car is always higher than the ones of the commuter vehicle. In both the OC and V2B strategy, this results in a relative lifetime improvement of about 6% for the supplementary car.

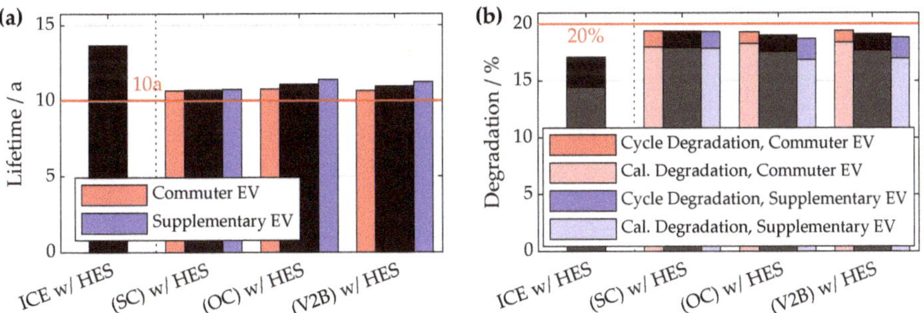

Figure 9. (**a**) Modelled lifetime of the lithium iron phosphate (LFP) home energy storage system (HES) with a nominal capacity of 9 kWh and (**b**) calendar and cyclic degradation during ten years of operation, with the end of life (EOL) condition of 80% remaining capacity. The dark-grey column represents the average value.

In Figure 9b, the relative calendar and cyclic degradation over the course of ten years of operation is illustrated. The results show that the 20% capacity fade is almost reached after ten years for the HES. In Figure 9a, it can be observed that a total lifetime of up to 13.6 years is reached. This can be explained by the initial intensity of degradation processes at the early stage of the battery's operation, which then decrease over time.

The fact that cells suffer particularly from SOC values in the lower and upper SOC range is reflected in the LFP model used for the simulations of this study [20]. Due to increased stress characteristics at these more extreme SOC regions, calendar degradation is accelerated. This, in turn leads to a reduced lifetime. At roughly 90%, calendar degradation processes are the main driver for the reduced battery lifetime. On the other hand, the cyclic degradation stress is fostered by high amounts of EFC. It should also be emphasized that the measured values shown are not the only drivers for battery degradation.

The battery's EFC are especially significant for cyclic degradation. The four HESs of the observed scenarios show annual EFC values of between 167 to 246, as shown in Figure 10a. Especially in the SC scheme, the EFC are significantly higher than those of the other scenarios. The lowest and almost equal amount of EFC is achieved in the settings of unidirectional (OC) and bidirectional (V2B) optimized charging.

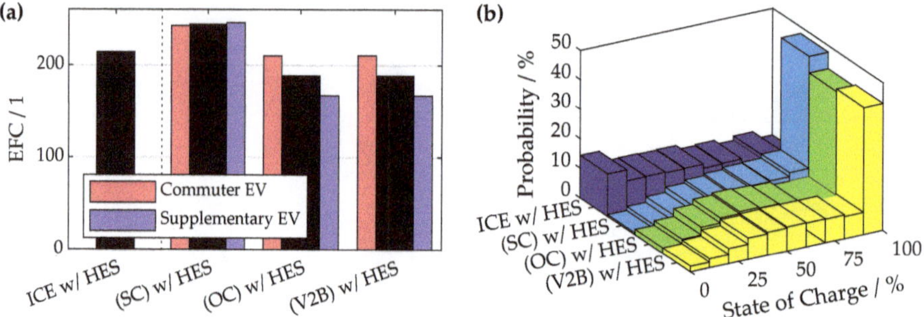

Figure 10. (**a**) Average amount of annual equivalent full cycles (EFC) of the home energy storage system and (**b**) probability distribution of the average state of charge (SOC) per scenario. The dark-grey column represents the average value.

Another degradation factor that is of importance for the lifetime of a LIB is the average SOC. This measure is illustrated in Figure 10b and gives insight into the probability distribution of the SOC for the four considered HESs. Here, a distinctive difference between the ICE vehicle and EV scenarios can be seen. While the SOC values of the HES have a rather homogeneous distribution in the ICE scenario, values in EV settings are much more heterogeneous. In all considered scenarios in which an EV and a HES are combined, it is shown that the SOC of the HES has a high probability density at high values. In the case of the simple charging (SC) scheme, the trend towards high SOC values is particularly strong. As with the number of EFC, here too, both scenarios OC and V2B show approximately the same, and better, results.

3.2.2. EV Battery

Like the evaluation of the HES's data, the battery of the EV is also examined with regard to degradation for the different scenarios. *SimSES* is used to model an isolated storage system behavior of the EV battery. Since the battery model used for the NMC cells is a generic model in comparison to the semi-empirical degradation model used for the LFP cells, results are shown in less detail for the EV battery.

A common standard for the expected lifetime and warranty period for EV batteries is seven to ten years [36]. Within this period, the remaining battery capacity should not fall below the defined EOL criteria of the battery. For the considered scenarios, it is shown that the EV battery has a lifetime of between 7.2 years and 11.8 years, as can be seen in Figure 11.

It is noticeable that its lifetime can be increased by an average of 19% if the EV battery works in conjunction with the stationary HES. The existence of a second storage system leads to a segmentation of the power flows, which results in a reduced stress level of the EV battery.

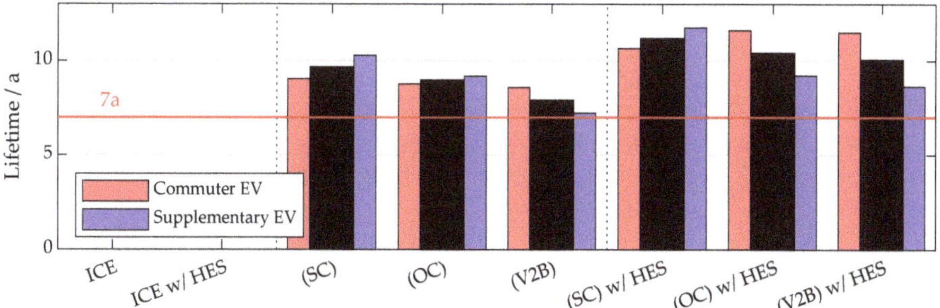

Figure 11. Lifetime calculation of the electric vehicle's nickel manganese cobalt (NMC) battery, based on a generic battery model, with the defined end of life (EOL) condition of 80% of the nominal capacity. The dark-grey column represents the average value.

The scenarios with SC and OC schemes show the same amount of EFC, due to the fact that in these unidirectional scenarios, only the power needed at a later time for driving is delivered from the building to the vehicle. Despite the same amount of EFC of the EV battery in the SC and OC scenarios, the lifetime of the optimized charging (OC) scheme is reduced by 7%. For better interpretation along with the degradation model used herein (based on Wöhler curves), the average absolute values of the DOC are shown in Table 4. Here, it can be seen that the average DOC in scenarios with a HES decreases by about 30% compared to the same settings without a HES.

Table 4. Annual amount of equivalent full cycles (EFC) and the absolute depth of cycle (DOC) (normalized to the amount of EFC) of the battery taken as an average from the commuter and supplementary electric vehicle (EV).

	ICE	ICE w/HES	SC	OC	V2B	SC w/HES	OC w/HES	V2B w/HES
EFC	n/a	n/a	85.5	85.5	119.3	85.5	85.5	89.5
|DOC|	n/a	n/a	1.00	0.98	0.98	0.58	0.76	0.76

The degradation in the case of V2B is significantly higher. Results show that the annual number of EFC at 119.3 increase by 40% when there is no additional BESS in the system besides the EV battery. This increase in EFC and the relatively high average DOC values result in a lifetime reduction of about 12% compared to the OC scheme.

For scenarios considering two BESSs, the V2B scenario again shows the highest battery degradation. Because of the permanently available HES, surplus PV power can also be stored in the stationary HES and therefore the number of EFC in the V2B scenario is only slightly higher than that of the unidirectional scenarios (SC and OC). However, the battery lifetime in the V2B case is shortened by about 3% compared to the same setting with OC scheme.

The commuter car battery in the V2B scenarios has a lower energy throughput and thus a lower number of EFC. The relatively higher plug-in time of the supplementary car allows more surplus energy to be charged into and discharged from the EV battery, resulting in a higher number of EFC and a reduced lifetime.

3.3. Storage System Efficiency

In addition to battery degradation, BESSs' round-trip efficiency values are also considered. For both BESS types, the stationary HES and the storage system of the EV, a round-trip efficiency of about 88% is achieved for all operational modes.

More detailed analysis reveals that the dominant source of storage losses comes from power-electronics. This is in line with efficiency analysis conducted on stationary storage systems [37].

Overall, between 8% and 10% efficiency losses are caused by the inverter. This emphasizes the relevance for optimizing the specifications of the technical components of a storage system.

Furthermore, storage losses are considered during the charging and discharging processes of the battery. Storage losses within the battery cells range from 2% to 4% in the considered simulations, which is in line with results from literature [17]. Self-discharge losses, which account for below 0.1% of the total energy throughput, play a subordinate role. This low percentage of storage losses is similar for both storage technologies in all scenarios.

4. Discussion and Conclusions

The following section summarizes the results derived from the simulations and discusses them in the context of previous literature. At the end of the section, related and future research fields are highlighted.

4.1. EV Versus ICE Vehicle

In the previously discussed results section it is shown that an EV can have a significant economic advantage compared to ICE powered vehicles when it comes to reducing electricity costs of a prosumer household. Considering a time span of ten years, it is shown that OPEX can be reduced by an average of 37% (without an additional stationary HES) and 42% (with HES). Even the least economically lucrative scenario with simple charging (SC) shows an average savings potential of 28% (without HES) and 37% (with HES) compared to the same scenarios with an ICE powered vehicle.

Looking at the average results of the individual EV scenarios, it can be said that the considered additional energy costs for the investigated ICE vehicle are about EUR 7400 higher than for its electric-powered counterpart, which may justify an investment in a higher priced EV. Of course, further cost components and economic and policy aspects must be taken into account in order to carry out a complete economic analysis [25,38]. Furthermore, at the moment, there is no consensus on when an EV is equivalent to an ICE vehicle in terms of investment costs.

In the context of battery lifetime, the simulations reveal a trend of stronger degradation when an EV is included in the consideration. The HES's battery reaches the defined EOL criterion earlier by 20%, on average, when an EV is connected to the household. Minimizing OPEX means that more self-generated energy is stored in the HES. In the EV scenarios, the effect leads to an increased occurrence of high SOC levels, which accelerates internal degradation processes of the LFP cells [10]. In order to compensate this effect, the developed charging strategies must be further optimized.

Furthermore, the share of automotive batteries that are used for further applications after their primary use as an EV battery is growing. Particularly, the installation and operation of such second-use batteries in stationary applications is increasing [39]. This use of second-use batteries allows an additional economic impact of the BESS, which makes it more lucrative for their stakeholders [40].

4.2. Impact of Vehicle Utilization Pattern

From the simulations it can be concluded that the supplementary vehicle type has a beneficial effect on electricity cost reductions. This is shown by the lower OPEX in all scenarios when compared to the commuter EV, which has less plug-in time at the building. This relation confirms the initial theory that a higher correlation coefficient between residual power and plug-in time leads to an economic improvement. It is expected that, from the perspective of an office building with PV generation, the connected EVs from commuting employees would have the same beneficial outcome. The underlying effect can also be explained by the household's increased self-consumption and self-sufficiency rate with the supplementary vehicle profile [7]. On average, OPEX in the commuter car scenarios are about 16% higher than those with the supplementary vehicle. This cost increase is particularly high when considering a bidirectional charging scheme (V2B).

In terms of battery degradation, on the other hand, it is shown that battery lifetime of the HES is slightly increased in the commuter car scenarios. However, the average battery lifetime for the

EV battery shows a favorable behavior in the supplementary scenarios, in particular during V2B charging schemes. DOC values and the underlying Wöhler curve for the EV battery degradation model represent the main drivers for this effect [27].

4.3. Impact of Considering an Additional HES

Previous literature has shown that it is still difficult to operate a HES in Germany in an economically lucrative way [4]. Although the results presented in this paper only relate to OPEX, it is noteworthy that a HES can reduce these costs by an average of 23% during the first year. When taking into account the rising electricity retail tariff estimated for the next ten years [16], the cost savings may rise by another few percentage points.

Due to the segmentation of power flows when considering a HES, both the energy throughput and relative DOC values of the EV battery can be reduced. The reduced stress level leads to an increase of the EV's battery lifetime by an average of 20%.

Whether and to what extent the advantages of the lifetime extension of the EV battery and OPEX reduction justify additional expenses of a HES depend, in turn, on the CAPEX. Taking into account the discussed prosumer and an operation period of ten years, HES investment costs below EUR 2305 (V2B scenario) and EUR 4437 (SC scenario) would be justified. The higher value in the SC scenario results from the fact that, here, an additional HES has a higher potential for OPEX improvement, which is further discussed in the subsequent paragraphs. Assuming steadily declining CAPEX for stationary battery packs [41], a HES can become increasingly interesting for residential buildings. If, in addition to the minimization of OPEX and self-consumption improvements, other applications are served, the economics of the HES can be increased even further [42].

4.4. Impact of Charging Scheme

Both in the scenarios with and without HES, the simple charging (SC) scheme resulted in the highest OPEX. The condition that the EV is charged as soon as it is connected to the building also results in overall low self-consumption and self-sufficiency rates of 33% and 26%, which are illustrated in Figure 6.

In the optimized charging (OC) scheme with a HES, the electricity costs can be reduced by 12% (about EUR 1300 for a ten-year operation period). This effect is even more pronounced when there is no additional HES and the EV battery is the only BESS in the setting. OPEX can be reduced by 15% (about EUR 2200) compared to the SC scheme when the EV battery is the only storage unit to decouple energy supply and demand.

By allowing a bidirectional power flow between the building and the EV (V2B) instead of the unidirectional power flow (OC), further cost savings can be achieved. Relative to the OC scheme, this results in a further OPEX reduction of 2% (with HES) and 11% (without HES). Analogous to the above comparison between the SC and OC schemes, there is an increased cost saving potential if the EV battery is the only storage unit in the system. When considering the absolute values of the savings potential, an OPEX reduction of EUR 160 (with HES) and EUR 1381 (without HES) results for an operation period of ten years. This comes at the cost for additional upfront investment costs: The low savings potential of the scenario with HES suggests that the additional investment costs for a power outlet with bidirectional power flow are difficult to compensate. On the other hand, in scenarios with a single EV battery, the V2B scheme could be economically lucrative in comparison to the OC scheme, if the additional investment costs are below the cost savings of EUR 1381.

In contrast to improved electricity expenditures, the lifetime of the EV battery decreases in the OC and V2B schemes. Due to increased energy throughput, particularly in the V2B scheme, the lifetime is reduced by up to 12% compared to the optimized unidirectional charging (OC). SC scenarios lead to the highest lifetime with a relative improvement of 7% compared to the OC scheme. One of the main drivers for the increased degradation are the relatively higher DOC values in the OC and V2B scheme. In addition, two more obstacles come into play: The prediction of power values is needed for effective OC and V2B schemes. Furthermore, automotive original equipment manufacturers (OEMs) provide warranties on the use of the EV battery for vehicle purposes. While this is maintained in the SC and OC schemes, the EV battery in the V2B scheme does not only function as an EV battery, but also as a buffer storage unit for the whole prosumer household. Thus this could pose a challenge to incentivizing V2B schemes.

The EV market shows a trend towards increasing battery capacity. EVs being manufactured currently are often more than twice as large in terms of nominal energy content than the 21.6 kWh EV battery that is considered in this work. It can be expected that the higher cost savings and the lower necessity of an additional HES due to the V2B scheme will be enhanced with these increased capacities.

4.5. Limitations and Future Research

The discussed simulations are conducted assuming perfect foresight of energy supply and demand, both for the household and the vehicle. In order to emphasize on limited foresight, other algorithms can be used. For instance, in [43] a fuzzy logic controller (FLC) is presented for a V2B environment.

Since the discussed RPF model is using constant efficiency values, it would be an improvement to implement non-linear relationships, as already implemented in *SimSES* [10]. This step would improve the RPF's validity, but would also lead to an increasing complexity of the optimization algorithm, resulting in an elevated computation time.

Although the generic battery model of the NMC cell provides values for the battery degradation [4,27], the quality of the battery model can be improved further. In comparison to the current model, more sophisticated degradation models could be implemented, as done for the semi-empirical degradation model of the LFP cell [20].

In order to generate a more profound insight into the economic results of the discussed settings, further research should take additional cost components into account. Although it is not clear how the CAPEX for batteries will develop in the future, there are estimations in recent literature that could be used [41]. Another cost component that is of relevance in that perspective are battery degradation costs [44]. The consideration of these factors would provide a more complete picture of the total cost of ownership (TCO), which in turn would allow for more precise conclusions.

Author Contributions: S.E. developed the conceptualization, including the generation of the optimization framework for the residential power-flow model, coordinated the study, and wrote the manuscript. H.H. significantly contributed to the conceptualization and methodology of the study. D.K. offered support in using simulation tool SimSES. A.J. provided overall guidance for the study and contributed to fruitful discussions on the methodology. All authors have read and approved the manuscript.

Funding: This research is funded by the Bavarian Ministry of Economic Affairs, Energy, and Technology via the research project StorageLink (grant number IUK-1711-0035//IUK551/002), and the Technical University of Munich (TUM) in the framework of the Open Access Publishing Program.

Acknowledgments: The authors thank Sebastian Fischhaber from the Forschungsstelle für Energiewirtschaft e. V. for providing electric-vehicle charging profiles.

Conflicts of Interest: The authors declare no conflict of interest. The funders had no role in the design of the study; in the collection, analyses, or interpretation of data; in the writing of the manuscript; or in the decision to publish the results.

Abbreviations

The following abbreviations are used in this manuscript:

AC	Alternating current
BESS	Battery energy storage system
CAGR	Compound annual growth rate
CAPEX	Capital expenditures
DC	Direct current
DOC	Depth of cycle
EFC	Equivalent full cycles

EOL	End of life
EV	Electric vehicle
HES	Home energy storage system
ICE	Internal combustion engine
LFP	Lithium iron phosphate
LIB	Lithium ion battery
LP	Linear programming
NMC	Nickel manganese cobalt
OC	Optimized charging
OPEX	Operating expenses
PV	Photovoltaic
RPF	Residential power flow
SC	Simple charging
SOC	State of charge
V2B	Vehicle-to-building

Appendix A

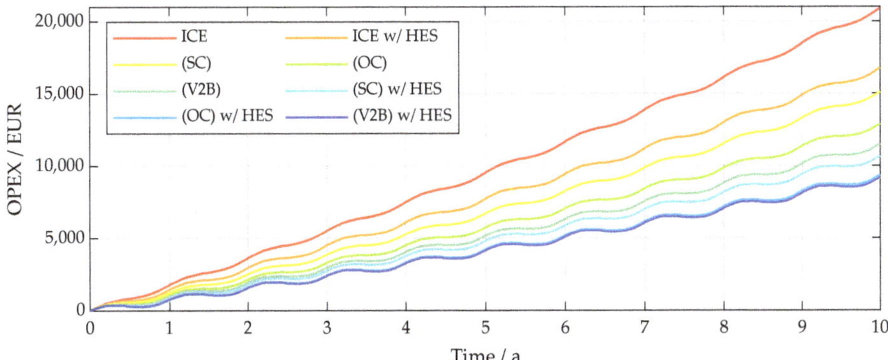

Figure A1. Operating expenses (OPEX) during ten years of operation showing a strong seasonal pattern.

References

1. Bunsen, T.; Cazzola, P.; Gorner, M.; Paoli, L.; Scheffer, S.; Schuitmaker, R.; Tattini, J.; Teter, J. *Global EV Outlook 2018: Towards Cross-Modal Electrification*; International Energy Agency: Paris, France, 2018.
2. Nykvist, B.; Nilsson, M. Rapidly falling costs of battery packs for electric vehicles. *Nat. Clim. Chang.* **2015**, *5*, 329–332, doi:10.1038/nclimate2564. [CrossRef]
3. Kästel, P.; Gilroy-Scott, B. Economics of pooling small local electricity prosumers—LCOE & self-consumption. *Renew. Sustain. Energy Rev.* **2015**, *51*, 718–729, doi:10.1016/j.rser.2015.06.057. [CrossRef]
4. Truong, C.N.; Naumann, M.; Karl, R.; Müller, M.; Jossen, A.; Hesse, H. Economics of Residential Photovoltaic Battery Systems in Germany: The Case of Tesla's Powerwall. *Batteries* **2016**, *2*, 14, doi:10.3390/batteries2020014. [CrossRef]
5. Naumann, M.; Karl, R.; Truong, C.N.; Jossen, A.; Hesse, H. Lithium-ion Battery Cost Analysis in PV-household Application. *Energy Procedia* **2015**, *73*, 37–47, doi:10.1016/j.egypro.2015.07.555. [CrossRef]
6. Tan, K.M.; Ramachandaramurthy, V.K.; Yong, J.Y. Integration of electric vehicles in smart grid: A review on vehicle to grid technologies and optimization techniques. *Renew. Sustain. Energy Rev.* **2016**, *53*, 720–732, doi:10.1016/j.rser.2015.09.012. [CrossRef]
7. Barone, G.; Buonomano, A.; Calise, F.; Forzano, C.; Palombo, A. Building to vehicle to building concept toward a novel zero energy paradigm: Modelling and case studies. *Renew. Sustain. Energy Rev.* **2019**, *101*, 625–648, doi:10.1016/j.rser.2018.11.003. [CrossRef]
8. Mwasilu, F.; Justo, J.J.; Kim, E.K.; Do, T.D.; Jung, J.W. Electric vehicles and smart grid interaction: A review on vehicle to grid and renewable energy sources integration. *Renew. Sustain. Energy Rev.* **2014**, *34*, 501–516, doi:10.1016/j.rser.2014.03.031. [CrossRef]

9. Infas Institut für angewandte Sozialwissenschaft GmbH. *Mobility in Germany 2017 (German: Mobilität in Deutschland 2017 (MiD 2017)): Federal Ministry of Transport and Digital Infrastructure*; Infas Institut für angewandte Sozialwissenschaft GmbH: Bonn, Germany, 2018.
10. Naumann, M.; Truong, C.N.; Schimpe, M.; Kucevic, D.; Jossen, A.; Hesse, H. (Eds.) SimSES: Software for techno-economic Simulation of Stationary Energy Storage Systems. In Proceedings of the International ETG Congress 2017, Bonn, Germany, 28–29 November 2017.
11. Gurobi Optimization. Gurobi Optimizer Version 8.1. 2018. Available online: http://www.gurobi.com/ (accessed on 14 December 2018).
12. Figgener, J.; Haberschusz, D.; Kairies, K.P.; Wessels, O.; Tepe, B.; Sauer, D.U. *Wissenschaftliches Mess- und Evaluierungsprogramm Solarstromspeicher 2.0: Jahresbericht 2018*; ISEA Institut für Stromrichtertechnik und Elektrische Antriebe RWTH Aachen: Aachen, Germany, 2018.
13. Dietrich, A.; Weber, C. What drives profitability of grid-connected residential PV storage systems? A closer look with focus on Germany. *Energy Econ.* **2018**, *74*, 399–416, doi:10.1016/j.eneco.2018.06.014. [CrossRef]
14. Tjaden, T.; Bergner, J.; Weniger, J.; Quaschning, V. *Representative Electrical Load Profiles of Residential Buildings in Germany with a Temporal Resolution of One Second*; HTW Berlin—University of Applied Sciences: Berlin, Germany, 2015.
15. KFW. KfW-Programm Erneuerbare Energien: Speicher. 2018. Available online: https://www.kfw.de (accessed on 15 October 2018).
16. Prognos, EWI, GWS. Entwicklung der Energiemärkte—Energiereferenzprognose. 2014. Available online: https://www.bmwi.de/Redaktion/DE/Publikationen/Studien/entwicklung-der-energiemaerkte-energiereferenzprognose-endbericht.pdf?__blob=publicationFile&v=7 (accessed on 15 October 2018).
17. Hesse, H.; Martins, R.; Musilek, P.; Naumann, M.; Truong, C.N.; Jossen, A. Economic Optimization of Component Sizing for Residential Battery Storage Systems. *Energies* **2017**, *10*, 835, doi:10.3390/en10070835. [CrossRef]
18. Stan, A.I.; Swierczynski, M.; Stroe, D.I.; Teodorescu, R.; Andreasen, S.J. Lithium ion battery chemistries from renewable energy storage to automotive and back-up power applications—An overview. In Proceedings of the 2014 International Conference on Optimization of Electrical and Electronic Equipment (OPTIM), Bran, Romania, 22–24 May 2014; IEEE: Piscataway, NJ, USA, 2014; pp. 713–720, doi:10.1109/OPTIM.2014.6850936. [CrossRef]
19. Vetter, J.; Novák, P.; Wagner, M.R.; Veit, C.; Möller, K.C.; Besenhard, J.O.; Winter, M.; Wohlfahrt-Mehrens, M.; Vogler, C.; Hammouche, A. Ageing mechanisms in lithium-ion batteries. *J. Power Sources* **2005**, *147*, 269–281, doi:10.1016/j.jpowsour.2005.01.006. [CrossRef]
20. Naumann, M. Techno-Economic Evaluation of Stationary Battery Energy Storage Systems with Special Consideration of Aging. Ph.D. Thesis, Technical University of Munich, Munich, Germany, 2018.
21. Hesse, H.; Schimpe, M.; Kucevic, D.; Jossen, A. Lithium-Ion Battery Storage for the Grid—A Review of Stationary Battery Storage System Design Tailored for Applications in Modern Power Grids. *Energies* **2017**, *10*, 2107, doi:10.3390/en10122107. [CrossRef]
22. Notton, G.; Lazarov, V.; Stoyanov, L. Optimal sizing of a grid-connected PV system for various PV module technologies and inclinations, inverter efficiency characteristics and locations. *Renew. Energy* **2010**, *35*, 541–554, doi:10.1016/j.renene.2009.07.013. [CrossRef]
23. Thiel, C.; Schmidt, J.; van Zyl, A.; Schmid, E. Cost and well-to-wheel implications of the vehicle fleet CO2 emission regulation in the European Union. *Transp. Res. Part A Policy Pract.* **2014**, *63*, 25–42, doi:10.1016/j.tra.2014.02.018. [CrossRef]
24. European Commission. *Merger Procedure Article 6(1)(b) of Council Regulation (EEC) No 4064/89—Decision on Case No IV/M.1406. HYUNDAI/KIA*; European Commission: Luxembourg, 1999.
25. Letmathe, P.; Suares, M. A consumer-oriented total cost of ownership model for different vehicle types in Germany. *Transp. Res. Part D Transp. Environ.* **2017**, *57*, 314–335, doi:10.1016/j.trd.2017.09.007. [CrossRef]
26. Jossen, A.; Weydanz, W. *Moderne Akkumulatoren Richtig Einsetzen*; Reichardt Verlag: Wiesbaden, Germany, 2006.
27. Rosenkranz, C.; Köhler, U.; Liska, J.L. Modern Battery Systems for Plug-In Hybrid Electric Vehicles. In Proceedings of the 23rd International Battery, Hybrid and Fuel Cell Electric Vehicle Symposium and Exhibition, Anaheim, CA, USA, 2–5 December 2007.

28. BMW Group. Der BMW i3. 2015. Available online: https://www.press.bmwgroup.com/austria/article/attachment/T0150664DE/299524 (accessed on 15 Dcember 2018).
29. Kraftfahrt-Bundesamt. Erneut mehr Gesamtkilometer bei Geringerer Jahresfahrleistung je Fahrzeug. 2018. Available online: https://www.kba.de/DE/Statistik/Kraftverkehr/VerkehrKilometer/verkehr_in_kilometern_node.html (accessed on 15 October 2018).
30. Tietge, U.; Mock, P.; Franco, V.; Zacharof, N. From laboratory to road: Modeling the divergence between official and real-world fuel consumption and CO2 emission values in the German passenger car market for the years 2001–2014. *Energy Policy* **2017**, *103*, 212–222, doi:10.1016/j.enpol.2017.01.021. [CrossRef]
31. Emission Test Cycle. US06 Supplemental Federal Test Procedure (SFTP). Available online at https://www.dieselnet.com/standards/cycles/ftp_us06.php (accessed on 7 December 2018).
32. Keil, P.; Schuster, S.F.; Wilhelm, J.; Travi, J.; Hauser, A.; Karl, R.; Jossen, A. Calendar Aging of Lithium-Ion Batteries: I. Impact of the Graphite Anode on Capacity Fade. *J. Electrochem. Soc.* **2016**, *163*, A1872–A1880, doi:10.1149/2.0411609jes. [CrossRef]
33. Englberger, S.; Hesse, H.; Truong, C.N.; Jossen, A. (Eds.) Autonomous Versus Coordinated Control of Residential Energy Storage Systems Monitoring Profit, Battery Aging, and System Efficiency: NEIS 2018. In Proceedings of the Conference on Sustainable Energy Supply and Energy Storage Systems, Hamburg, Germany, 20–21 September 2018.
34. Lunz, B.; Yan, Z.; Gerschler, J.B.; Sauer, D.U. Influence of plug-in hybrid electric vehicle charging strategies on charging and battery degradation costs. *Energy Policy* **2012**, *46*, 511–519, doi:10.1016/j.enpol.2012.04.017. [CrossRef]
35. Pistoia, G.; Liaw, B.Y. *Behaviour of Lithium-Ion Batteries in Electric Vehicles: Battery Health, Performance, Safety, and Cost/Gianfranco Pistoia, Boryann Liaw*; Green Energy and Technology; Springer: Cham, Switzerland, 2018.
36. Fischhaber, S.; Regett, A.; Schuster, S.F.; Hesse, H. *Studie: Second-Life-Konzepte für Lithium-Ionen-Batterien aus Elektrofahrzeugen: Analyse von Nachnutzungsanwendungen, ökonomischen und ökologischen Potenzialen*; Begleit- und Wirkungsforschung Schaufenster Elektromobilität (BuW): Frankfurt, Germany 2016.
37. Schimpe, M.; Naumann, M.; Truong, C.N.; Hesse, H.; Santhanagopalan, S.; Saxon, A.; Jossen, A. Energy efficiency evaluation of a stationary lithium-ion battery container storage system via electro-thermal modeling and detailed component analysis. *Appl. Energy* **2018**, *210*, 211–229, doi:10.1016/j.apenergy.2017.10.129. [CrossRef]
38. Massiani, J. Cost-Benefit Analysis of policies for the development of electric vehicles in Germany: Methods and results. *Transp. Policy* **2015**, *38*, 19–26, doi:10.1016/j.tranpol.2014.10.005. [CrossRef]
39. Neubauer, J.; Simpson, M. *Deployment of Behind-the-Meter Energy Storage for Demand Charge Reduction*; Tech. Rep. NREL/TP-5400-63162; National Renewable Energy Laboratory: Golden, CO, USA, 2015.
40. Heymans, C.; Walker, S.B.; Young, S.B.; Fowler, M. Economic analysis of second use electric vehicle batteries for residential energy storage and load-levelling. *Energy Policy* **2014**, *71*, 22–30, doi:10.1016/j.enpol.2014.04.016. [CrossRef]
41. Tsiropoulos, I.; Tarvydas, D.; Lebedeva, N. *Li-ion Batteries for Mobility and Stationary Storage Applications Scenarios for Costs and Market Growth*; Publications Office of the European Union: Luxembourg, 2018; doi:10.2760/87175.
42. Metz, D.; Saraiva, J.T. Simultaneous co-integration of multiple electrical storage applications in a consumer setting. *Energy* **2018**, *143*, 202–211, doi:10.1016/j.energy.2017.10.098. [CrossRef]
43. Singh, M.; Kumar, P.; Kar, I. Implementation of Vehicle to Grid Infrastructure Using Fuzzy Logic Controller. *IEEE Trans. Smart Grid* **2012**, *3*, 565–577, doi:10.1109/TSG.2011.2172697. [CrossRef]
44. Koller, M.; Borsche, T.; Ulbig, A.; Andersson, G. Defining a degradation cost function for optimal control of a battery energy storage system. In Proceedings of the 2013 IEEE Grenoble Conference, Grenoble, France, 16–20 June 2013; IEEE: Piscataway, NJ, USA, 2013; pp. 1–6, doi:10.1109/PTC.2013.6652329. [CrossRef]

Article

A General Parameter Identification Procedure Used for the Comparative Study of Supercapacitors Models

Henry Miniguano *, Andrés Barrado, Cristina Fernández, Pablo Zumel and Antonio Lázaro

Power Electronics Systems Group; Universidad Carlos III de Madrid, 28911 Leganés, Spain; barrado@ing.uc3m.es (A.B.); cfernand@ing.uc3m.es (C.F.); pzumel@ing.uc3m.es (P.Z.); alazaro@ing.uc3m.es (A.L.)

* Correspondence: hminiguano@gmail.com; Tel.: +34-916-24-9188

Received: 22 March 2019; Accepted: 7 May 2019; Published: 10 May 2019

Abstract: Supercapacitors with characteristics such as high power density, long cycling life, fast charge, and discharge response are used in different applications like hybrid and electric vehicles, grid integration of renewable energies, or medical equipment. The parametric identification and the supercapacitor model selection are two complex processes, which have a critical impact on the system design process. This paper shows a comparison of the six commonly used supercapacitor models, as well as a general and straightforward identification parameter procedure based on Simulink or Simscape and the Optimization Toolbox of Matlab®. The proposed procedure allows for estimating the different parameters of every model using a different identification current profile. Once the parameters have been obtained, the performance of each supercapacitor model is evaluated through two current profiles applied to hybrid electric vehicles, the urban driving cycle (ECE-15 or UDC) and the hybrid pulse power characterization (HPPC). The experimental results show that the model accuracy depends on the identification profile, as well as the robustness of each supercapacitor model. Finally, some model and identification current profile recommendations are detailed.

Keywords: supercapacitor models; parameter estimation; ECE15; HPPC; Simulink; Simscape; Matlab; Identification

---

1. Introduction

Energy storage systems are essential in the industrial, medical, renewable or transportation sectors, as well as other sectors. Some characteristics like high power density, reliability and safety are critical in those sectors, this is why the electrochemical double layer capacitor or the supercapacitor play an important role [1].

Many application areas in which supercapacitors are used can be mentioned like magnetic resonance imaging (MRI) that needs very short pulses with high current [2] or fuel cell supercapacitor hybrid bus, where the supercapacitor satisfy the dynamic power demand [3]. In addition, the supercapacitor can be used for the integration of a photovoltaic power plant [4], grid integration of renewable energies [5] and the improvement of energy utilization for mine hoist applications [6]. However, many applications are limited by the self-discharge behavior in wireless sensor network applications [7], where the new techniques of chemical modification to suppress this phenomenon are shown in reference [8] and reference [9].

In general, the supercapacitors models classify into three categories: electrochemical, mathematical, and electrical. Electrochemical models consist of a set of partial differential-algebraic equations with many parameters. The estimation of the electrochemical model is very accurate [10]. However, the simulation of these models consumes many resources. Mathematical models are an alternative based on three dimensional ordered structures [11]. It can get a good fitting with experimental data but with a complex process to get the different parameters. Finally, circuit-based or electrical models are able to reproduce the electrical behavior of supercapacitors with equivalent circuits [12].

In the literature, there are some studies comparing supercapacitor models. Reference [13] reviews three types of equivalent circuits with linear components, with only an identification current profile and several verification current profiles. These models are the classic model, the multi-stage ladder model, and the dynamic model, which are used in electric vehicle applications. In this case, a genetic algorithm (GA) is used to estimate the different constant parameters of the resistors and capacitors (*RC*) circuits. Reference [14] analyzes three basic constant parameters RC networks models showing the relationship among them. However, as shown in reference [15], the model accuracy can be improved with a nonlinear equivalent circuit model. In reference [16], the authors compared three circuits models (Miller Model, Zubieta Model, and Thevenin Model) with a specific identification current profile for every model. In general, the papers found in the state-of-the-art compare some of the known supercapacitor models, applying different identification current profiles, and using different parameters identification procedures, as it is difficult to obtain reliable conclusions to identify the best model for every application.

The main contribution of this paper is the proposal of a general, practical and effective parameters identification procedure applied to supercapacitors models and obtained in offline mode. The parameters of this model can also be used as an initial estimation of the parameters in online supercapacitor models [17]. The numeric optimization is developed by means of the interactive interface provided by the Identification Toolbox of Matlab (Version R2018b, MathWorks, Natick, MA, USA), once the equivalent models are built in Simulink or Simscape. In addition, the paper shows the comparison of different identification current profiles applied to six kinds of models in order to obtain the best features of each model, as well as the best accurate vs. complexity model.

The next sections are organized as follows: Section 2 shows the six supercapacitor models selected to make the comparative study, as well as their circuits implemented in Simulink or Simscape. Section 3 describes the parameters estimation procedure. Section 4 depicted the different current profiles and the experimental setup to get the supercapacitor voltage and current responses. Section 5 shows the obtained statistical metrics using ECE15 and HPPC dynamic driving cycles, and the discussion about the experimental vs. simulation results. Finally, in Section 6, the main conclusions are presented.

2. Supercapacitors Models

In this section, six representative supercapacitor models are selected from the literature, which cover most of the typical applications. All of them are nonlinear models since this kind of models obtains better accuracy. The selected models are the Stern-Tafel Model [18], Zubieta Model [19], Series Model [20], Parallel Model [21], Transmission Line Model [22] and Thevenin Model [23]. In this section, the electrical equivalent circuit and the parameters of each model are reviewed.

2.1. Stern-Tafel Model

The supercapacitor proposed in reference [24] and reference [25] uses the Stern-Tafel model to describe the nonlinear capacitance. This electrochemical model reproduces the double layer capacitance (C_T) related to the nonlinear diffusion dynamics. To do this, the supercapacitor model combines both the Helmholtz's capacitance (C_H) and Gouy-Chapman's capacitance (C_{GC}) [26],

$$C_T = \frac{N_p}{N_s} \cdot \left(\frac{1}{C_H} + \frac{1}{C_{GC}} \right)^{-1} \quad (1)$$

Being

$$C_H = \frac{N_e \cdot \varepsilon \cdot \varepsilon_0 \cdot A_i}{d} \quad (2)$$

$$C_{GC} = \frac{F \cdot Q_T}{2 \cdot N_e \cdot R \cdot T} sinh\left(\frac{Q_T}{N_e^2 \cdot A_i \sqrt{8 \cdot R \cdot T \cdot \varepsilon \cdot \varepsilon_0 \cdot c}}\right) \tag{3}$$

where N_p is the number of parallel supercapacitor cells, N_s is the number of series of supercapacitor cells, N_e is the number of layers of electrodes, d the molecular radius (m), c the molar concentration (mol.m^{-3}), A_i is the interfacial area between electrode and electrolyte (m^2), T is the operating temperature (K), F_c is the Faraday constant (C/mol), R is the ideal gas constant (J/(K·mol)), ε is the relative permittivity of the electrolyte material (F/m), and ε_0 is the free space permittivity (F/m) [18].

The model equivalent circuit has a controlled voltage source and an internal resistance, as shown in Figure 1a. This model depends on several parameters where C_n is the nominal capacitance (F), V_{max} is the maximum supercapacitor voltage (V), R_{dc} is the internal resistance (Ω), V_T is the total voltage (V), and i_{sd} is the self-discharge current (A) which is determined by the Tafel Equation (4) described in reference [27] as:

$$i_{sd}(t) = N_e \cdot I_f \cdot e^{\left(\frac{\alpha \cdot F_c \cdot \left(\frac{V_{init}}{N_s} - \frac{V_{max}}{N_s} - \Delta V\right)}{R \cdot T}\right)} \tag{4}$$

where I_f is the leakage current (A), V_{init} is the initial voltage (V), α is the charge transfer coefficient and ΔV is the over-potential (V). The capacitance of the electrochemical model requires only a few data from manufacturer datasheet and universal constant as described in reference [28]. The Simulink implementation is shown in Figure 1b.

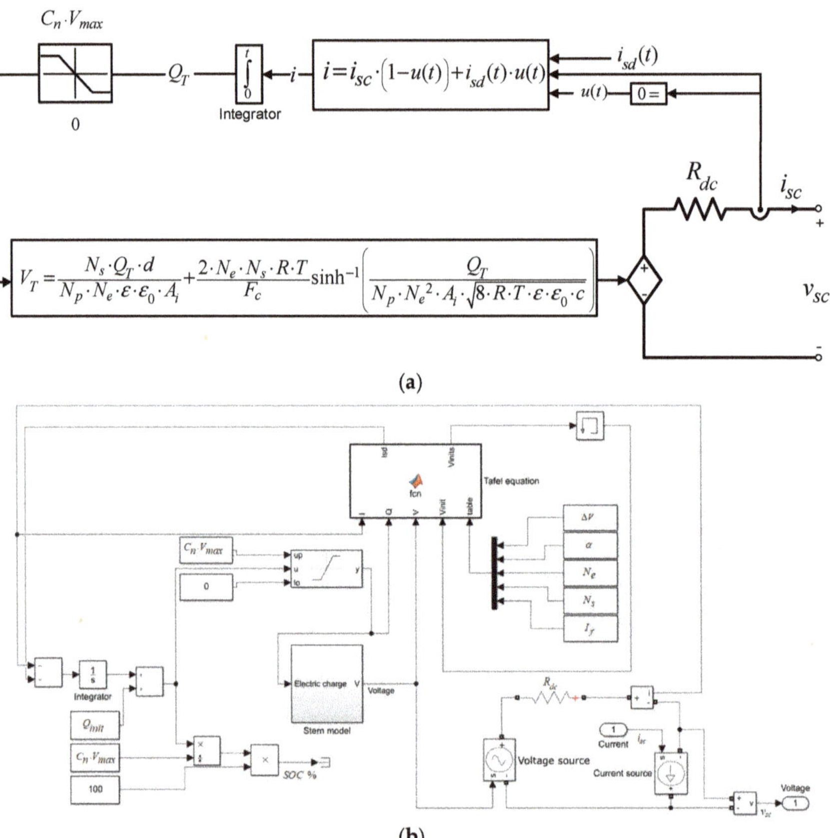

Figure 1. Stern-Tafel model: (**a**) Electric circuit; (**b**) Simulink implementation.

2.2. Zubieta Model

The proposed model in reference [19] includes a circuit with three parallel *RC* time constant, Figure 2a. The first branch, with the elements R_0C_0, and the voltage-dependent $k_c{\cdot}v_c$ defines the response in seconds. The second branch R_1C_1 provides the response in the range of minutes. The branch R_2C_2 represents the response for a time longer than minutes. Finally, a resistor R_{lk} reproduces the leakage resistance.

A simplified equivalent circuit with two branches is shown in reference [29], with a simplified parameter identification procedure through the differential equation of the circuit. Similar studies are proposed in reference [30] in which the model parameters are easily obtained when the supercapacitor is discharged with constant power. In addition, reference [31] proposes a multivariable minimization function to find the parameters, they are validated with a current profile of a hybrid electric vehicle.

The total capacitance and current of the voltage-controlled capacitance implemented in Simscape are shown in Figure 2b, which are defined by (5) and (6):

$$C(v_c) = C_0 + k_c{\cdot}v_c \tag{5}$$

$$i_c = \frac{dQ}{dt} = \frac{d(C(v_c){\cdot}v_c)}{dt} = (C_0 + 2k_c{\cdot}v_c)\frac{dv_c}{dt} \tag{6}$$

where C_0 is the initial linear capacitance which represents the electrostatic capacitance of the capacitor, and k_c a positive coefficient which represents the effects of the diffused layer of the supercapacitor.

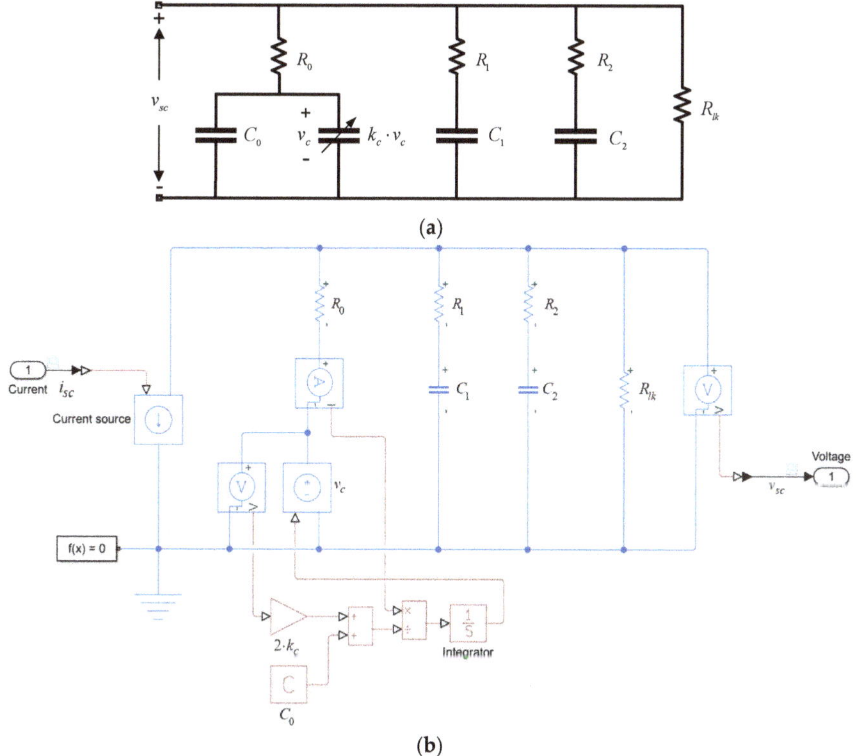

Figure 2. Zubieta model: (**a**) Electric circuit; (**b**) Simscape implementation.

2.3. Series Model

The series model is an equivalent circuit obtained through the AC impedance approach, which consists of two parallel *RC* circuit compound by $R_1(vsc)$, $C_1(vsc)$, $R_2(vsc)$, $C_2(vsc)$, connected in series with another *RC* circuit compound by R_s and $C_s(vs)$, as described in references [20,32,33]. This equivalent circuit shows in the first branch of Figure 3a. In reference [34] a modified version of this circuit was presented, which includes the model proposed by Buller and Zubieta, in order to represent a complete model for a full frequency range. This complete model includes three branches in a parallel compound by R_3 and C_3, R_4 and C_4, and the leakage resistance R_{lk}, as shown in Figure 3a. Figure 3b shows the Simscape implementation of the modified series model.

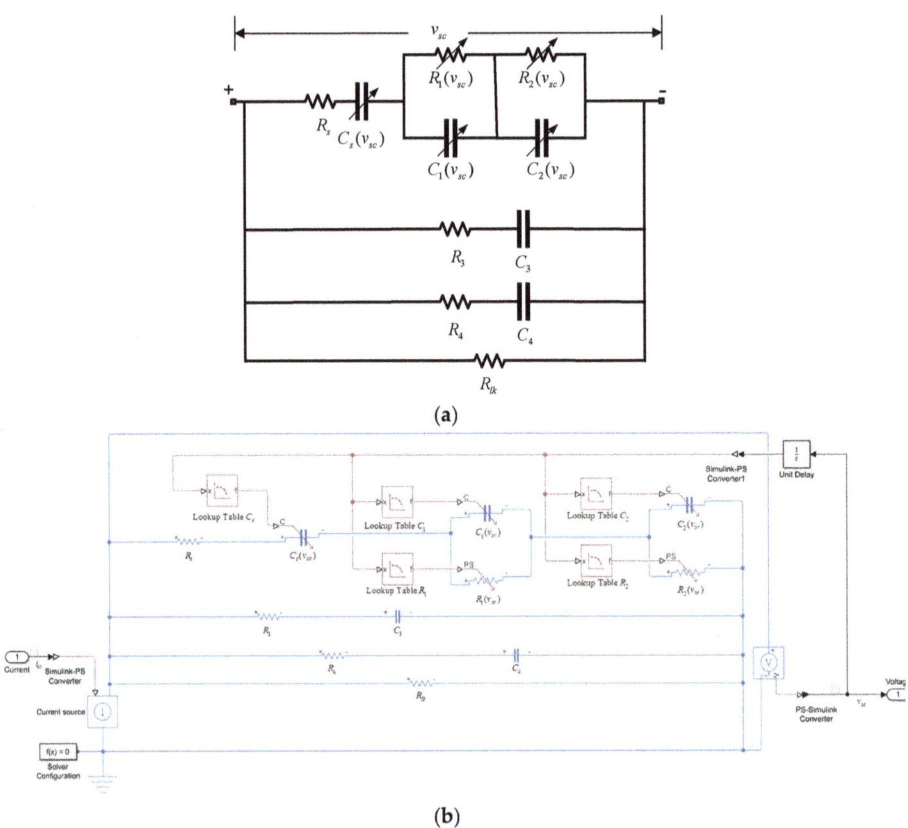

Figure 3. Series model: (**a**) Electric circuit; (**b**) Simscape implementation.

2.4. Parallel Model

The basic parallel model with constant values is described in reference [35] and reference [36]. Reference [37] describes an approximation to calculate the parameters without data acquisition, only using the information provided by a supercapacitor datasheet, as well as the main basic equations to obtain the constant parameters using this information. A modified four parallel *RC* networks with voltage-dependent parameters are presented in reference [21], and it is shown in Figure 4a. This model is more complex, but it achieves better accuracy. Figure 4b shows the implementations of the modified parallel model in Simscape.

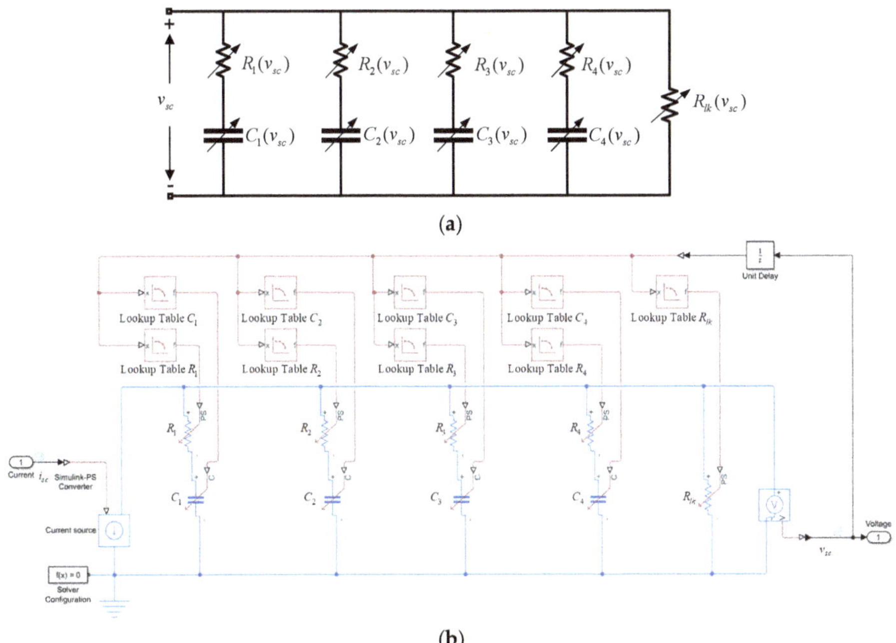

(a)

(b)

Figure 4. Series model: (**a**) Electric circuit; (**b**) Simscape implementation.

2.5. Transmission Line Model

Transmission line model is composed of *nRC* branches in order to reproduce the supercapacitor frequency response from 10 mHz to 1 kHz. This model was proposed for hybrid and electric vehicles, and it was described in reference [38] and reference [39]. This model consists of four parallel networks based on R_1, $C_1(v_1)$, R_2, $C_2(v_2)$, R_3, $C_3(v_3)$ and R_4, $C_4(v_4)$, and a parallel leakage resistance R_{lk}, as shown in Figure 5a. Reference [22] describes a procedure to estimate the parameters through time response and the equations of the circuit. Also, this model is used to evaluate the supercapacitor physical aging process [40], by estimating the uncertainties of the parameters. Reference [41] uses a different number of networks according to the simulation time step.

Figure 5b shows the model implemented in Simscape with the described Equations (5) and (6).

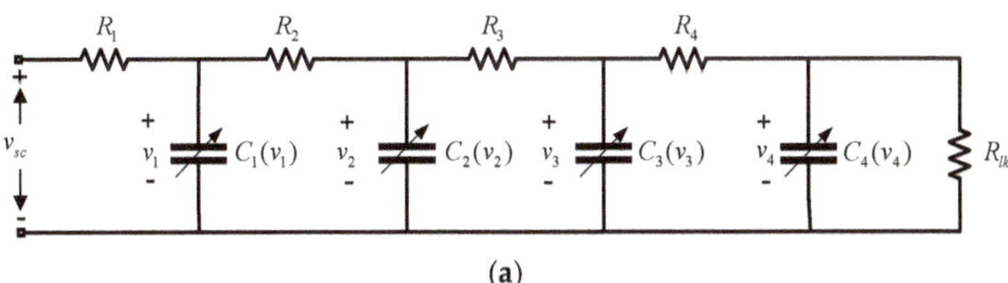

(a)

Figure 5. *Cont.*

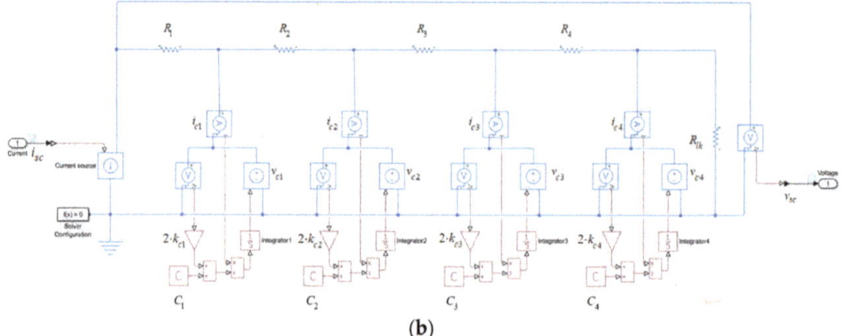

(**b**)

Figure 5. Series model: (**a**) Electric circuit; (**b**) Simscape implementation.

2.6. Thevenin Model

The equivalent electric circuit of the Thevenin model, which includes several parallel RC and a nonlinear state-of-charge (*SOC*) voltage-dependent source is described in reference [42]. The *SOC* is calculated by coulomb counting using (7):

$$SOC = \frac{Q_{init} - \int_0^t i(\tau)d\tau}{Q_T} \tag{7}$$

with Q_{init} being the initial supercapacitor charge, Q_T being the total supercapacitor charge and $i(\tau)$ as the supercapacitor current.

In this paper, three *RC* branches are used to get a better accuracy, where *OCV* represents the open circuit voltage, R_0 represents the internal resistance, and three parallel networks based on R_1, C_1, R_2, C_2, R_3, and C_3 reproduce the supercapacitor dynamic, as shown in Figure 6a. All parameters are state-of-charge dependent. The proposed model applied to a hybrid storage system for an electric vehicle gives a better agreement for a simulated vs. experimental response when 3-branches are used in the model [23]. Figure 6b shows the Simscape implementation.

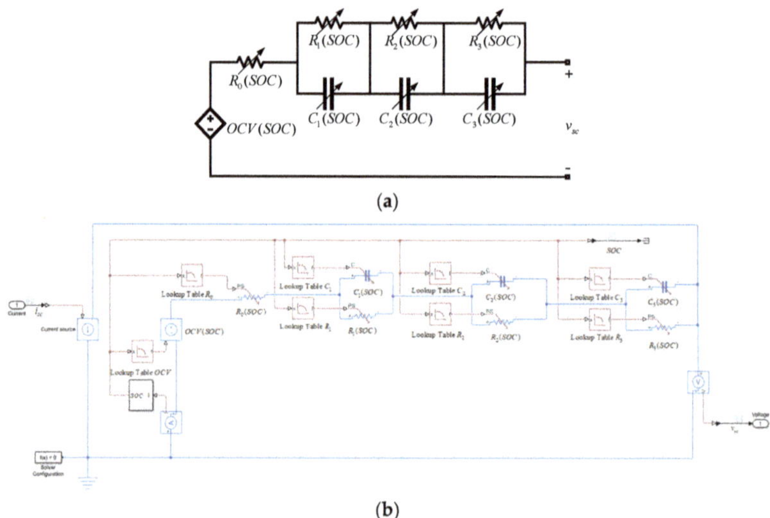

(**a**)

(**b**)

Figure 6. Series model: (**a**) Electric circuit; (**b**) Simscape implementation.

3. Parameters Estimation Procedure

Parametric models explicitly contain differential equations, transfer function or block diagrams. The parameters update could be offline or online. For obtaining the parameters, in the offline mode, the data are stored to later process, on the other hand, in the online mode, the procedure is executed in parallel to the experiment [43]. In the literature, there are many proposed procedures to obtain the model parameters such as e.g., the unscented Kalman filter [44] or the Luenberger-style technique [17].

Taking into account the literature, this paper focuses on the proposal of a practical, interactive, simple and enough general offline procedures for estimating the model parameters.

Figure 7a shows the proposed identification procedure block diagram. This procedure can be divided into several steps, shown and described in Table 1.

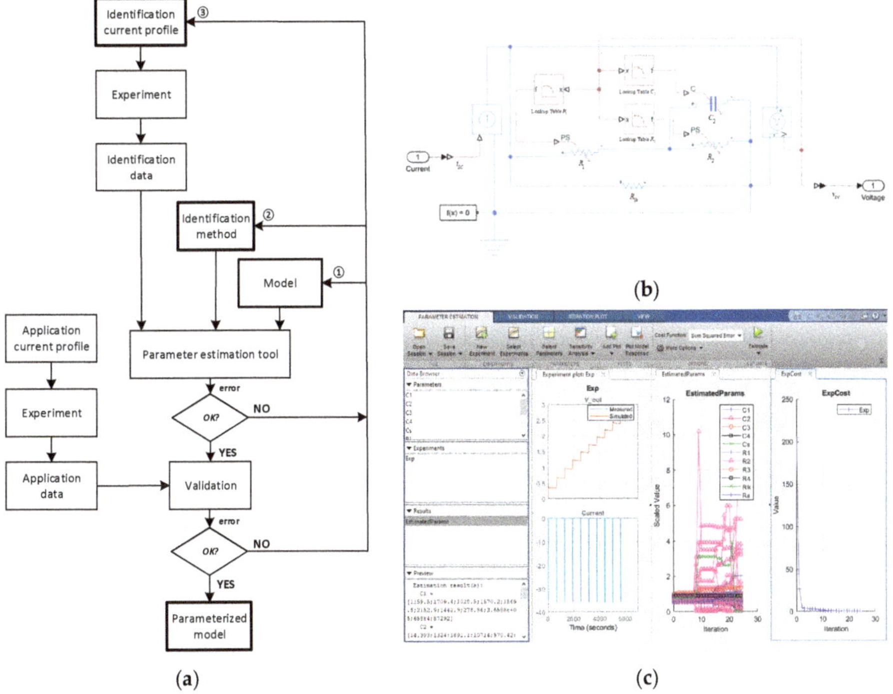

Figure 7. Parameter estimation procedure: (**a**) Identification block diagram process; (**b**) Simscape model; (**c**) Interactive interface by Simulink [45].

Table 1. Parameter Estimation Steps.

Steps	Description
1	Apply the identification current profile to obtain supercapacitor current and voltage waveforms (identification data) from the experimental test. E.g., as shown in Section 4.2: current profiles and supercapacitor voltage response (a), (b) and (c).
2	Select and build the equivalent circuit model in Simulink or Simscape through a block diagram or circuit. E.g., as shown in Section 3: Figure 7b.
3	Create a new experiment in Simulink and to import the identification data. Simulate the model with the initial parameters and the identification current profile to obtain the simulation data. E.g., as shown in Section 3: Figure 7c.
4	Choose the variables and their limits to estimate their value. E.g., as shown in Section 3: Figure 7c.

Table 1. *Cont.*

Steps	Description
5	Set up optimization options (optimization method, algorithm, and parameter and function tolerance). E.g., as shown in Section 3: Figure 8.
6	Run the parameter estimation process applying the selected optimization solver (E.g., sum-squared error) to match the identification data with the simulation data. E.g., as shown in Section 3: Figure 7c. If the error is not small enough, return to step 1 (①); or change the identification method and return to step 3 (②); or modify the current profile and return to step 2 (③), Figure 7a.
7	Once the model parameters have been obtained from the identification data, the next step is to verify the model response using the application current profile and the application data. For that, it is necessary to compare the application data with the new simulated data, using the obtained parameters in step 6, E.g., as shown in Section 4.2: Figure 9d,e. If the error is not small enough, return to step 1 (①); or change the identification method and return to step 3 (②); or modify the current profile and return to step 2 (③), Figure 7a.

In step 5, the optimization method has to be selected. This paper uses an offline parameters estimation based on the error minimization between the measured and simulated supercapacitor voltage. The iterative procedure tunes the supercapacitors model parameters (p) to get a simulated response (V_s) that tracks the measured response (V_m), with a finite number of samples (n). To do that, the solver minimizes the next cost function for each current profile:

$$F(p) = \min_{p} \sum_{i=0}^{n} \left[V_{m_i} - V_{s_i}(p) \right]^2 \tag{8}$$

where p varies between zero and infinity (e.g., 0 to 10^{10}).

The minimization problem is carried out with Simulink® Design Optimization™ of Matlab (Version R2018b, MathWorks, Natick, MA, USA). This toolbox provides an interactive interface that helps to minimize the square of the error between the measured and simulated supercapacitor voltage, using the nonlinear least squares method for parameters estimation. This method is selected in the user interface as shown in Figure 8.

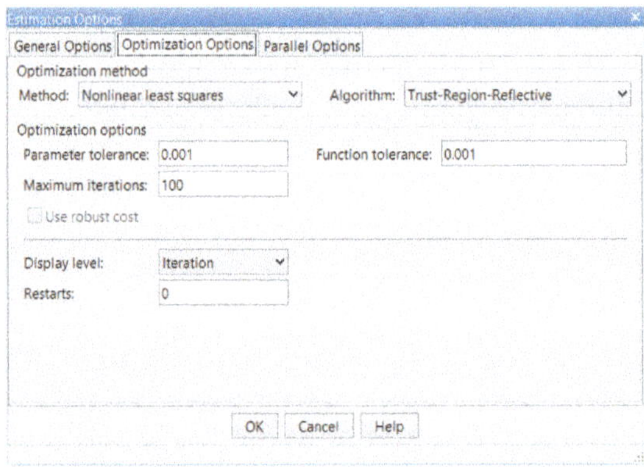

Figure 8. The optimization option user interface for parameter estimation.

This method uses the Simulink function named as lsqnonlin, that requires at least $(2k + 1)$ simulations per iteration, where k is the number of parameters to be estimated [46]. The required CPU time and memory increase as a function of the numbers of parameters and their initial values. The offline runtime estimation is in the order of minutes.

If runtime estimation has to be reduced, other techniques based on the layered technique to break the global optimization into a smaller task [47], or based on differential mutation strategy [48], or based on genetic programming [49], among others, could be used, although the flexibility and simplicity provided by the Simulink user interface could be affected.

On the other hand, the algorithm selected is the Trust-Region-Reflective, which is based on a gradient process with a trial step by solving a trust region. Specific details of the algorithm can be found in reference [45]. Additional information is detailed in reference [50], in which the process of how to import, analyze, prepare and estimate model parameters in Simulink is described.

Using the proposed procedure, based on Simulink® Design Optimization™ of Matlab, the most model can be built, from a practical point of view. Nevertheless, this procedure is limited by the optimization methods and algorithms included in Simulink.

4. Experimental Setup

4.1. Supercapacitor Testing System

The experimental setup includes a supercapacitor, a data acquisition system, a power source, and an electronic load, as shown in Figure 9. The supercapacitor used to develop the test has been the Maxwell BCAP3000. An equivalent bidirectional current source compound of the electronic load and the power source, connected in parallel, emulates the current profile. This equivalent current source includes the typical regenerative breaking present in automotive applications. The experimental current profile and the data acquisition system are conducted using the following set of equipment listed in Table 2:

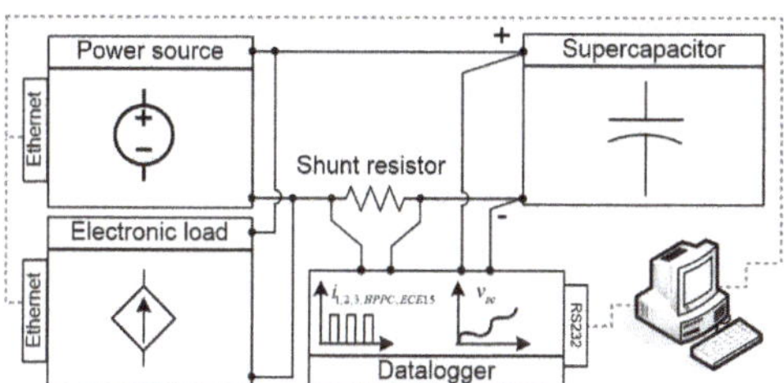

Figure 9. Experimental setup.

Table 2. Equipment and Components used in the Experimental Setup.

Component	Specifications	Use
Supercapacitor: Maxwell BCAP3000	2.7 V/3000 F	Cell under test
Datalogger: Agilent 34970A	100 nV–1000 V/500 kHz	Measure voltage
Power source: Sorensen SGI400/38	400 V/38 A	Current charge
Electronic load: Chroma 63206A-600	600 V/420 A	Current discharge
Shunt resistor: Newtons4th HF200	0.5m Ω/200 mA → 200 Arms	Measure current

All these elements have been synchronized with a computer running to manage the data logging and supervisory control using LabVIEW® software.

4.2. Supercapacitor Test Schedule

The parameter identification procedure uses three different current profiles. The current profile i_1 is a current step, Figure 10a; the current profile i_2 are repetitive charging current steps applied until to reach the maximum supercapacitor voltage, Figure 10b; and the current profile i_3 is a dynamic charge-discharge current step modulated in amplitude and time applied until the middle value of the supercapacitor voltage range, [51], Figure 10c. From the modeling perspective, the validation current profile must be more dynamic in amplitude and frequency than the identification current profile, as shown in Figure 10d,e.

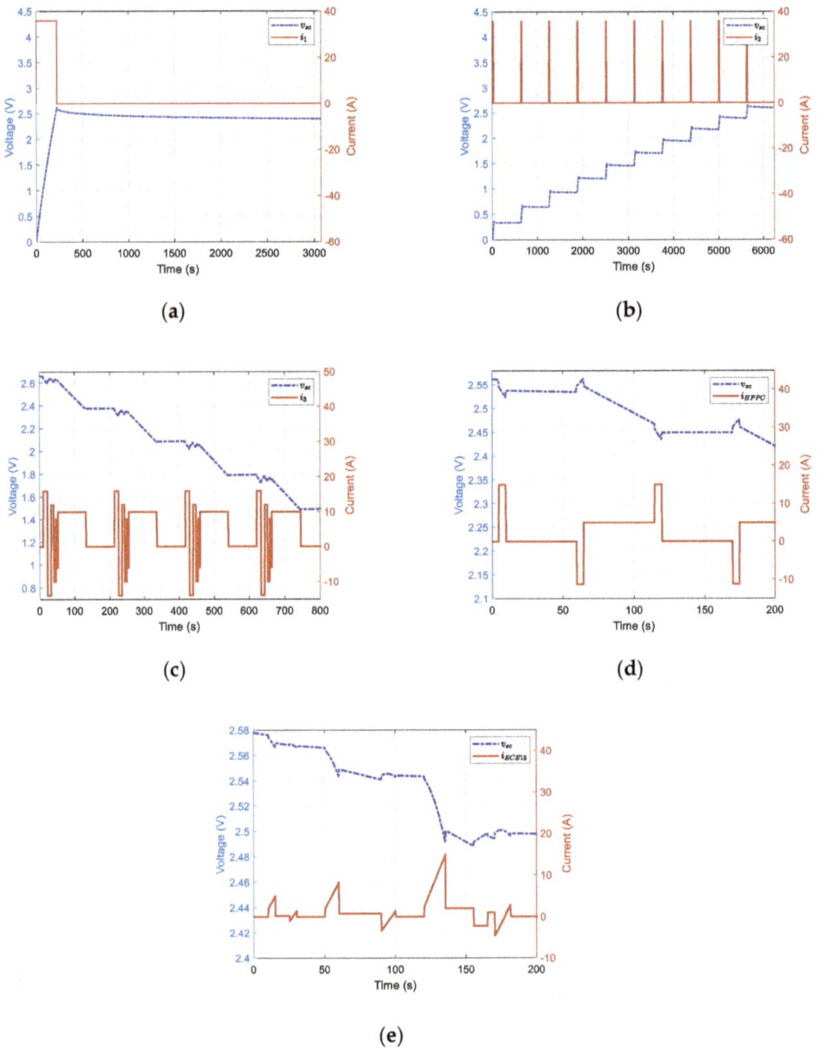

Figure 10. Current profiles and supercapacitor voltage response: **(a)** i_1; **(b)** i_2; **(c)** i_3; **(d)** Verification HPPC profile; **(e)** Verification ECE15 profile.

These identification current profiles apply to those models aforementioned in Section 2 to obtain their parameters. The current profile applied in every model is shown in Table 3.

The robustness and accuracy of the supercapacitor models are evaluated by means of different standardized test profiles, which include the Hybrid Pulse Power Characterization (HPPC) test and European Urban Driving Cycle (ECE15) for long-time responses. Figure 9d shows the HPPC test that is described in the Freedom Car Battery Manual [52]. The ECE15 test, described in reference [53], is a more dynamic current profile, as shown in Figure 10e.

Table 3. Identification Current Profiles Used to Supercapacitor Parameters Estimation.

Model	Current Profile		
	i_1	i_2	i_3
Stern-Tafel	✓	–	–
Zubieta	✓	✓	✓
Series	–	✓	✓
Parallel	–	✓	✓
Transmission line	–	✓	✓
Thevenin	–	✓	✓

✓= Applicable; – = Not applicable.

5. Experimental Results, Comparison, and Discussion

After obtaining the parameters for each model, detailed in Appendix A in Tables A1–A9, using the procedure described in Section 3 and identification current profiles described in Section 4, the output voltage accuracy and robustness analysis for the six supercapacitor models described in Section 2 is performed based on statistical metrics, such as relative error and root-mean-square (RMS) error.

Comparative results with identification current profile i_1 are illustrated in Figure 11a–d for the HPPC test and Figure 11e–h for the ECE15 test. Figure 11a,e show the experimental supercapacitor voltage and the voltages provided by the Stern-Tafel and Zubieta models. Figure 11b,f show the relative error between these models and the experimental data.

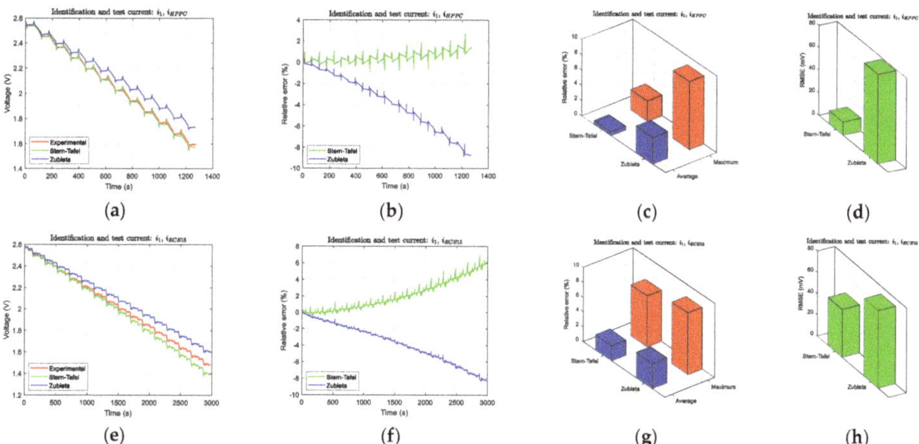

Figure 11. Experimental data and supercapacitor models time response, relative error (%), RMS error (mV): **(a)**–**(d)** current profile HPPC for i_1, **(e)**–**(h)** current profile ECE15 for i_1.

Figure 11c,g represent the relative error in percentage. Figure 11d,h show the RMS error in mV. It shows that the Stern-Tafel model has lower error values in comparison with the Zubieta model. In any case, the relative error tendency with the time increase in both models, therefore the accuracy of both models identified with the i_1 current profile is not proper.

Similar information is shown when current profile i_2 is used to obtain the model parameters. Figure 12a–d depicted the obtained result for the HPPC test and Figure 12e–h for the ECE15 test. This current profile is applied to five out of the six models, with the exception of the Stern-Tafel model. In this case, the Series model is the best one, since it presents a reduced relative error that maintained with the time.

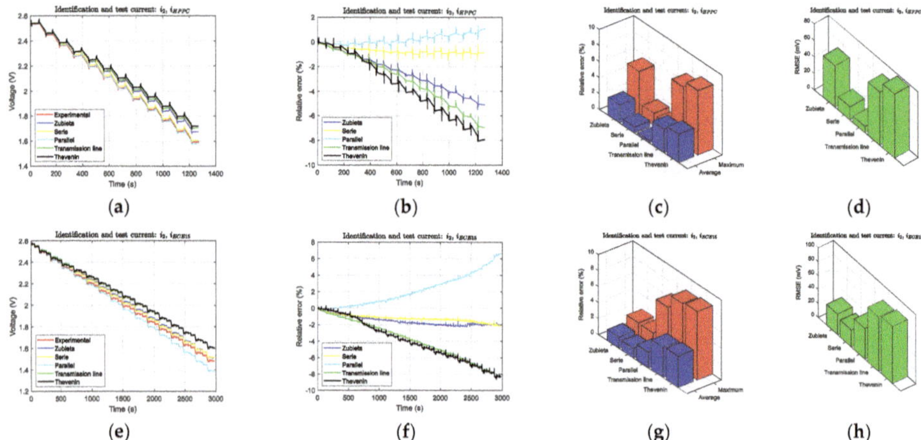

Figure 12. Experimental data and supercapacitor models time response, relative error (%), RMS error (mV): (**a**)–(**d**) current profile HPPC for i_2, (**e**)–(**h**) current profile ECE15 for i_2.

Finally, the result obtained with the current profile i_3, which is the most dynamic current profile, is depicted in Figure 13a–d for the HPPC test and Figure 13e–h for the ECE15 test. This current profile has been applied to the same models as current profile i_2. Again, the Serie Model has the best performance, and even the obtained relative error is lower than using the previous current profiles. Nevertheless, the Parallel model, Transmission Line model and Thevenin model get good behaviors.

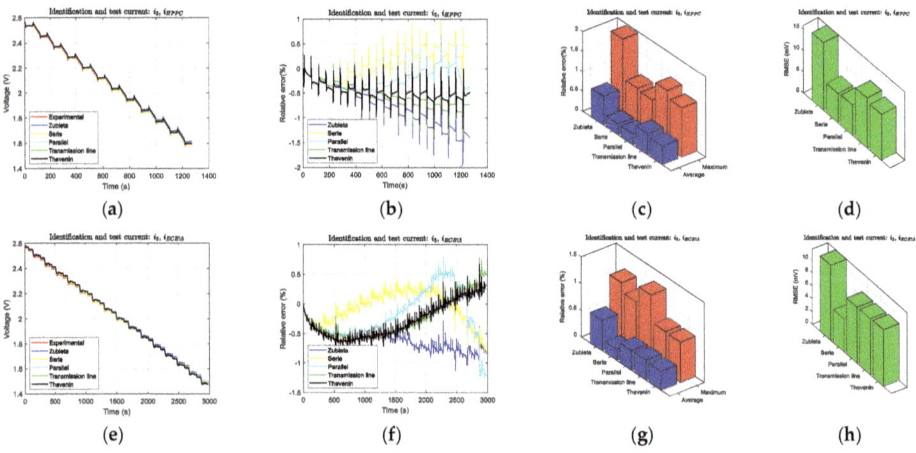

Figure 13. Experimental data and supercapacitor models time response, relative error (%), RMS error (mV): (**a**)–(**d**) current profile HPPC for i_3, (**e**)–(**h**) current profile ECE15 for i_3.

The main conclusions obtained from these results are the following:

- The greater complex identification current profile i_3 gets greater accuracy for every model in which it can be applied.
- In most cases, the Series model provides the minimum relative error.
- If a simple and basic supercapacitor model has to be used, the best option is to use Zubieta model identified with the current profile i_3.

Tables 4–6 include the numeric values for different current profiles identification and the response of each model for HPPC and ECE15 test. These values are those shown in Figures 10–12.

Table 4. The statistical metric with Current Profile i_1.

Model	HPPC			ECE15		
	Maximum Relative Error (%)	Average Relative Error (%)	Root Mean Square Error (mV)	Maximum Relative Error (%)	Average Relative Error (%)	Root Mean Square Error (mV)
Stern-Tafel	2.6624	0.4975	11.6037	6.9044	2.0402	45.1645
Zubieta	8.8702	3.5055	79.0721	8.2994	3.3861	72.6719

Table 5. The statistical metric with Current Profile i_2.

Model	HPPC			ECE15		
	Maximum Relative Error (%)	Average Relative Error (%)	Root Mean Square Error (mV)	Maximum Relative Error (%)	Average Relative Error (%)	Root Mean Square Error (mV)
Zubieta	5.5291	2.2657	50.5217	2.2409	1.4013	29.2705
Serie	1.5156	0.7592	15.7086	2.1871	1.2212	24.2654
Parallel	1.2932	0.2971	7.3963	6.6324	1.9509	43.9770
Transmission line	7.2856	2.9581	65.0135	8.3307	3.8223	78.3289
Thevenin	8.0835	3.5226	77.9992	8.5904	3.9647	82.2516

Table 6. The statistical metric with Current Profile i_3.

Model	HPPC			ECE15		
	Maximum Relative Error (%)	Average Relative Error (%)	Root Mean Square Error (mV)	Maximum Relative Error (%)	Average Relative Error (%)	Root Mean Square Error (mV)
Zubieta	1.9602	0.7170	15.8874	1.1287	0.5673	11.6797
Serie	0.9718	0.2728	6.8365	0.9562	0.2119	5.1513
Parallel	0.9263	0.2714	7.0952	1.2706	0.3724	8.7621
Transmission line	1.4099	0.5227	11.6305	0.7700	0.3722	8.7803
Thevenin	1.2297	0.4617	10.7354	0.7593	0.3493	8.8829

6. Conclusions

This paper describes a parameter identification general procedure with a flexible and interactive interface used to build supercapacitor models in Simulink or Simscape. This procedure enables estimating the different models parameters based on the use of the Optimization Toolbox of Matlab®. Once, the procedure steps are explained, the procedure is used to develop a comparative study of six commonly used supercapacitor models. In addition, the procedure enables using different identification current profiles, providing the possibility of analyzing the influence of three different identification current profiles in the accuracy and robustness of every model.

The experimental results obtained from the six models and three different identification current profiles, used to develop the study, show that both the model and the identification current profile are critical to obtaining good accuracy and robustness, which must be maintained over time.

From the comparison between the experimental results and the simulation results obtained using the model, it can be concluded that the greater complexity of the current identification profile, the greater accuracy and robustness of the model. In this case, the most complex identification current profile i_3 gets the best accuracy for every model in which it can be applied.

In a short simulation period, most models provide enough accuracy results. However, in a long simulation period the differences among models as well as among the current identification profiles increase, and models responses cumulate voltage errors and, in some cases, they cannot correctly represent the voltage of the supercapacitor. The Stern-Tafel model is proper for a short simulation and as a first approximation. However, in a long-time simulation, the Series Model represents a good performance, followed by the Parallel Model. In most cases, the Series model provides the minimum relative error. However, the Zubieta model provides a good compromise between complexity and accuracy. Then, if a simple and basic supercapacitor model has to be used, the best option is to use a Zubieta model identified by means of the current profile i_3.

Author Contributions: Conceptualization, H.M. and A.B.; methodology, A.B.; software, H.M.; validation, A.B., P.Z. and A.L.; formal analysis, C.F.; resources, A.B.; writing—original draft preparation, H.M.; writing—review and editing, P.Z. and A.L.; supervision, C.F.; funding acquisition, A.B.

Funding: This research was funded by the Spanish Ministry of Economy and Competitiveness and ERDF, grant number DPI2014-53685-C2-1-R.

Acknowledgments: This work has been supported by the Ministry of Economy and Competitiveness and FEDER funds through the research project "Storage and Energy Management for Hybrid Electric Vehicles based on Fuel Cell, Battery and Supercapacitors"—ELECTRICAR-AG-(DPI2014-53685-C2-1-R).

Conflicts of Interest: The authors declare no conflicts of interest.

Appendix A. Tables of Supercapacitors Parameter

Table A1. Stern-Tafel Model Parameters with current profile i_1.

Parameter	Value
C (F)	3000
R_{dc} (mΩ)	2.1000
I_f (mA)	5.2000
V_n (V)	2.7000
N_s	1
N_p	1
N_e *	2
d (nm) *	1.0115
α *	0.3200 ($0 < \alpha < 1$)
ΔV *	0.4100

* = Estimated parameters.

Table A2. Zubieta Model Parameters.

Parameter	Current Profile		
	i_1	i_2	i_3
R_0 (mΩ) *	1.1080	0.8653	0.6504
C_0 (F) *	2290.3000	2172.6000	2081.7000
k_c (F/V) *	244.4400	240.5600	220.1800
R_1 (Ω) *	16.9130	19.9280	10.5170
C_1 (F) *	471.1500	368.1700	111.1500
R_2 (Ω) *	0.6729	0.3773	3.5770
C_2 (F) *	292.0100	176.3200	382.0700
R_{lk} (kΩ) *	171210	11023	51.4030

* = Estimated parameters.

Table A3. Series Model Parameters with current profile i_2.

Parameter	Voltage (V)										
	0	**0.330**	**0.6417**	**0.9295**	**1.1977**	**1.4522**	**1.6973**	**1.934**	**2.1622**	**2.3830**	**2.5948**
C_s (kΩ) *	2.4655	2.3629	2.5419	2.6699	2.8260	2.8941	2.9450	2.962	3.0666	3.1518	3.5999
R_1 (μΩ) *	581.02	470.26	563.94	546.61	540.87	507.120	544.94	238.4	0.0080	1.3731	1.6410
C_1 (kF) *	1.1595	1.7094	1.0285	1.5702	1.5695	2.1529	1.4429	0.277	265.980	65.884	87.292
R_2 (μΩ) *	0.120	58.126	14.810	25.842	43.970	108.870	0.0164	6.128	707.200	692.73	501.860
C_2 (F) *	14.393	1324	1891.1	10714	570.42	34.5400	58.919	13.51	43.7690	161.84	54.3090
R_3 (Ω) *						10.6610					
C_3 (kF) *						1.0877					
R_4 (Ω) *						14.4660					
C_4 (F) *						32.5750					
R_{lk} (kΩ) *						5.5436					

* = Estimated parameters.

Table A4. Series Model Parameters with current profile i_3.

Parameter	Voltage (V)				
	1.4932	**1.7931**	**2.0874**	**2.3761**	**2.6618**
C_s (kΩ) *	2.8890	2.9626	3.0489	3.1745	3.1935
R_1 (μΩ) *	192.3800	460.1300	428.0600	182.1800	437.5600
C_1 (F) *	15.3320	923.8500	30.6020	9.3121	229.6200
R_2 (μΩ) *	166.3700	1.4114	32.8880	303.2600	99.7610
C_2 (kF) *	0.0309	0.2120	1.9380	1.9937	0.1050
R_3 (Ω) *			7.6440		
C_3 (kF) *			1.2486		
R_4 (Ω) *			14.0350		
C_4 (F) *			110.7000		
R_{lk} (kΩ) *			1584.4000		

* = Estimated parameters.

Table A5. Parallel Model Parameters with current profile i_2.

Parameter	Voltage (V)										
	0.0002	**0.3301**	**0.6417**	**0.9295**	**1.1977**	**1.4522**	**1.6973**	**1.9337**	**2.1622**	**2.3830**	**2.5948**
C_1 (kF) *	11.494	56.47	86.826	0.2254	82.203	0.00353	100.120	219.83	125.880	182.96	170
R_1 (Ω) *	18502	91.543	1980.4	15.8680	421.90	2545.4	51657	857.68	398.850	254.02	279.950
C_2 (F) *	561.28	258.20	362.59	239.440	57.1310	65.969	56.8820	112.78	134.450	46.676	50.050
R_2 (mΩ) *	0.00358	11.937	0.3757	1.9524	155.180	137.89	117.090	14.308	13.746	4622.1	500.020
C_3 (F) *	1068.9	1780.7	0.3458	43.1960	47.3320	0.1059	665.260	8.367	12.691	30.798	25.003
R_3 (mΩ) *	2.9311	5.5832	0.00014	0.6223	16.946	3.0161	5.6190	1.2579	0.01369	369.01	100
C_4 (kF) *	1.8016	0.0452	2.4350	2.4355	2.7939	2.7622	2.4572	3.013	3.0641	3.3026	3.119
R_4 (mΩ) *	6.4737	8.4052	0.1441	1.0450	0.5837	0.6355	0.6429	0.4083	0.3057	0.5758	0.199
R_{lk} (Ω) *	1.6786	15.397	13.829	25.6290	11.714	13.701	17.801	17.821	16.065	16.649	18.947

* = Estimated parameters.

Table A6. Parallel Model Parameters with current profile i_3.

Parameter	Voltage (V)				
	1.4932	**1.7931**	**2.0874**	**2.3761**	**2.6618**
C_1 (kF) *	21.1090	652.920	119.440	503.490	28.4650
R_1 (Ω) *	0.3961	765.50	364.790	112.370	14.1460
C_2 (F) *	59.1990	123.90	110.42	126.430	26.3940
R_2 (Ω) *	9.1786	0.12605	0.11993	0.09205	0.01963
C_3 (F) *	657.710	15.920	21.9840	17.6280	15.6860
R_3 (mΩ) *	16.0070	4.7105	11.4340	0.0615	12.3470
C_4 (kF) *	3.0746	2.867	2.9347	3.0351	3.1868
R_4 (uΩ) *	507.890	657.90	689.490	470.030	850.340
R_{lk} (Ω) *	120.930	768.810	6150.30	585.340	303.670

* = Estimated parameters.

Table A7. Transmission Line Model Parameters.

Parameter	Current Profile	
	i_2	i_3
R_1 (μΩ) *	826.72	575.75
R_2 (mΩ) *	115.67	2.7568
R_3 (Ω) *	1.3558	0.0116
R_4 (Ω) *	3.9621	1.5723
k_{c1} (F/V) *	209.20	85.715
k_{c2} (F/V) *	26.6080	21.482
k_{c3} (F/V) *	47.1510	13.4470
k_{c4} (F/V) *	11.8540	15.2340
C_1 (F) *	2303.30	2408.90
C_2 (F) *	0.1000	5.2926
C_3 (F) *	21.2820	120.67
C_4 (F) *	37.9010	55.4780
R_{lk} (kΩ) *	111.01	13.8080

* = Estimated parameters.

Table A8. Thevenin Model Parameters with i_2 Current Profile.

Parameter	*SOC* (%)										
	0.01	**10**	**20.04**	**29.96**	**39.95**	**49.86**	**59.89**	**69.77**	**70.77**	**89.67**	**99.64**
OCV (V)	0.0002	0.330	0.6417	0.9295	1.1977	1.4522	1.6973	1.9337	2.162	2.3830	2.5953
C_1 (F) *	17046	551.64	236.58	303.76	363.06	509.86	434.30	401.51	227.47	411.24	147.11
C_2 (kF) *	119.28	29.58	35.427	30.323	39.527	22.579	20.577	771.57	11.614	94.522	19.695
C_3 (kF) *	2943.8	498.12	248.36	205.07	541.22	449.09	844.16	97.252	2507.5	703.25	3.772
R_0 (μΩ) *	723.2	431.02	0.2476	13.381	104.41	17.20	3.7923	213.26	115.96	23.528	372.15
R_1 (mΩ) *	1.2436	0.9416	1.2504	1.1690	1.1998	1.3676	1.3495	1.2395	1.0854	1.1086	1.2487
R_2 (mΩ) *	68.622	10.261	1.1819	35.648	3.3491	0.32405	0.4529	7.7754	14.9250	1.6529	186.44
R_3 (Ω) *	33.592	22.765	0.9059	2.143	2.6865	1.9369	432.480	548140	29422	558.260	555

* = Estimated parameters.

Table A9. Thevenin Model Parameters with i_3 Current Profile.

Parameter	*SOC* (%)				
	53.37	**64.68**	**75.97**	**87.25**	**98.59**
OCV (V)	1.4786	1.7794	2.0742	2.365	2.6499
C_1 (F) *	0.070731	11.1610	11.1610	12.30	123.90
C_2 (kF) *	7.2709	1.3362	0.7907	16.107	28.906
C_3 (MF) *	1929.30	0.00769	0.00638	76.5800	9.8952
R_0 (μΩ) *	6.5167	13.9100	39.4740	75.8540	53.527
R_1 (μΩ) *	204.02	168.97	35.226	339.940	565.850
R_2 (μΩ) *	684.340	486.780	512.330	352.520	284.940
R_3 (μΩ) *	24.6710	4.8491	7.8730	74.0560	0.25154

* = Estimated parameters.

References

1. Zhang, L.; Hu, X.; Wang, Z.; Sun, F.; Dorrell, D.G. A review of supercapacitor modeling, estimation, and applications: A control/management perspective. *Renew. Sustain. Energy Rev.* **2018**, *81*, 1868–1878. [CrossRef]
2. Ristic, M.; Gryska, Y.; McGinley, J.V.; Yufit, V. Supercapacitor energy storage for magnetic resonance imaging systems. *IEEE Trans. Ind. Electron.* **2014**, *61*, 4255–4264. [CrossRef]
3. Wu, W.; Partridge, J.; Bucknall, R. Development and Evaluation of a Degree of Hybridisation Identification Strategy for a Fuel Cell Supercapacitor Hybrid Bus. *Energies* **2019**, *12*, 142. [CrossRef]
4. Ciccarelli, F.; Di Noia, L.; Rizzo, R. Integration of Photovoltaic Plants and Supercapacitors in Tramway Power Systems. *Energies* **2018**, *11*, 410. [CrossRef]

5. Vazquez, S.; Lukic, S.M.; Galvan, E.; Franquelo, L.G.; Carrasco, J.M. Energy storage systems for transport and grid applications. *IEEE Trans. Ind. Electron.* **2010**, *57*, 3881–3895. [CrossRef]
6. Yang, X.; Wen, P.; Xue, Y.; Zheng, T.; Wang, Y. Supercapacitor energy storage based MMC for energy harvesting in mine hoist application. *Energies* **2017**, *10*, 1428. [CrossRef]
7. Yang, H.; Zhang, Y. A study of supercapacitor charge redistribution for applications in environmentally powered wireless sensor nodes. *J. Power Sources* **2015**, *273*, 223–236. [CrossRef]
8. Wang, H.; Zhou, Q.; Yao, B.; Ma, H.; Zhang, M.; Li, C.; Shi, G. Suppressing the Self-Discharge of Supercapacitors by Modifying Separators with an Ionic Polyelectrolyte. *Adv. Mater. Interfaces* **2018**, *5*, 1701547. [CrossRef]
9. Xia, M.; Nie, J.; Zhang, Z.; Lu, X.; Wang, Z.L. Suppressing self-discharge of supercapacitors via electrorheological effect of liquid crystals. *Nano Energy* **2018**, *47*, 43–50. [CrossRef]
10. Sarwar, W.; Marinescu, M.; Green, N.; Taylor, N.; Offer, G. Electrochemical double layer capacitor electro-thermal modelling. *J. Energy Storage* **2016**, *5*, 10–24. [CrossRef]
11. Wang, H.; Pilon, L. Mesoscale modeling of electric double layer capacitors with three-dimensional ordered structures. *J. Power Sources* **2013**, *221*, 252–260. [CrossRef]
12. Lee, J.; Yi, J.; Kim, D.; Shin, C.; Min, K.S.; Choi, J.; Lee, H.Y. Modeling of the Electrical and Thermal Behaviors of an Ultracapacitor. *Energies* **2014**, *7*, 8264–8278. [CrossRef]
13. Zhang, L.; Wang, Z.; Hu, X.; Sun, F.; Dorrell, D.G. A comparative study of equivalent circuit models of ultracapacitors for electric vehicles. *J. Power Sources* **2015**, *274*, 899–906. [CrossRef]
14. Shi, L.; Crow, M. Comparison of ultracapacitor electric circuit models. In Proceedings of the IEEE Power and Energy Society General Meeting—Conversion and Delivery of Electrical Energy in the 21st Century, Pittsburgh, PA, USA, 20–24 July 2008; pp. 1–6.
15. Xie, C.; Liu, X.; Huang, L.; Fang, W. Modeling of supercapacitor modules in the hybrid powertrain with a nonlinear 3-branch equivalent circuit. *Int. J. Energy Res.* **2018**, *42*, 3524–3534. [CrossRef]
16. Miniguano, H.; Raga, C.; Barrado, A.; Lázaro, A.; Zumel, P.; Olías, E. A comparative study and parameterization of electrical battery models applied to hybrid electric vehicles. In Proceedings of the International Conference on Electrical Systems for Aircraft, Railway, Ship Propulsion and Road Vehicles & International Transportation Electrification Conference (ESARS-ITEC), Toulouse, France, 2–4 November 2016; pp. 1–6.
17. Ceraolo, M.; Lutzemberger, G.; Poli, D. State-Of-Charge Evaluation Of Supercapacitors. *J. Energy Storage* **2017**, *11*, 211–218. [CrossRef]
18. Motapon, S.N.; Dessaint, L.A.; Al-Haddad, K. A comparative study of energy management schemes for a fuel-cell hybrid emergency power system of more-electric aircraft. *IEEE Trans. Ind. Electron.* **2014**, *61*, 1320–1334. [CrossRef]
19. Zubieta, L.; Bonert, R. Characterization of double-layer capacitors for power electronics applications. *IEEE Trans. Ind. Appl.* **2000**, *36*, 199–205. [CrossRef]
20. Wu, C.; Hung, Y.; Hong, C. On-line supercapacitor dynamic models for energy conversion and management. *Energy Convers. Manag.* **2012**, *53*, 337–345. [CrossRef]
21. Quintáns, C.; Iglesias, R.; Lago, A.; Acevedo, J.M.; Martínez-Peñalver, C. Methodology to obtain the voltage-dependent parameters of a fourth-order supercapacitor model with the transient response to current pulses. *IEEE Trans. Power Electron.* **2017**, *32*, 3868–3878. [CrossRef]
22. Noh, S.; Choi, J.; Kim, H.C.; Lee, E.K. PSIM Based electric modeling of supercapacitors for line voltage regulation of electric train system. In Proceedings of the EEE 2nd International Power and Energy Conference, Johor Bahru, Malaysia, 1–3 December 2008; pp. 855–859.
23. Michalczuk, M.; Grzesiak, L.M.; Ufnalski, B. Experimental parameter identification of battery-ultracapacitor energy storage system. In Proceedings of the IEEE 24th International Symposium on Industrial Electronics (ISIE), Buzios, Brazil, 3–5 June 2015; pp. 1260–1265.
24. Kang, J.; Wen, J.; Jayaram, S.H.; Yu, A.; Wang, X. Development of an equivalent circuit model for electrochemical double layer capacitors (EDLCs) with distinct electrolytes. *Electrochim. Acta* **2014**, *115*, 587–598. [CrossRef]
25. Xu, N.; Riley, J. Nonlinear analysis of a classical system: The double-layer capacitor. *Electrochem. Commun.* **2011**, *13*, 1077–1081. [CrossRef]
26. Oldham, K.B. A Gouy–Chapman–Stern model of the double layer at a (metal)/(ionic liquid) interface. *J. Electroanal. Chem.* **2008**, *613*, 131–138. [CrossRef]

27. Conway, B.E.; Pell, W.; Liu, T. Diagnostic analyses for mechanisms of self-discharge of electrochemical capacitors and batteries. *J. Power Sources* **1997**, *65*, 53–59. [CrossRef]
28. Ji, H.; Zhao, X.; Qiao, Z.; Jung, J.; Zhu, Y.; Lu, Y.; Zhang, L.L.; MacDonald, A.H.; Ruoff, R.S. Capacitance of carbon-based electrical double-layer capacitors. *Nat. Commun.* **2014**, *5*, 3317. [CrossRef]
29. Faranda, R. A new parameters identification procedure for simplified double layer capacitor two-branch model. *Electr. Power Syst. Res.* **2010**, *80*, 363–371. [CrossRef]
30. Yang, H.; Zhang, Y. Characterization of supercapacitor models for analyzing supercapacitors connected to constant power elements. *J. Power Sources* **2016**, *312*, 165–171. [CrossRef]
31. Solano, J.; Hissel, D.; Pera, M.C. Modeling and parameter identification of ultracapacitors for hybrid electrical vehicles. In Proceedings of the IEEE Vehicle Power and Propulsion Conference (VPPC), Beijing, China, 15–18 October 2013; pp. 1–4.
32. Kim, S.H.; Choi, W.; Lee, K.B.; Choi, S. Advanced dynamic simulation of supercapacitors considering parameter variation and self-discharge. *IEEE Trans. Power Electron.* **2011**, *26*, 3377–3385.
33. Buller, S.; Karden, E.; Kok, D.; De Doncker, R. Modeling the dynamic behavior of supercapacitors using impedance spectroscopy. In Proceedings of the Conference Record of the 2001 IEEE Industry Applications Conference. 36th IAS Annual Meeting (Cat. No.01CH37248), Chicago, IL, USA, 30 September–4 October 2001; Volume 4, pp. 2500–2504.
34. Musolino, V.; Piegari, L.; Tironi, E. New Full-Frequency-Range Supercapacitor Model With Easy Identification Procedure. *IEEE Trans. Ind. Electron.* **2013**, *60*, 112–120. [CrossRef]
35. Signorelli, R.; Ku, D.C.; Kassakian, J.G.; Schindall, J.E. Electrochemical Double-Layer Capacitors Using Carbon Nanotube Electrode Structures. *Proc. IEEE* **2009**, *97*, 1837–1847. [CrossRef]
36. Wei, T.; Qi, X.; Qi, Z. An improved ultracapacitor equivalent circuit model for the design of energy storage power systems. In Proceedings of the International Conference on Electrical Machines and Systems (ICEMS), Seoul, South Korea, 8–11 October 2007; pp. 69–73.
37. Miller, J.M.; McCleer, P.J.; Cohen, M. *Ultracapacitors as Energy Buffers in a Multiple Zone Electrical Distribution System*; Maxwell Technologies: San Diego, CA, USA, 2004.
38. Lajnef, W.; Vinassa, J.M.; Briat, O.; Azzopardi, S.; Woirgard, E. Characterization methods and modelling of ultracapacitors for use as peak power sources. *J. Power Sources* **2007**, *168*, 553–560. [CrossRef]
39. Yang, H. Estimation of Supercapacitor Charge Capacity Bounds Considering Charge Redistribution. *IEEE Trans. Power Electron.* **2018**, *33*, 6980–6993. [CrossRef]
40. Schaeffer, E.; Auger, F.; Shi, Z.; Guillemet, P.; Loron, L. Comparative analysis of some parametric model structures dedicated to EDLC diagnosis. *IEEE Trans. Ind. Electron.* **2016**, *63*, 387–396. [CrossRef]
41. Dougal, R.; Gao, L.; Liu, S. Ultracapacitor model with automatic order selection and capacity scaling for dynamic system simulation. *J. Power Sources* **2004**, *126*, 250–257. [CrossRef]
42. Parvini, Y.; Siegel, J.B.; Stefanopoulou, A.G.; Vahidi, A. Supercapacitor electrical and thermal modeling, identification, and validation for a wide range of temperature and power applications. *IEEE Trans. Ind. Electron.* **2016**, *63*, 1574–1585. [CrossRef]
43. Isermann, R.; Münchhof, M. *Identification of Dynamic Systems: An Introduction with Applications*; Springer Science & Business Media: Berlin, Germany, 2010.
44. Wang, Y.; Liu, C.; Pan, R.; Chen, Z. Modeling and state-of-charge prediction of lithium-ion battery and ultracapacitor hybrids with a co-estimator. *Energy* **2017**, *121*, 739–750. [CrossRef]
45. MathWorks. *Simulink® Design Optimization User 's Guide*; MathWorks: Natick, MA, USA, 2018.
46. Jackey, R.A.; Plett, G.L.; Klein, M.J. *Parameterization of a Battery Simulation Model Using Numerical Optimization Methods*; Technical Report; SAE: Warrendale, PA, USA, 2009.
47. Jackey, R.; Saginaw, M.; Sanghvi, P.; Gazzarri, J.; Huria, T.; Ceraolo, M. *Battery Model Parameter Estimation Using a Layered Technique: An Example Using a Lithium Iron Phosphate Cell*; Technical Report; SAE: Warrendale, PA, USA, 2013.
48. Chakraborty, U.K.; Abbott, T.E.; Das, S.K. PEM fuel cell modeling using differential evolution. *Energy* **2012**, *40*, 387–399. [CrossRef]
49. Chakraborty, U.K. Static and dynamic modeling of solid oxide fuel cell using genetic programming. *Energy* **2009**, *34*, 740–751. [CrossRef]
50. MathWorks. *Simulink® Design Optimization Getting Started Guide*; MathWorks: Natick, MA, USA, 2019.

51. Burke, A. Ultracapacitor technologies and application in hybrid and electric vehicles. *Int. J. Energy Res.* **2010**, *34*, 133–151. [CrossRef]
52. He, H.; Xiong, R.; Zhang, X.; Sun, F.; Fan, J. State-of-charge estimation of the lithium-ion battery using an adaptive extended Kalman filter based on an improved Thevenin model. *IEEE Trans. Veh. Technol.* **2011**, *60*, 1461–1469.
53. Raga, C.; Barrado, A.; Lázaro, A.; Quesada, I.; Sanz, M.; Zumel, P. Driving profile and fuel cell minimum power analysis impact over the size and cost of fuel cell based propulsion systems. In Proceedings of the 9th International Conference on Compatibility and Power Electronics (CPE), Costa da Caparica, Portugal, 24–26 June 2015; pp. 390–395.

Article

A Dual-Objective Substation Energy Consumption Optimization Problem in Subway Systems

Hongjie Liu [1,2], Tao Tang [1,2], Jidong Lv [1,2] and Ming Chai [1,2,*]

[1] School of Electronics and Information Engineering, Beijing Jiaotong University, Beijing 100044, China; hjliu2@bjtu.edu.cn (H.L.); ttang@bjtu.edu.cn (T.T.); jdlv@bjtu.edu.cn (J.L.)

[2] State Key Laboratory of Rail Traffic Control and Safety, Beijing Jiaotong University, Beijing 100044, China

[*] Correspondence: chaiming@bjtu.edu.cn; Tel.: +86-10-5168-5741

Received: 12 April 2019; Accepted: 13 May 2019; Published: 17 May 2019

Abstract: Maximizing regenerative energy utilization is an important way to reduce substation energy consumption in subway systems. Timetable optimization and energy storage systems are two main ways to improve improve regenerative energy utilization, but they were studied separately in the past. To further improve energy conservation while maintaining a low cost, this paper presents a strategy to improve regenerative energy utilization by an integration of them, which determines the capacity of each Wayside Energy Storage System (WESS) and correspondingly optimizes the timetable at the same time. We first propose a dual-objective optimization problem to simultaneously minimize substation energy consumption and the total cost of WESS. Then, a mathematical model is formulated with the decision variables as the configuration of WESS and timetable. Afterwards, we design an ϵ-constraint method to transform the dual-objective optimization problem into several single-objective optimization problems, and accordingly design an improved artificial bee colony algorithm to solve them sequentially. Finally, numerical examples based on the actual data from a subway system in China are conducted to show the effectiveness of the proposed method. Experimental results indicate that substation energy consumption is effectively reduced by using WESS together with a correspondingly optimized timetable. Note that substation energy consumption becomes lower when the total size of WESS is larger, and timetable optimization further reduces it. A set of Pareto optimal solutions is obtained for the experimental subway line—based on which, decision makers can make a sensible trade-off between energy conservation and WESS investment accordingly to their preferences.

Keywords: regenerative energy; timetable optimization; energy storage system; ϵ-constraint method; improved artificial bee colony

---

1. Introduction

Energy conservation in subway systems has attracted great attention in recent years. As a great proportion of total energy is consumed by train traction systems [1–4], many measures have been taken to reduce traction energy consumption. The adoption of energy-efficient train operations has been a focus in the past decades [5]. It aims to find a driving strategy which consumes the least energy. However, Regenerative Energy (RE) is not considered in these studies, which limits the effect of this method on energy conservation [3,6]. Regenerative energy is the electrical energy converted from kinetic energy by a regenerative braking system during the braking of trains. As a potential energy supply, regenerative energy can be used to accelerate trains. Regenerative energy takes a great amount of the total energy consumption in a subway system [7]. Substation energy consumption can be reduced dramatically if the driving strategies of trains remain unchanged and the regenerative energy is fully utilized. Thus, with regenerative braking systems being widely applied in subway systems, optimizing Regenerative Energy Utilization (REU) has become a hot topic in recent years [8]. Generally, there are three ways to do so [9–12].

The first one is to adopt Timetable Optimization (TO). It aims to coordinate the accelerating and braking of trains, thus the regenerative energy from braking trains can be utilized by the accelerating trains immediately. It is the most preferable way as its cost is the lowest according to [8,13–26]. However, RE cannot be fully utilized in this way in general, and the surplus regenerative energy is dissipated into heat via resistors.

The second way is to store regenerative energy temporarily by using wayside and/or on-board Energy Storage Systems (ESS), e.g., super-capacitors, and to reuse it later. Due to the advancement of power electronics and energy storage technologies, ESS can be integrated into subway systems to utilize regenerative energy more sufficiently. For example, Wayside Energy Storage Systems (WESSs) can store the surplus regenerative energy temporarily and deliver it back to accelerate trains in the same Electricity Supply Interval (ESI) when needed. Thus, Substation Energy Consumption (SEC) can be reduced. Furthermore, the stored energy in WESS can contribute to shaving power peaks during the acceleration of trains, and WESS can be used as a temporary electricity supplier in case of power grid failure. Thus, not only can WESS improve the efficient energy management, but it can also stabilize the power network [10]. It becomes the second choice as it needs to pay for the cost of ESS and the rapid development of ESS that has happened in recent years.

The last way is to feed regenerative energy back into a utility grid network in a city through reversible power substations. Thus, the surplus regenerative energy can be used by other electricity facilities outside of a subway system. However, this is not yet diffused, as it needs to modify the substations greatly, which is complex and comes with high costs [11].

The management of timetables and ESS belong to different departments in subway systems. Thus, they were developed and optimized separately in the past. Regenerative energy from braking trains can either be utilized by traction trains immediately or be stored in ESS for later use. Thus, when ESS is used in a subway subway system, the utilization of regenerative energy is different from without it. It is obvious that the configuration of ESS affects its effect on energy saving. In addition, note that the effect of ESS is also affected by the timetable used. As a timetable defines the schedules of all the trains, it determines the synchronization of traction and braking trains, which affects the regenerative energy that can be absorbed by ESS. Thus, when a timetable is changed, the effect of energy conservation is changed accordingly, even if the ESS remains unchanged. On the other hand, by absorbing and/or releasing electrical energy, ESS affects currents on the power supply line, thus the utilization of regenerative energy between traction and braking trains is also affected. Therefore, both timetable optimization and the configuration of ESS affect substation energy consumption, and their effects on energy saving interact with each other. However, the integration of timetable optimization and ESS has seldom been studied in the past literature. As will be shown in Section 2, most of the studies on REU improvement focus on timetable optimization, a few talk about the configuration of ESS, and only very few of them study the integration of these two methods.

Motivated by the above, we present in this paper an integration optimization problem to reduce energy consumption in a subway system, which simultaneously uses timetable optimization and the application of WESS. To reduce the financial cost caused by WESS, we formulated a dual-objective optimization problem. An ε constraint method together with an Improved Artificial Bee Colony (IABC) algorithm are designed to the problem, and numerical examples are conducted to show the effectiveness of the proposed methods.

The main ideas of this work are shown in Figure 1. The main contributions are concluded as follows:

1. An integration of timetable optimization and WESS is proposed to maximize regenerative energy utilization, thus to minimize substation energy consumption in a subway system.
2. To maximize energy saving with the least cost, a dual-objective optimization model is thereby formulated to simultaneously minimize substation energy consumption and the total investment cost of WESSs. Note that a subway line is divided into several electricity supply intervals. One WESS is installed in each electricity supply interval. The size of each WESS can be different from each other, and both of them are greater than or equal to 0. Their total size determines the total cost of WESSs.
3. To solve the proposed dual-objective optimization problem, an ϵ-constraint method is first designed to transform it into several single-objective optimization problems. Then, an improved artificial bee colony algorithm is designed to solve these single-objective optimization problems sequentially.
4. Numerical examples are constructed based on the actual data obtained from a subway system in China to show the effectiveness of the proposed resolution methods. A set of Pareto optimal solutions is obtained. i.e., for each value of the total size of WESSs, the minimal substation energy consumption is determined, and the optimal configuration of each WESS and the correspondingly optimized timetable are obtained to reach the maximum energy saving.

The remainder of this paper is organized as follows. We review the related work on improvement of regenerative energy utilization in Section 2. In Section 3, a dual-objective optimization problem is proposed to simultaneously minimize substation energy consumption and the corresponding WESS costs, and a mathematical model for the proposed problem is formulated. To solve the dual-objective optimization problem, we design both an ϵ-constraint method and an improved artificial bee colony algorithm in Section 4. Then, numerical experiments are conducted in Section 5 to show the effectiveness of the proposed method. Finally, Section 6 concludes this paper and points out some future research directions.

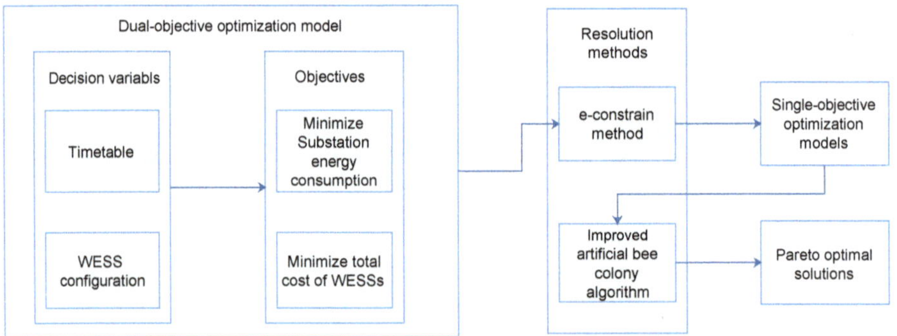

Figure 1. Main ideas of this work.

2. Literature Review

In this section, we review the studies on improvement of regenerative energy utilization. As two main measures, timetable optimization and the application of energy storage systems are studied separately in the literature. However, their integration was not considered until recently. Consequently, a dual-objective optimization problem to minimize energy consumption and investment cost was also seldom studied. The main related publications are reviewed in the following.

An optimized timetable can improve regenerative energy utilization between traction and braking trains, hence reduce substation energy consumption in a subway system. In addition, the cost of timetable optimization is relatively low. Therefore, timetable optimization has been studied by many researchers to save energy [7,8,13–27]. However, ESS was seldom considered in these studies, which limits their effects on energy saving. Furthermore, most of the studies are single-objective optimization problems and only a few of them are dual-objective optimization ones, which limits the resolution methods to be applied in our work.

In a single-objective timetable optimization problem to save energy, the direct objective is to maximize regenerative energy utilization, which may be represented by energy or the overlap time between trains' traction and braking phases. Mathematical models of timetable and regenerative energy utilization are formulated in these studies, and the problems are usually solved by an analytical method or an intelligent algorithm, which are helpful for further research. For example, Ramos et al. [13] present a timetabling problem to maximize the overlap time of the speed-up and slow-down actions of trains. Kim et al. [15] propose a multi-criteria mixed integer program to minimize the peak energy consumed and to maximize regenerative energy utilization through timetable optimization. Pena et al. [16] propose a timetable optimization model for an underground rail system to maximize regenerative energy utilization. Fournier et al. [17] develop an optimization model to maximize regenerative energy utilization by subtly modifying dwell time for trains at stations, and a hybrid genetic/linear programming algorithm is implemented to tackle this problem. Yang et al. [18] propose a cooperative scheduling model to maximize the overlap time of accelerating and braking processes of adjacent trains. Yang et al. [21] formulate an integer programming model with real-world speed profiles to minimize traction energy consumption by adjusting dwell time. They coordinate the arrivals and departures of trains in the same electricity supply interval (ESI), such that regenerative energy is effectively utilized. A genetic algorithm (GA) is designed to solve their problems in both [18,21]. Zhao et al. [20] develop a nonlinear integer program to maximize regenerative energy utilization, which searches for the optimal headway and dwell time at each station. Gong et al. [23] present a timetable optimization model to maximize regenerative energy utilization with dwell time control, and GA is used to find a near-optimal solution.

Few studies on timetable optimization are dual-objective optimization problems, where energy conservation and passenger time are usually minimized at the same time. The way to formulate a dual-objective optimization model and the possible resolving method are referrable. e.g., Yang et al. [8] propose a dual-objective timetable optimization model to coordinate up and down trains at the same station to improve regenerative energy utilization and reduce passenger waiting time. Two objectives are combined into one by a weighted-sum method, and it is solved by GA. Zhao et al. [19] propose a dual-objective optimization problem to maximize regenerative energy utilization measured by the overlap time and to shorten total passenger time. Two objectives are combined into one through weighting, and a simulated annealing (SA) method is designed to solve it. Xu et al. [24] propose a dual-objective optimization problem to minimize both passenger time and traction energy, by controlling running time at each section and dwell time at each platform. They adopt a weighted-sum method to combine two objectives into one, and designed a genetic algorithm to solve it.

Installing ESS also can improve regenerative energy utilization, thereby reducing substation energy consumption in subway systems. Therefore, few researchers have studied the application of ESS in subway systems. Although timetable optimization is seldom considered in these studies, the modelling method of ESS is referable to our work. Ceraolo et al. [11] develop models to analyze the impact of regenerative braking in high-speed railway systems, where ESS is used. Feasibility of using wayside and on-board ESS is analyzed, respectively. They evaluate the cost-effectiveness of different solutions by taking into account the capital cost of the investment and annual energy saving. The results prove the effectiveness on improving regenerative energy utilization through ESS application. Ciccarelli et al. [12] propose a control strategy for on-board super-capacitors integrated with motor drive control. Simulation results show its effectiveness on energy saving and reducing the voltage surge at the overhead contact line during train braking. Liu et al. [28] propose a single-objective timetable optimization problem with application of WESS to minimize total energy consumption. An algorithm integrating tabu search and simulation is designed to solve it. Experimental results prove the effectiveness of WESS on energy saving. The model of WESS energy management is relatively simple in their work. Gao et al. [29] propose a control strategy to use super-capacitors as WESS, which is helpful for WESS energy management. Huang et al. [30] propose an energy-saving model to optimize trains' speed profiles in a subway system. On-board ESS was used as a basis of their optimization problem and the running time is allowed to be optimized within a predefined window. A dynamic programming method is used to solve their problems. Kampeerawat and Koseki [10] propose a single-objective optimization problem to reduce substation energy consumption, which raised the studies on integration of timetable optimization and WESS. A mathematical model is formulated to minimize a linear weighted sum of substation energy consumption and the energy capacity of WESS; then, GA is designed to solve their problem. Although different weight factors can be assigned in theory, it is hard to find all the non-dominated solutions for the original dual-objective optimization problem. Ahmadi et al. [31] propose to reduce substation energy consumption in subway systems by simultaneous application of WESS and speed profile optimization. To demonstrate the validity of the proposed method, they first optimized the configuration of WESS together with the real world (i.e., non-optimized) speed profiles to reduce substation energy consumption. Then, the speed profiles and the configuration of WESS were simultaneously optimized. GA was used to solve their problems. Experimental results prove that substantial reduction in substation energy consumption was achieved and total size of WESS is decreased when speed profiles were optimized in comparison with the non-optimized speed profiles.

From the above we can see that the integration of timetable optimization and WESS is very few in the existing studies, which highlights the first feature of our work. i.e., the energy saving strategy proposed in this work, which combines timetable optimization and WESS, is challenging and different from the existing studies. Consequent on the integration of the two different methods, to simultaneously minimize substation energy consumption and the financial cost of WESS, a dual-objective optimization problem is proposed, which is also innovative. Furthermore, to solve a dual-objective optimization problem, a weighted-sum method is often adopted in the previous studies to transform it into a single-objective optimization problem, where the optimal solutions are limited by the weights adopted. While in this work we design an ϵ-constraint method to transform the original dual-objective optimization problem into several single-objective ones, then the Pareto front is able to be obtained in theory [32,33]. Finally, instead of applying the most frequently used genetic algorithm to solve the single-objective optimization problems, an improved artificial bee colony algorithm is designed and experimental results prove its better performance over a genetic algorithm.

3. Mathematical Model

3.1. Problem Assumptions

According to the characteristics of the dual-objective substation energy-consumption optimization problem, we formulate a mathematical model based on the following assumptions.

1. There is reserve time for dwell time at each platform.
2. The subway line is divided into several electricity supply intervals. Each electricity supply interval provides energy for several stations and sections. Electricity can be transmitted within an an electricity supply interval.
3. Each electricity supply interval includes one substation, one WESS and a bank of resistors. Each WESS can be installed near the place of a substation. Each WESS consists of several parallel Basic Energy Storage Modules (BESMs). The total cost of all WESSs grows linearly with their total energy capacity (represented by the number of BESMs in this work).

4. The transmission loss coefficient of regenerative energy is a constant.
5. The other electrical utilities (e.g., lights and air conditioners) are not considered during the simulation, i.e., electrical energy can only be consumed by traction trains, resistors and charged into WESS.
6. Regenerative energy is fed back into the power supply line and can be immediately used by traction trains in the same electricity supply interval. If regenerative energy cannot be used fully by traction trains, the surplus is charged into WESS and consumed by resistors if needed.
7. The condition of a charge/discharge state transition for a WESS is defined as follows:
    - When trains are braking and the power of surplus regenerative energy exceeds a predefined limit, WESS in the same electricity supply interval begins to charge; When the power of surplus regenerative energy gets smaller, the corresponding WESS stops charging.
    - When trains are in a traction phase and the power of surplus traction energy exceeds a predefined limit, WESS begins to discharge; when the surplus traction energy gets smaller, WESS stops discharging (the surplus traction energy demand is satisfied by a substation).

3.2. Train Movement and Timetable Modeling

As shown in Figure 2, we label each station, platform and section with indices $y \in \{1, 2, \cdots, Y\}$, $n \in \{1, 2, \cdots, N\}$ and $l \in \{1, 2, \cdots, L\}$ in this work, respectively. Note that total numbers of stations, platforms and sections are Y, N and L, respectively. There are two platforms at each station, except the terminal station Y, since we dismiss the detailed turnaround process in this work. Thus, we have $N = 2Y - 1$. The platforms in station $y \in \{1, 2, \cdots, Y - 1\}$ are labeled as platforms n and $N - n + 1$ respectively, by noting $y = n$. In addition, the platform in terminal station Y is labeled as platform $\frac{N+1}{2}$. A section is used to connect any two successive platforms, thus $L = N - 1 = 2Y - 2$. The section connecting platforms n and $n + 1$ is labeled as section l, by noting $l = n$.

In a subway system, every train travels at the same closed path in a subway system with a time interval. Every train $i \in \{1, 2, \cdots, I\}$ begins its travel from platform 1. It runs along the down direction, through each section and platform sequentially, until it arrives at the terminal station. Then, it turns around to the up direction and repeats a similar process until it arrives the last platform N and finishes its travel. In the process, it stays at each platform $n \in \{1, 2, \cdots, N - 1\}$ for some time period $x_{i,n}$ for passengers to get on and off the train.

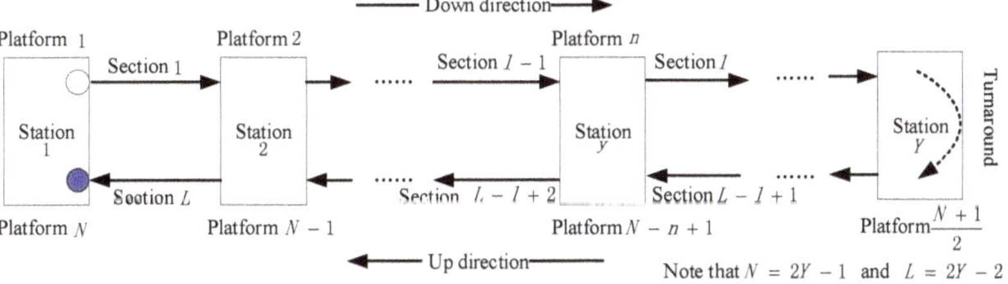

Figure 2. Train travel process.

A timetable is often used to describe the key time points of trains in a subway system. Given a timetable, we can easily obtain all the key time points of trains, including the time instant when each train arrives at and departs from any platform. Dwell time and headway time are two of the most important factors in a timetable, and they are to be optimized in this work. The running process at a section of each train is also important for calculating energy consumption. It is often divided into three phases, i.e., traction, coasting and braking phases, in the timetable optimization studies [18,21]. Detailed process at any section l and the related parameters are illustrated in Figure 3. A red dotted line stands for a traction phase, a green solid line for a braking phase and a blue dashed line for a coasting phase. $t_{i,l}^u$ is the time instant when train i is at time point u and section l, where $u \in \{1,2,3,4\}$ is the key time point index of the running phases at a section, and its value from 1 to 4 represents the start time of a traction phase (train leaves a platform), switch time from traction to a coasting phase, switch time from coasting to a braking phase and the end time of a braking phase (train arrives at a platform), respectively. $v_{i,l}^u$ is the train speed when train i is at time point u and section l. r_l^γ is the time duration of phase γ at section l, where $\gamma \in \{1,2,3\}$ is the train running phase index at a section, and its value from 1 to 3 represents a traction, coasting and braking phase, respectively. a_l^γ is the train acceleration in phase γ at section l. $x_{i,n}$ is the dwell time for train $i \in \{1,2,\cdots,I\}$ at platform $n \in \{1,2,\cdots,N-1\}$, which is the time interval from train i arrives at to it leaves from platform n.

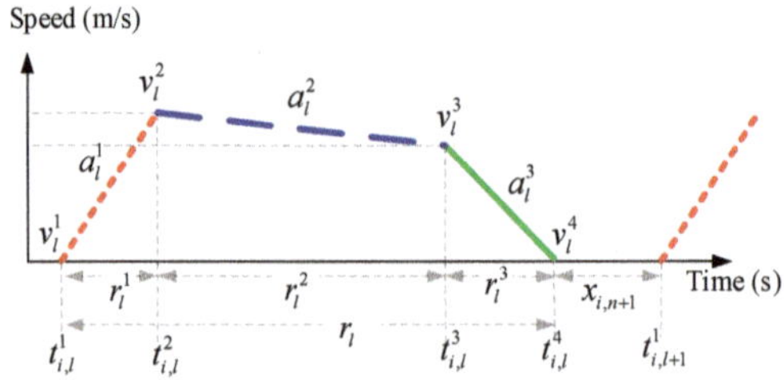

Figure 3. General process of a subway train running at a section.

Based on the above, the time of all the key points for all trains is determined by the time instant when the first train starts its travel ($t_{1,0}^4$) via the following procedures:

1. Determine the time of the other key points for train i at section l by that train's start time at that section:

$$t_{i,l}^u = t_{i,l}^{u-1} + r_l^\gamma, \quad u \in \{2,3,4\}, \gamma = u-1. \tag{1}$$

2. Determine $t_{i,l}^1$ by the time instant when train i starts its travel:

$$t_{i,l}^1 = \begin{cases} t_{i,l-1}^4 + x_{i,n} & n \in \{1,2,\cdots,N-1\} \backslash \{\frac{N+1}{2}\}, \\ t_{i,l-1}^4 + x_{i,n} + \Phi & n = \frac{N+1}{2}, \end{cases} \tag{2}$$

3. Determine $t_{i,0}^4$ by using $t_{1,0}^4$:

$$t_{i,0}^4 = t_{i-1,0}^4 + h_{i-1} = t_{1,0}^4 + \sum_{j=1}^{i-1} h_j, \; i \in \{2,\cdots,I\}, \tag{3}$$

where $t_{1,0}^4$ is a given constant which represents the time instant that the first train ends its braking phase at a virtual section 0 (also means arrival time at platform 1); h_i is the headway time between trains i and $i+1$ and it is represented by the time difference of their arrival time at platform 1 in this work, i.e., $h_i = t_{i+1,0}^4 - t_{i,0}^4$.

Travel time of train i is obtained as:

$$\tau_i = \sum_{n=1}^{N-1} x_{i,n} + \sum_{l=1}^{L} r_l + \Phi, \tag{4}$$

where τ_i is the travel time duration of train i, i.e., $\tau_i = t^4_{i,L} - t^4_{i,0}$; Φ is the turnaround time duration at terminal station Y.

Operation time of a subway system is obtained as:

$$\Lambda = t^4_{I,0} - t^4_{1,0} = \sum_{i=1}^{I-1} h_i, \tag{5}$$

where Λ is the operation time duration of a subway system. It is represented by the first platform in this work.

Train speed is 0 when a train stops at a platform, and it obeys uniformly accelerated/decelerated motion in every phase at a section. Thus, train speed is determined as follows:

$$v_i(t) = \begin{cases} 0 & t \in [t^4_{i,l-1}, t^1_{i,l}], \\ a^1_l \left(t - t^1_{i,l}\right) & t \in [t^1_{i,l}, t^2_{i,l}], \\ a^1_l r^1_l - a^2_l \left(t - t^2_{i,l}\right) & t \in [t^2_{i,l}, t^3_{i,l}], \\ a^3_l \left(t^4_{i,l} - t\right) & t \in [t^3_{i,l}, t^4_{i,l}], \end{cases} \tag{6}$$

where $i \in \{1, 2, \cdots, I\}$ and $l \in \{1, 2, \cdots, L\}$; $v_i(t)$ is the train speed at any time t. Note that a^1_l and a^3_l are the maximum traction and braking acceleration of a train at section l respectively, which are both given constants. In addition, it is easy to obtain a^2_l as $a^2_l = \dfrac{a^1_l r^1_l - a^3_l r^3_l}{r^2_l}$. Thus, the kinetic energy of each train at any time is determined actually.

3.3. Energy Consumption Calculation

Each electricity supply interval $z \in \{1, 2, \cdots, Z\}$ has several energy suppliers and consumers, where the energy suppliers include a substation, braking trains and WESS in a discharging state, and energy consumers include traction trains, a charging WESS and resistors installed in this electricity supply interval, as shown in Figure 4. $P^a_z(t)$ is the power of traction energy demand in electricity supply interval z at time t; $P^b_z(t)$ is the power of available regenerative energy in electricity supply interval z at time t; $P^c_z(t)$ is the charging power of WESS z at time t; $P^d_z(t)$ is the discharging power of WESS z at time t; $P^r_z(t)$ is the power of energy consumed by the resistors in electricity supply interval z at time t; $P^s_z(t)$ is the power of energy supplied by substation z at time t.

According to the energy conservation law, the total power of energy suppliers should be the same as that of energy consumers, i.e.,

$$P^b_z(t) + P^d_z(t) + P^s_z(t) = P^a_z(t) + P^c_z(t) + P^r_z(t) \tag{7}$$

Note that traction trains consume electrical energy and convert it into kinetic energy, and braking trains supply regenerative energy to the power supply line by converting from kinetic energy. The electrical energy demand from traction trains are satisfied by regenerative energy first. If regenerative energy is not enough, the surplus traction energy demand are satisfied by the discharging energy of WESS and the energy from a substation. Otherwise, if there is surplus regenerative energy, it can be consumed by resistors into heat and be charged into WESS for later use.

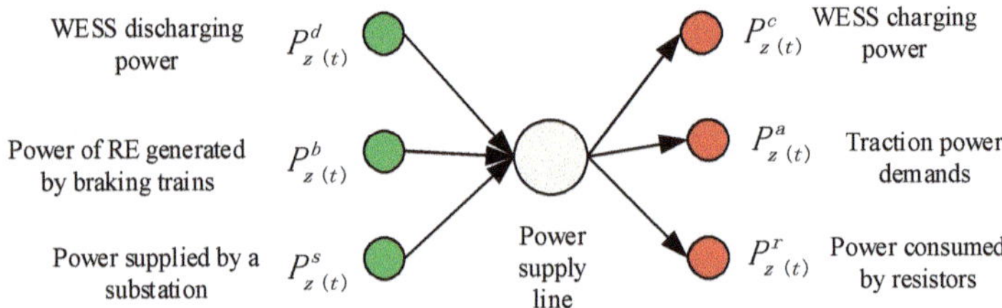

Figure 4. Energy suppliers and consumers in an electricity supply interval.

3.3.1. Total Traction Energy Demand

The required electrical energy for accelerating train i at section l (denoted as $E^a_{i,l}$) is determined by the kinetic energy increased in its traction phase, and the required electrical power for accelerating train i at section l at time t (denoted as $P^a_{i,l}(t)$) is its derivative. They are determined as follows:

$$E^a_{i,l}(t) = \begin{cases} \psi_1\left((v_{i,l}(t))^2 - (v^1_l)^2\right), & t \in [t^1_{i,l}, t^2_{i,l}], \\ 0, & else, \end{cases} \tag{8}$$

$$P^a_{i,l}(t) = \frac{dE^a_{i,l}(t)}{dt} = \begin{cases} \psi_2\left(t - t^1_{i,l}\right), & t \in [t^1_{i,l}, t^2_{i,l}], \\ 0, & else, \end{cases} \tag{9}$$

where ψ_1 and ψ_2 are constants satisfying $\psi_1 = m/2\eta_1$ and $\psi_2 = m\left(a^1_l\right)^2/\eta_1$, respectively, where m is train mass and η_1 is the conversion efficiency from electrical energy to a train's kinetic energy. Thus, the total traction energy demand of all trains (denoted as E^a) in a subway system is determined as

$$E^a = \sum_{i=1}^{I}\sum_{l=1}^{L} E^a_{i,l}(t). \tag{10}$$

The power of traction energy demand in electricity supply interval z at time t is determined by all the traction trains in that electricity supply interval, i.e.,

$$P^a_z(t) = \sum_{i=1}^{I}\sum_{l\in A_z} P^a_{i,l}(t). \tag{11}$$

Note that the total traction energy demand E^a is a constant in this work, as it is not affected by the change of headway time and dwell time.

3.3.2. Available Regenerative Energy from Braking Trains

The available regenerative energy from train i at section l (denoted as $E^b_{i,l}$) is determined by the kinetic energy decreased in its braking phase, and the power of regenerative energy at time t (denoted as $P^b_{i,l}(t)$) is the derivative of regenerative energy. They are determined as follows:

$$E_{i,l}^{b}(t) = \begin{cases} \psi_3\left(\left(v_l^3\right)^2 - \left(v_{i,l}(t)\right)^2\right), & t \in [t_{i,l}^3, t_{i,l}^4], \\ 0, & else, \end{cases} \tag{12}$$

$$P_{i,l}^{b}(t) = \frac{dE_{i,l}^{b}(t)}{dt} = \begin{cases} \psi_4\left(t_{i,l}^4 - t\right), & t \in [t_{i,l}^3, t_{i,l}^4], \\ 0, & else, \end{cases} \tag{13}$$

where ψ_3 and ψ_4 are constants satisfying $\psi_3 = m\eta_2\left(1-\zeta_1\right)/2$ and $\psi_4 = m(a_l^3)^2\eta_2(1-\zeta_1)$, respectively; η_2 is the conversion efficiency from kinetic energy to regenerative energy; ζ_1 is the transmission loss coefficient of regenerative energy.

The total available regenerative energy from all braking trains in a subway system (E^b) is

$$E^b = \sum_{i=1}^{I}\sum_{l=1}^{L} E_{i,l}^{b}(t). \tag{14}$$

The total power of regenerative energy from braking trains in electricity supply interval z is

$$P_z^b(t) = \sum_{i=1}^{I}\sum_{l \in A_z} P_{i,l}^{b}(t). \tag{15}$$

3.3.3. Energy Dynamics of Wess

As described in Section 3.1, one WESS is installed in each electricity supply interval, and it is comprised of several BESMs in parallel. Thus, the energy capacity and maximum power of WESS z are determined by the number of BESMs κ_z and BESM specification. i.e.,

$$\begin{cases} C_z^e = \kappa_z \times C^M, \\ P_z^e = P^M, \end{cases} \tag{16}$$

where C_z^e and P_z^e is the maximum power and energy capacity of WESS z respectively; C^M and P^M are given constants, represents the maximum power and energy capacity of a BESM respectively; κ_z is the number of BESMs in WESS $z \in \{1, 2, \cdots, Z\}$. Thus, the total energy capacity of all WESSs (denoted as C^e) in a subway system is:

$$C^e = \sum_{z=1}^{Z} C_z^e = C^M \times K, \tag{17}$$

where K is the total number of BESMs in a subway system, satisfying $K = \sum_{z=1}^{Z}\kappa_z$.

As for the control strategy design and modelling of ESS, there are several good references in smart home domains. For example, Carli and Dotoli [34,35] present a distributed technique to control the charging/discharging of ESS, controllable electrical appliances and renewable energy sources in a smart city scenario. Sperstad et al. [36] formulate the charging/discharging of ESS as a multi-period optimal power flow problem, and propose a framework of methods and models to handle the uncertainties in energy exchange due to distributed wind and solar photovoltaic power generation. These studies are very valuable for the control strategy and explicit modelling of ESS. Nevertheless, besides timetable optimization, this paper mainly focuses on the optimization of WESS configuration in a subway system, i.e., to determine the capacity of the WESS in each electrical supply interval. Thus, the control strategies of WESS are reasonably simplified as in [11,29]. For example, we assume that each WESS can only be in a charging, discharging or idle state at a time, the abnormal case that charging and discharging commands are simultaneously sent to a ESS is not considered.

Note that, according to [11,29], the maximum charging and discharging power of WESS are both variables changing with its state-of-charge, as shown in Figure 5. The horizontal axes are state-of-charge, and the vertical axes are maximum charging/discharging power of a WESS. Note that WESS cannot be thoroughly discharged. For example, the discharging process of WESS stops when its state-of-charge decreases to β_0.

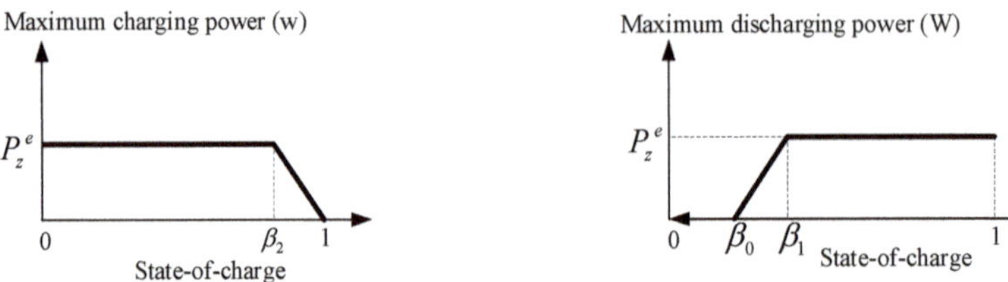

Figure 5. Maximum charging and discharging power of WESS.

The state-of-charge of WESS z at time t is defined as:

$$S_z(t) = \frac{Q_z^e(t)}{C_z^e} \times 100\%, \tag{18}$$

where $s_z(t)$ and $Q_z^e(t)$ are the state-of-charge and energy stored in WESS z at time t, respectively.

The maximum charging and discharging power of WESS z at time t (denoted as $\overline{P_z^c}(t)$ and $\overline{P_z^d}(t)$ respectively) are determined as follows:

$$\overline{P_z^c}\left(S_z(t)\right) = \begin{cases} P_z^e, & 0 \leq S_z(t) \leq \beta_2, \\ P_z^e \times \dfrac{1 - S_z(t)}{1 - \beta_2}, & \beta_2 < S_z(t) \leq 1, \end{cases} \tag{19}$$

$$\overline{P_z^d}\left(S_z(t)\right) = \begin{cases} P_z^e, & \beta_1 \leq S_z(t), \leq 1, \\ P_z^e \times \dfrac{S_z(t) - \beta_0}{\beta_1 - \beta_0}, & \beta_0 < S_z(t) < \beta_1, \\ 0, & 0 \leq S_z(t) \leq \beta_0, \end{cases} \tag{20}$$

where β_2 is the point where the maximum charging power of WESS begins to decrease; β_1 is the point where the maximum discharging power of WESS begins to decrease, and β_0 is the point where the maximum discharging power of WESS decreases to 0.

The energy stored in WESS z at time t is determined by its initial state, charging and discharging energy, i.e.,

$$Q_z^e(t) = Q_z^e(t = 0) + \int_0^t \left(P_z^c(t) \times \eta_3 - P_z^d(t)/\eta_4\right) dt, \tag{21}$$

where $Q_z^e(t = 0)$ is the energy stored in WESS z at the start time; η_3 and η_4 are the energy conversion efficiencies when a WESS is in a charging and discharging state, respectively.

Real-time charging power of a WESS

WESS charges when the regenerative energy from braking trains is more than the energy demand from traction trains, and the surplus regenerative energy exceeds the threshold of WESS charging. When

WESS charges, its real-time charging power (denoted at $P_z^c(t)$) is determined by the surplus regenerative energy and maximum charging power of WESS, i.e.,

$$P_z^c(t) = \begin{cases} \min\left\{P_z^+(t) \times \zeta_2, \overline{P_z^c}(t)\right\}, & P_z^+(t) \geq \underleftarrow{P_z^c}, \\ 0, & P_z^+(t) < \underleftarrow{P_z^c}, \end{cases} \tag{22}$$

where $P_z^+(t)$ is the power of surplus regenerative energy in electricity supply interval z , i.e., $P_z^+(t) = \max\left\{P_z^b(t) - P_z^a(t), 0\right\}$; ζ_2 is the proportion coefficient of surplus regenerative energy charging into WESS. Note that the surplus regenerative energy is partially charged into WESS and partially consumed via resistors into heat if needed.

Real-time discharging power of a WESS

WESS discharges when the power demand from traction trains is more than regenerative energy from braking trains, and the surplus traction energy demand is greater than the threshold of WESS discharging. When WESS discharges, its real-time discharging power (denoted at $P_z^d(t)$) is determined by the surplus traction energy and maximum discharging power of WESS. i.e.,

$$P_z^d(t) = \begin{cases} \min\left\{P_z^-(t) \times \zeta_3, \overline{P_z^d}(t)\right\}, & P_z^-(t) \geq \underleftarrow{P_z^d}, \\ 0, & P_z^-(t) < \underleftarrow{P_z^d}, \end{cases} \tag{23}$$

where $P_z^-(t)$ is the power of surplus traction energy in electricity supply interval z, i.e., $P_z^-(t) = \max\left\{P_z^a(t) - P_z^b(t), 0\right\}$; ζ_3 is the proportion coefficient of electrical energy supplied by WESS when it is discharging. Note that the surplus traction energy is partially provided by the discharging power of WESS and partially by a substation if needed.

3.3.4. Substation Energy Consumption

The realtime power of energy consumption from substation z is determined by the surplus traction energy in electricity supply interval z and discharging power of WESS z:

$$P_z^s(t) = \max\left\{P_z^a(t) - P_z^b(t) - P_z^d(t), 0\right\}. \tag{24}$$

Thus, the whole day total energy consumption from all substations (denoted as E^s) in a subway system is obtained as:

$$E^s = \sum_{z=1}^{Z} \int_{t_{1,0}^4}^{t_{I,L}^4} P_z^s(t)dt. \tag{25}$$

3.4. Dual-Objective Substation Energy Consumption Optimization Model

For energy optimization problems, it is necessary to consider both profit and costs paid for that. Note that total traction energy demand is a constant in this work, thereby the profit is maximized if substation energy consumption is minimized. The cost is determined by the energy capacity of WESS in this work. It is correspondingly determined by the total number of BESMs. Thus, a mathematical model of the dual-objective substation energy-consumption optimization problem is formulated as follows:

- **Problem P:**

$$\min E^s = \sum_{z=1}^{Z} \int_{t_{1,0}^4}^{t_{I,L}^4} P_z^s(t)dt = f_1(\boldsymbol{H}, \boldsymbol{X}, \boldsymbol{\kappa}), \tag{26}$$

$$\min K = \sum_{z=1}^{Z} \kappa_z = f_2(\boldsymbol{\kappa}), \tag{27}$$

s.t.

$$\underline{h} \leq h_i \leq \overline{h},\ i \in \{1, 2, \cdots, I-1\}, \tag{28}$$

$$\underline{x_n} \leq x_{i,n} \leq \overline{x_n},\ i \in \{1, 2, \cdots, I\}, n \in \{1, 2, \cdots, N-1\}, \tag{29}$$

$$\sum_{i=1}^{I-1} h_i = C_1, \tag{30}$$

$$\underline{\tau} \leq \sum_{n=1}^{N-1} x_{i,n} + \sum_{l=1}^{L} r_l + \Phi \leq \overline{\tau},\ i \in \{1, 2, \cdots, I\}, \tag{31}$$

$$h_i \in Z^+,\ i \in \{1, 2, \cdots, I-1\}, \tag{32}$$

$$x_{i,n} \in Z^+,\ i \in \{1, 2, \cdots, I\}, n \in \{1, 2, \cdots, N-1\}, \tag{33}$$

$$\kappa_z \in \{0, Z^+\},\ z \in \{1, 2, \cdots, Z\}, \tag{34}$$

where $\boldsymbol{H} = (h_1, h_2, \cdots h_i, \cdots, h_{I-1})$, $\boldsymbol{X} = (\boldsymbol{X_1}, \boldsymbol{X_2}, \cdots, \boldsymbol{X_i}, \cdots, \boldsymbol{X_I})$, where $\boldsymbol{X_i} = (x_{i,1}, x_{i,2}, \cdots, x_{i,N-1})$ and $\boldsymbol{\kappa} = (\kappa_1, \kappa_2, \cdots, \kappa_Z)$ in (26) are the decision vectors of headway time, dwell time and number of BESMs, respectively; $\underline{h}$ and $\overline{h}$ are the lower and upper limits of headway time, respectively; $\underline{x_n}$ and $\overline{x_n}$ are the lower and upper limits of dwell time at platform n, respectively; C_1 is a constant satisfying $C_1 = t_{I,0}^4 - t_{1,0}^4$, where $t_{1,0}^4$ and $t_{I,0}^4$ are the time instants when the earliest train and the latest train go into service, respectively; $\underline{\tau}$ and $\overline{\tau}$ are the lower and upper limits of train travel time, respectively; Z^+ represents the set of non-negative integers.

Objective (26) aims to minimize substation energy consumption. Objective (27) aims to minimize the total number of BESMs, which is a measurement of the cost paid for energy conservation. Constraint (28) guarantees the lower and upper limits of headway time. The lower limit is determined by safety requirements of a train signaling system and operation cost, and the upper one is determined by required service quality. Constraint (29) ensures the lower and upper limits of dwell time for a train at a platform. They are determined by the passenger flow and its potential variance. Constraint (30) guarantees that the operation time duration of a subway system is fixed. (31) guarantees the punctuality of each train travel, which is an important measurement of service quality in a subway system. (32) and (33) ensure that headway time and dwell time are positive integers, respectively. (34) ensures that the number of BESMs in a WESS is non-negative.

4. Resolution Method

There are several techniques to solve a multi-objective optimization problem [32,37–40]. The most popular and straightforward one is a weighted-sum method. It converts the former into a single-objective optimization problem by using a linear weighted sum formulation that combines all the objectives. Then, the single-objective optimization problems can be solved by using an analytical method, or commercial software (e.g., CPLEX), or an intelligent optimization algorithm. A weighted-sum method together with GA are frequently used in solving dual-objective problems in subway systems [8,24,41]. However, the weights are hard to decide sometimes. Moreover, this method is inappropriate if not all objectives can be represented via a linear combination, and it is ill-suited for a multi-objective problem with non-convex objective space [32,42].

Another well-known technique to solve a multi-objective optimization problem is an ε-constraint method. This method is first introduced by Haimes et al. [43], some good application examples can also be found in [38,44]. In solving a multi-objective optimization problem, it aims to optimize only one primary objective each time, and the other objectives are transformed into constraints. By gradually changing the constraints with a step length of ε, a series of optimal results can be obtained. Thus, a set of Pareto optimal solutions can be found for the original multi-objective optimization problem. It is very convenient to be applied in the resolution of a dual-objective optimization problem. Since its first use to solve a dual-objective shortest path problem [45], it has been successfully applied to solve many problems [32,38,45–49]. Note that the dual-objective substation energy-consumption optimization problem presented in this work is a dual-objective problem with discrete decision variables (integers), and objective (27) is obviously linear. Thus, its Pareto front is able to be obtained by using an ϵ-constraint method in theory [32,33]. Therefore, an ϵ-constraint method is designed to transform the dual-objective substation energy-consumption optimization problem into several single-objective optimization problems.

To solve these single-objective optimization problems, an improved artificial bee colony algorithm is designed. By combining the ϵ-constraint method and the improved artificial bee colony algorithm, the non-dominated solutions of the dual-objective substation energy-consumption optimization problem are obtained.

4.1. ϵ-Constraint Method

As our main objective is to minimize substation energy consumption in a subway system, we treat objective (26) in problem P as the primary objective, and objective (27) is converted into a constraint. In this way, the initial dual-objective substation energy-consumption optimization problem can be transformed into the following single-objective optimization problems by applying an ϵ-constraint method. The detailed procedure of applying the proposed ϵ-constraint method is shown in Algorithm 1, and the obtained single-objective optimization problems are listed as follows. Note that only one objective in the original dual-objective optimization problem is kept in each single-objective optimization problem, the other objective is either dismissed or converted into a constraint. The order to solve the single-objective optimization problems is also important, as the solution of the previous problem may be used in the following problems.

- **Problem P_1:**

$$\min E^s = \sum_{z=1}^{Z} \int_{t_{1,0}^4}^{t_{I,L}^4} P_z^s(t)dt = f_1(\boldsymbol{H}, \boldsymbol{X}, \boldsymbol{\kappa}),$$

s.t.

Constraints (28)–(34).

Note that the objective $f_1()$ in this problem is corresponding to objective (26) in problem P, and objective (27) in problem P is not considered in this problem. By solving problem P_1, we obtain the minimal value of E^s denoted as f_1^0. It represents the minimal substation energy consumption when sufficient WESS is installed.

- **Problem P_2:**

$$\min K = \sum_{z=1}^{Z} \kappa_z = f_2(\boldsymbol{\kappa}),$$

s.t.

$$\text{Constraints (28)–(34).}$$

Note that the objective $f_2()$ in this problem is corresponding to objective (27) in problem P, and objective (26) in problem P is not considered in this problem. By solving problem P_2, we obtain the minimal value of K denoted as ϵ_Ω. By using an analytical method, it is easy to obtain that $\epsilon_\Omega = 0$. It represents the minimal number of BESM installed in a subway system, when substation energy consumption is not considered. Note that objective vector (f_1^0, ϵ_Ω) is the ideal point of problem P. It represents the lower limit of the Pareto front, which is unreachable.

- **Problem P_3:**

$$\min K = \sum_{z=1}^{Z} \kappa_z = f_2(\boldsymbol{\kappa}),$$

s.t.

$$E^s = f_1(\boldsymbol{H}, \boldsymbol{X}, \boldsymbol{\kappa}) \le f_1^0, \tag{35}$$

$$\text{Constraints (28)–(34).}$$

Note that objective (26) in problem P is transformed into constraint (35) in this problem. By solving this problem, we obtain its optimal result denoted as ϵ_0. It represents the minimal number of BESMs needed while substation energy consumption is optimally minimized. Thus, (f_1^0, ϵ_0) is a non-dominated point of problem P. Substation energy consumption cannot be reduced by further increasing the number of BESMs.

- **Problem P_4:**

$$\min E^s = \sum_{z=1}^{Z} \int_{t_{1,0}^4}^{t_{I,L}^4} P_z^s(t)dt = f_1\left(\boldsymbol{H}, \boldsymbol{X}, \boldsymbol{\kappa}\right),$$

s.t.

$$K = f_2(\boldsymbol{\kappa}) \le \epsilon_\Omega, \tag{36}$$

$$\text{Constraints (28)–(34).}$$

Note that objective (27) in problem P is transformed into constraint (36) in this problem. By solving this problem, we obtain the optimal result of E^s denoted as f_1^Ω. Note that $(f_1^\Omega, \epsilon_\Omega)$ is also a non-dominated point of problem P. It represents the minimal substation energy consumption when a minimal number of WESS is installed. As $\epsilon_\Omega = 0$, f_1^Ω is the minimal substation energy consumption when a timetable is optimized and no WESS is installed.

- **Problem P_5:**

$$\min E^s = \sum_{z=1}^{Z} \int_{t_{1,0}^4}^{t_{I,L}^4} P_z^s(t)dt = f_1\left(\boldsymbol{H}, \boldsymbol{X}, \boldsymbol{\kappa}\right),$$

s.t.

$$K = f_2(\boldsymbol{\kappa}) \le \epsilon_\omega, \tag{37}$$

Constraints (28)–(34).

Note that objective (27) in problem P is transformed into constraint (37) in this problem. Problem P_5 aims to minimize substation energy consumption with a limited number of BESMs. By solving it, we obtain the optimal result denoted f_1^ω. As ϵ_ω can be assigned with different values, problem P_5 represents a series of problems. With ϵ_ω decreasing from ϵ_0 to ϵ_Ω, we obtain a series of optimal results of problem P_5. Each pair of $(f_1^\omega, \epsilon_\omega)$ is a non-dominated point of problem P. Thus, $\left\{ (f_1^\omega, \epsilon_\omega) \,|\omega \in \{0, 1, \cdots, \Omega\} \right\}$ constitutes the Pareto front of problem P. Note that $\epsilon_\Omega = 0$ and K is a non-negative integer in this work.

Algorithm 1 Procedure of applying the ϵ-constraint method

Input: Mathematical model of problem P

Output: $\left\{ (f_1^\omega, \epsilon_\omega) \right\}$ %Non-dominated points on Pareto front

```
Transform problem P into problems P1, P2, P3, P4 and P5;
Solve problems P1, P2, P3 and P4 sequentially to obtain the optimal results f1^0, εΩ, ε0 and
  f1^Ω, respectively;
Set Ω = ε0;
for ω = 1 to Ω − 1 do
    Set εω = εω−1 − 1;
    Solve problem P5 to obtain the optimal result f1^ω;
end for
Remove dominated points from {(f1^ω, εω) |ω ∈ {0, 1, ··· , Ω}} and output the remaining as the Pareto
  front of problem P;
```

According to [50], one drawback of an ϵ-constraint method is that an improper selection of ϵ may result in a formulation with no feasible solution in general. Fortunately, the second constraint in our dual-objective optimization problem, i.e., (27) in problem P is a linear function, and its decision variables are all non-negative integers. Thus, its objective value K is definitely a non-negative integer too, i.e., the changing step length of ϵ_ω can be set to 1 naturally and the complete set of Pareto optimal solutions is able to be obtained correspondingly, providing the exact solution for each single-objective optimization problem can be obtained. In addition, for any non-negative value of K, there is always at least one feasible solution for each κ_z satisfying $K = \sum_{z=1}^{Z} \kappa_z$. e.g., one possible solution is $\kappa_1 = K$ and $\kappa_z = 0, \forall z \in \{2, 3, \cdots, Z\}$. Furthermore, whatever value is K, the timetable currently used in a subway system is always a feasible solution to (26). Thus, the drawback of potentially no feasible solution in an ϵ-constraint method does not exist for our dual-objective optimization problem.

The results obtained by applying an ϵ-constraint method are illustrated in Figure 6. Point A is the ideal point and B is the nadir one of the dual-objective optimization problem. Points C and E are the end points on the Pareto front. Note that point D represents a typical point on the Pareto front. With ϵ_ω changes from ϵ_0 to ϵ_Ω, f_1^ω is changed from f_1^0 to f_1^Ω correspondingly, thus point D can be any point on the green curve in Figure 6.

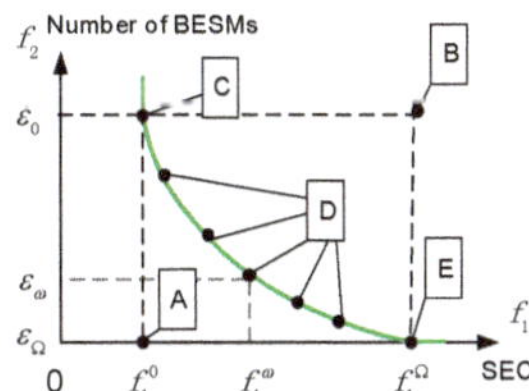

Figure 6. Points obtained by an ϵ-constraint method.

The procedure of applying an ϵ-constraint method to transform a dual-objective optimization problem P into several single-objective ones (i.e., problems P_1 to P_5) is detailed as follows. Note that the first objective in problem P (i.e., substation energy consumption) is treated as the primary objective.

1. Obtain problems P_1 and P_2 respectively, where problem P_1 dismisses the second objective in problem P and problem P_1 dismisses the other one. The constraints in problem P are kept unchanged in both problems P_1 and P_2.
2. By solving problem P_1, we obtain the minimal value of the first objective (i.e., substation energy consumption) without consideration of the other objective (WESS cost). The result is illustrated as f_1^0 in Figure 6. Similarly, by solving problem P_2, we obtain the minimal value of the second objective without consideration of the other one. The result is illustrated as ϵ_Ω.
3. Obtain problem P_3 by adding constraint (35) into obtaining problem P_2. Note that the constraint is the first objective function in problem P and f_1^0 is the optimized result of problem P_1.
4. By solving problem P_3, we obtain the minimal value of WESS size needed (denoted as ϵ_0) whilst maintaining minimum substation energy consumption. Note that (f_1^0, ϵ_0) is a non-dominated point of problem P, shown as point C in Figure 6.
5. Obtain problem P_4 by adding constraint constraint (36) into obtain problem P_1. Note that the constraint is the second objective function in problem P and ϵ_Ω is the optimized result of problem P_2.
6. By solving problem P_4, we obtain the minimal value of substation energy consumption (denoted as f_1^Ω) whilst the least size of WESS is installed. Note that $(f_1^\Omega, \epsilon_\Omega)$ is also a non-dominated point of problem P, shown as point E in Figure 6.
7. Obtain problem P_5 by adding constraint constraint (37) into obtain problem P_1. Note that the constraint is the second objective function in problem P and ϵ_ω is a variable which changes from ϵ_0 to ϵ_Ω with a predefined step length. For different values of ϵ_ω, it becomes a different problem. Thus, problem P_5 represents a series of optimization problems actually.
8. By solving problem P_5 with each value of ϵ_ω, we obtain a set of minimal values of substation energy consumption (denoted as $f_1^{(\omega)}$) whilst certain size of WESS (i.e., totally ϵ_ω number of BESMs) is installed. Note that each $(f_1^\omega, \epsilon_\omega)$ is a non-dominated point of problem P, shown as point D in Figure 6. Also note that point D actually represents a set of points in the green curve, as ϵ_ω is a changing from ϵ_0 to ϵ_Ω.

4.2. Improved Artificial Bee Colony Algorithm

For the single-objective optimization problems transformed from problem P, except P_2, all the other problems (i.e., P_1, P_3, P_4 and P_5) are nonlinear. It is hard to find an optimal solution by using commercial software (e.g., CPLEX) within an acceptable time. Thus, heuristic algorithms are often used to find near-optimal solutions for these kinds of problems [8,10,17–25].

An Artificial Bee Colony (ABC) algorithm is a swarm intelligence algorithm. Since 2005 [51], it has been successfully used to handle many complicated optimization problems [52–58]. It has been compared with differential evolution (DE) [59,60], GA [25,53,61], particle swarm optimization (PSO) [62,63] and evolutionary algorithm (EA) [64] for multi-dimensional numeric problems. Its performance is better than or similar to these algorithms. Therefore, an improved artificial bee colony algorithm is designed to solve the single-objective optimization problems in this work.

The proposed IABC starts with an initial set of feasible solutions, where one of them is the currently used timetable with no WESS, and the others are randomly generated. There are three kinds of bees, i.e., employed, onlooker and scout bees, used in an optimization process. In each iteration, each employed bee is employed at a particular solution, and finds a neighbor one via a local search operator, which is randomly chosen from swap, insertion, mutation and crossover operators; each onlooker bee chooses a solution by spinning a roulette wheel, and then finds a neighbor one with the same procedure as an employed bee; each scout bee randomly generates a feasible solution, to enhance IABC's global search ability. The best solution found by all bees is kept as an initial one in the next iteration. When the iteration count reaches a predefined threshold, the algorithm restarts. Note that all the solutions, except the best one found, are replaced by randomly generated ones. IABC terminates when the restart count reaches a predefined limit. Then, the best solution found is output as an optimal result. The main procedure of IABC is shown in Algorithm 2. The detailed procedure of IABC can also be found in our previous work in [7]. n_1, n_2 and n_3 are the numbers of employed, onlooker and scout bees, respectively; n_4 is the total number of all bees, i.e., $n_4 = n_1 + n_2 + n_3$; p_1, p_2, p_3 and p_4 are the probability of swap, insertion, mutation and crossover operators been chosen, respectively; M_1 and M_2 are predefined iteration count limits; S^* is a specified initial solution; $S_1, S_2, \cdots$, and S_{n_4+1} are one of $n_4 + 1$ solutions, respectively; S' and F' are the optimal solution found and its fitness value, respectively.

Algorithm 2 Procedure of IABC

Input: $n_1, n_2, n_3, n_4, p_s, p_i, p_m, p_c, M_1, M_2, S^*$
Output: S', F'

Set $S_{n_4+1} = S^*$; %init the best solution found
for $j = 1$ to $M_1 \times M_2$ **do**
  if $(j \mod M_2) = 1$ **then**
    for $k = 1$ to n_4 **do**
      Randomly generate a feasible solution S_k;
    end for
  end if
  for $k = 1$ to $n_4 + 1$ **do**
    Calculate the fitness value F_k for each solution S_k;
  end for
  Sort $F_1, F_2, \cdots, F_{n_4+1}$ in 'descend' order and pick out the first n_1 number of elements to generate a list F;
  Generate a list $\bar{S}$ containing the n_1 number of solutions corresponding to F;
  Set $S_{n_4+1} = \bar{S}_{(1)}$; %record the best solution
  Set $F_{n_4+1} = F_{(1)}$; %Fitness value of the best solution
  %employed bee phase
  for $k = 1$ to n_1 **do**
    Set solution $B = \bar{S}_{(k)}$;
    Generate S_k from B via a local search operator randomly chosen from swap, insertion, mutation and crossover;
  end for
  %onlooker bee phase
  for $k = 1$ to n_2 **do**
    Set B as a solution in $\bar{S}$ by roulette wheel selection;
    Generate S_{n_1+k} from B via a local search operator randomly chosen from swap, insertion, mutation and crossover;
  end for
  %scout bee phase
  for $k = 1$ to n_3 **do**
    Randomly generate a feasible solution $S_{n_1+n_2+k}$;
  end for
end for
Output $S' = S_{n_4+1}$ and $F' = F_{n_4+1}$;

The currently used timetable without WESS is used as an initial solution and the idea of elitism is adopted in IABC, to ensure the optimized solution is not getting worse than it. In each iteration, all solutions are evaluated and sorted based on their fitness values. A list containing n_1 best solutions is generated accordingly. Then, each employed bee is assigned to a specific solution in the list sequentially, and a new solution is generated by applying a local search operator. The local search operator is randomly chosen from swap, insertion, mutation and crossover operators with the probability of p_1, p_2, p_3 and p_4, respectively. To ensure each newly generated solution is feasible, i.e, satisfying constraints (28)–(34), its feasibility is checked immediately and repair is applied if it is infeasible. Similarly, each onlooker bee generates a new solution based on an assigned solution in the same way. The difference between employed and onlooker bees is as follows: the former is associated with a particular solution in the best solution list, and the latter randomly chooses one in the list by spinning a roulette wheel. A scout bee randomly generates a feasible solution in each iteration to ensure the global search ability of IABC. IABC converges to a local optimum in M_2 iterations. Then, the best solution found so far is kept and the other solutions are replaced by randomly generated feasible ones, to reduce the chance of IABC being trapped in a local optimum. Thereafter, a similar process repeats until the total iteration count reaches $M_1 \times M_2$. Finally, a near-optimal solution and its fitness value are obtained.

5. Experimental Results and Analysis

To show the effectiveness of the proposed method, numerical examples are conducted based on the actual data obtained from Yanfang Line in Beijing, China. The actual dwell time at each platform and running time at each section are shown in Table 1. Note that x_n is dwell time at a departure platform with index n, $\underline{x_n}$ and $\overline{x_n}$ being its lower and upper bounds in the optimization process, and r_l is the running time from platform x_n to x_{n+1}. They are the same for all trains. The headway time and other parameters of Yanfang Line are listed in Table 2. Note that headway time between any two successive trains is identical in the current timetable. The sections in each electricity supply interval are given in Table 3. A super-capacitor is selected as a BESM in this work. Parameters about WESS are shown in Table 4.

Table 1. Actual dwell time and running time of Yanfang Line.

Departure Platform Index n	x_n (s)	$\underline{x_n}$ (s)	$\overline{x_n}$ (s)	Arrival Platform Index $n+1$	r_l (s)
1	30	25	35	2	129
2	30	25	35	3	98
3	30	25	35	4	117
4	30	25	35	5	135
5	25	20	30	6	139
6	30	25	35	7	84
7	30	25	35	8	128
8	30	25	35	9	141
9	30	25	35	10	136
10	30	25	35	11	124
11	30	25	35	12	83
12	30	25	35	13	140
13	25	20	30	14	132
14	30	25	35	15	117
15	30	25	35	16	96
16	30	25	35	17	119

Table 2. Other Parameters of Yanfang Line.

Parameter	Value	Parameter	Value	Parameter	Value
I	131	N	9	m (kg)	287,080
h (s)	482	$\underline{h}$ (s)	422	$\overline{h}$ (s)	542
η_1	0.7	η_2	0.8	ζ_1	0.05
τ (s)	2576	$\underline{\tau}$ (s)	2516	$\overline{\tau}$ (s)	2636
Φ (s)	188	a_l^1	0.8	a_l^3 (m/s^2)	1
C_1 (s)	65236	r_l^1 (s)	27	r_l^3 (s)	21

Table 3. Electricity supply intervals distribution of Yanfang Line.

Electricity Supply Interval Index	Section Indices					
1	8	9	-	-	-	-
2	5	6	7	10	11	12
3	3	4	13	14	-	-
4	1	2	15	16	-	-

Table 4. Parameters about WESS.

Parameter	Value	Parameter	Value	Parameter	Value
P^m (kW)	2000	C^m (kWh)	1	ζ_2	1
$\overleftarrow{P_z^c}$ (kW)	300	$\overleftarrow{P_z^d}$ (kW)	300	ζ_3	0.1
β_2 (%)	90	β_1 (%)	30	β_0 (%)	20

Parameters of IABC are set as shown in Table 5. GA is also applied to solve the single-objective optimization problems as a baseline method. The detailed processes of GA can be found in [21]. To compare IABC and GA fairly, the population size of GA is set to be 40 and its maximum iteration count is 300. The local search operators used in GA include selection, crossover and mutation. The probability of crossover and mutation are 0.8 and 0.2, respectively.

Table 5. Parameters of IABC.

n_e	n_o	n_s	M_1	M_2	p_s	p_i	p_m	p_c
10	10	20	6	50	0.1	0.1	0.2	0.6

Compared with identical dwell time, more energy can be saved when dwell time at a platform varied for different trains. But regenerative energy utilization improvement is relatively limited, and it increases the solution space of an optimization problem dramatically [25]. Thus, in order to simplify the problem and keep consistent with the actual timetable, we keep the dwell time for different trains i and j at platform n identical in the experiments. i.e., $x_{i,n} = x_{j,n}, \forall i,j \in \{1,2,...,I\}$ and $n \in \{1,2,...,N-1\}$.

The experiments are implemented in MATLAB 2014 and runs on a notebook with Intel(R) Core(TM) i5-3210M CPU @2.50 GHz, 12 GB RAM and a Windows 7 Operating System.

The simulation results are are shown in Figure 7. Note that there are several WESS in a subway line, different WESS may have different number of BESMs. The horizontal axis in Figure 7 is the sum of the BESM numbers in each WESS. The number of BESMs equals 0 means there are no WESS installed at all. When it is greater than 0, the detailed configuration in each WESS should be obtained in the optimized solution. The vertical axis is the substation energy consumption. A black asterisk in Figure 7 denotes substation energy consumption with the current timetable (i.e., the timetable is not optimized). Note that the configuration of each WESS is set as the optimized result, as there does not exist an actual configuration of WESS for each case. A blue cross denotes an optimized result obtained by GA where timetable and WESS are simultaneously optimized, a green circle denotes an optimized result of our dual-objective optimization problem obtained by IABC, and a red "+ " in a green circle denotes a non-dominated point on the Pareto front of problem P.

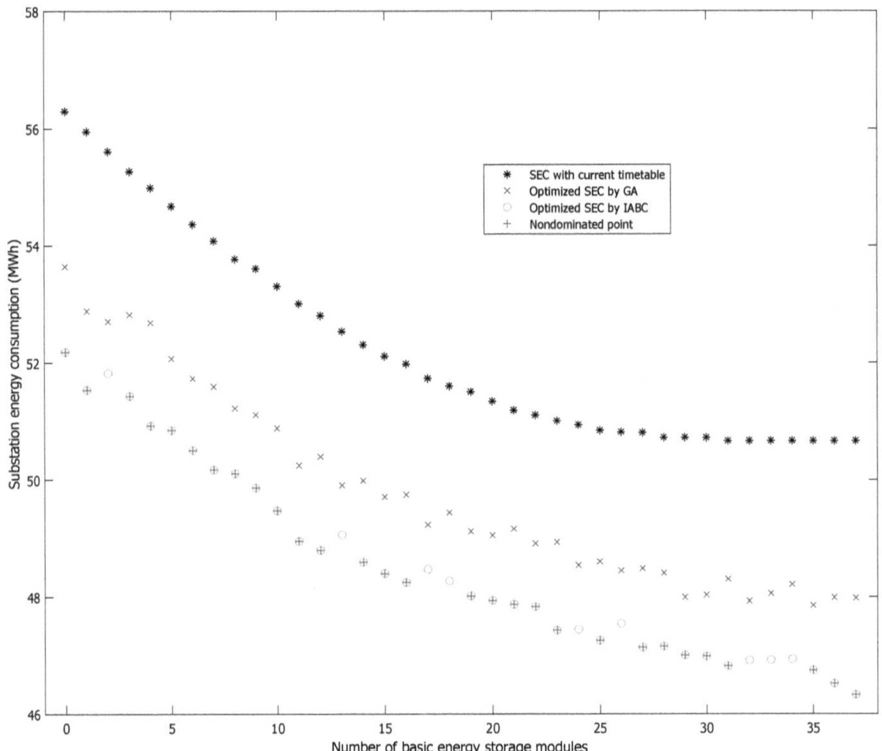

Figure 7. Traction energy saved in different numbers of basic energy storage modules.

From the experimental results, it is easy to see that: (1) Substation energy consumption decreases with the total size of WESS increases, and the decreasing speed gradually slows down. (2) With the same size of WESS installed, timetable optimization further reduces substation energy consumption, which shows the effectiveness of the integration optimization of timetable and WESS. (3) IABC and GA can both to solve the single-objective optimization problems, and IABC performs better than GA, as the results obtained by IABC dominate those obtained by GA. (4) A set of Pareto optimal solutions of the dual-objective optimization problem is obtained by applying the proposed ϵ-constraint method and IABC. The non-dominated points are diverse and well distributed over the Pareto front.

Set the current timetable with no WESS installed, which is the actual conditions in the experimental subway line, as a basis for comparison, substation energy consumption reduces by $(56.293 - 52.180)/56.293 \times 100\% = 7.31\%$ when timetable is optimized without WESS invested, optimization of WESS with total size as 37 BESMs reduces substation energy consumption by $(56.293 - 50.663)/56.293 \times$

100% = 10.00% when timetable is optimized, and the decrement comes to (56.293 − 46.331)/56.293 × 100% = 17.70% when timetable is optimized together with 37 BESMs installed as WESS. Thus, timetable optimization and WESS installation can both improve energy conservation, and the integration of them reaches the greatest extent. Note that the results we obtained is a set of Pareto optimal solutions, each element of them reduces substation energy consumption to different degrees with different WESS investment cost. The experimental results are helpful for optimal decision making. Based on relationship between energy conservation profit and WESS cost, decision makers can easily choose a preferred solution according to their particular needs, e.g., the solution with the lowest cost, the one with the greatest profit, or that with the best cost-effectiveness.

As an illustration of one optimized solution, the optimized timetable and configuration of each WESS when the total number of BESMs as $K = 37$ are given in Figure 8. The upper part shows the difference between the optimized headway time and that in the current timetable; the lower left part shows the optimized dwell time and the current dwell time; in addition, the lower right part shows the numbers of BESMs in each WESS, note that there are four electricity supply intervals in this subway line, which means four WESSs are installed. It is seen that the headway time and dwell time are slightly changed from the current timetable, and the configuration for each WESS is proposed. Note that there are several electricity supply intervals in a subway system, and one WESS is installed in an electricity supply interval. Thus, there are several WESSs in a subway system (e.g., WESS 1 to z). As each WESS is independent from others, they can have different capacities (i.e., number of BESMs) in theory.

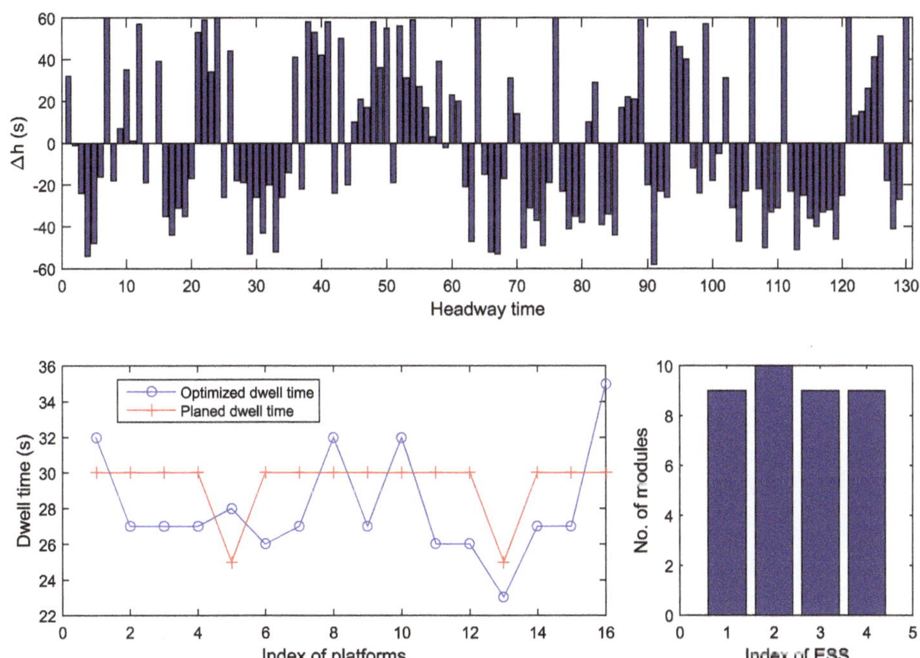

Figure 8. One optimized timetable and WESS configuration.

The profiles of the state-of-charge of each WESS are shown in Figure 9. The peak values of all WESS are between 90–100%. There is some residual capacity, but it is very small. Thus, this kind of WESS configuration is reasonable for regenerative energy utilization, i.e., if the total number of BESMs is reduced, some regenerative energy may not be able to be utilized; and if the number was increased, the increased energy capacity would be a waste of money.

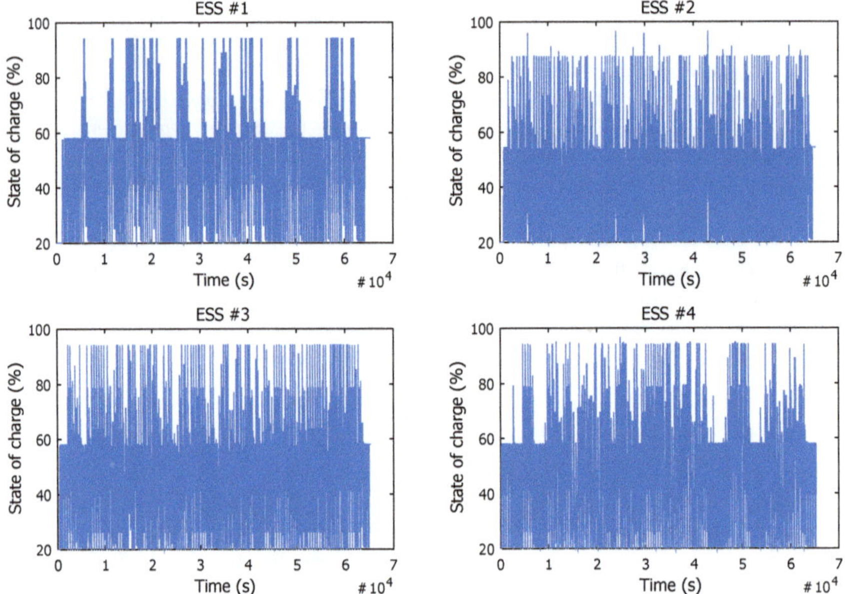

Figure 9. State-of-charge of each WESS.

A real-world subway line is often subject to minor real-time perturbations which affects the adherence to the timetable [7,17]. The presence of unexpected disturbances brings uncertainties to scheduling optimization problems [65,66], which may affect the effectiveness of the optimized results. To make it clear, the problem and method proposed in this work are intended to be used offline indeed, which means the optimization problem is solved only once before trains go into operation and the potential uncertainties are not particularly considered. However, the offline method is still of great significance, as the size of each WESS is hard to be changed over time; it should be decided at the first stage of design.

To solve the potential uncertainties in real-world application, there are two alternative ways. One is to adopt a real-time optimization method [65,66], e.g., using a receding horizon principle to optimize the timetable iteratively; the other is to evaluate the robustness of the optimized results on disturbances and subject to it if acceptable (i.e., the optimized results are not too sensitive to disturbances). It is relatively complicated to introduce an online optimization method in this work, so we leave it as our future work direction and adopt the second method here.

To evaluate the robustness of the optimized timetable, we conduct the following experiments by adding random noise to it. In the experiment, we traverse the headway time and dwell time vector and add a random number to each element in it. The random number is randomly chosen from set $\{-\delta, 0, \delta\}$ (s), which represents an actual noise. As a comparison, we add this king of noise both on the current timetable and the optimized one, and δ is assigned with different values to compare their effects on energy conservation. To be justified, we run the experiment 100 times with each value of δ and their average substation energy consumption is obtained, respectively. The experimental results with δ varies from 0 to 3 are given in Table 6.

Table 6. Substation energy consumption (SEC) for timetable with noise.

Noise δ (s)	SEC with Current Timetable (kWh)	SEC with Optimized Timetable (kWh)	Energy Saving Ratio
0	56,293	46,331	17.70%
1	56,302	46,418	17.55%
2	56,313	46,592	17.26%
3	56,233	46,925	16.55%

Note that $\delta = 0$ represents no noise added on the timetable. It is seen that the energy saving ratio slightly decreases when the disturbance on train operations becomes larger. However, even with 3-s of noise, the energy saving ratio is still around 17% for the optimized timetable over the current one. Thus, the optimized result is robust enough for energy conservation under disturbances. The proposed offline optimization method is obviously acceptable to this particular problem.

6. Conclusions

We propose an integration of timetable optimization and WESS to improve regenerative energy utilization, thus to reduce substation energy consumption in a subway system. A dual-objective optimization model is formulated to simultaneously minimize substation energy consumption and WESS investment cost accordingly. To solve the dual-objective optimization problem, an ϵ-constraint method is first designed to transform it into several single-objective optimization problems; then, an improved artificial bee colony algorithm is designed to solve them sequentially. By combining the proposed ϵ-constraint method and improved artificial bee colony algorithm, a set of Pareto optimized solutions for the original dual-objective optimization can be obtained. Experiments based on the actual data from a subway system in China are conducted to illustrate the procedure to use them and their effectiveness. The results are useful for decision makers in a subway system, based on which they can easily make a sensible trade-off between energy saving profit and WESS investment cost according to their practical needs. In addition, the proposed improved artificial bee colony algorithm is also compared with a commonly used genetic algorithm during the experiments to prove its effectiveness.

The proposed model and resolution method can be used in any subway system which is equipped with regenerative braking systems, to maximize energy saving with the minimum investment cost paid for that. Although the optimized results are robust enough to random disturbances, one limit of the proposed method is that it is mainly applicable for offline optimization now. Thus, part of our future work will be the study of an online optimization method, to better handle the parameter uncertainty problems caused by unexpected disturbances to train operations. In addition, train speed profiles at sections also affect regenerative energy utilization and substation energy consumption. Thus, in our future research, we also plan to integrate energy-efficient train operations into our model to further reduce substation energy consumption.

Author Contributions: H.L. and T.T. contributed to the conception of the study. H.L. and M.C. formulated the problem and designed the resolution algorithms. J.L. helped program the algorithm and proposed several constructive suggestions for this work. H.L. analyzed the experimental results and drafted this paper.

Funding: The work was funded by the Fundamental Research Funds for the Central Universities under grant 2019JBM006.

Conflicts of Interest: The authors declare no conflict of interest.

References

1. González-Gil, A.; Palacin, R.; Batty, P.; Powell, J. A systems approach to reduce urban rail energy consumption. *Energy Convers. Manag.* **2014**, *80*, 509–524. [CrossRef]
2. González-Gil, A.; Palacin, R.; Batty, P. Optimal energy management of urban rail systems: Key performance indicators. *Energy Convers. Manag.* **2015**, *90*, 282–291. [CrossRef]
3. Yang, X.; Li, X.; Ning, B.; Tang, T. A survey on energy-efficient train operation for urban rail transit. *IEEE Trans. Intell. Transp. Syst.* **2016**, *17*, 2–13. [CrossRef]
4. Su, S.; Tang, T.; Wang, Y. Evaluation of strategies to reducing traction energy consumption of metro systems using an optimal train control simulation model. *Energies* **2016**, *9*, 105. [CrossRef]
5. Liu, R.R.; Golovitcher, I.M. Energy-efficient operation of rail vehicles. *Transp. Res. Part A Policy Pract.* **2003**, *37*, 917–932. [CrossRef]
6. Su, S.; Tang, T.; Roberts, C. A cooperative train control model for energy saving. *IEEE Trans. Intell. Transp. Syst.* **2015**, *16*, 622–631. [CrossRef]
7. Liu, H.; Zhou, M.; Guo, X.; Zhang, Z.; Ning, B.; Tang, T. Timetable optimization for regenerative energy utilization in subway systems. *IEEE Trans. Intell. Transp. Syst.* **2018**, 1–11. [CrossRef]
8. Yang, X.; Ning, B.; Li, X.; Tang, T. A two-objective timetable optimization model in subway systems. *IEEE Trans. Intell. Transp.* **2014**, *15*, 1913–1921. [CrossRef]
9. González-Gil, A.; Palacin, R.; Batty, P. Sustainable urban rail systems: Strategies and technologies for optimal management of regenerative braking energy. *Energy Convers. Manag.* **2013**, *75*, 374–388. [CrossRef]
10. Kampeerawat, W.; Koseki, T. A strategy for utilization of regenerative energy in urban railway system by application of smart train scheduling and wayside energy storage system. *Energy Procedia* **2017**, *138*, 795–800. [CrossRef]
11. Ceraolo, M.; Lutzemberger, G.; Meli, E.; Pugi, L.; Rindi, A.; Pancari, G. Energy storage systems to exploit regenerative braking in dc railway systems: Different approaches to improve efficiency of modern high-speed trains. *J. Energy Storage* **2018**, *16*, 269–279. [CrossRef]
12. Ciccarelli, F.; Iannuzzi, D.; Tricoli, P. Control of metro-trains equipped with onboard supercapacitors for energy saving and reduction of power peak demand. *Transp. Res. Part C Emerg. Technol.* **2012**, *24*, 36–49. [CrossRef]
13. Ramos, A.; Pena, M.T.; Fernandez, A.; Cucala, P. Mathematical programming approach to underground timetabling problem for maximizing time synchronization. In Proceedings of the International Conference on Industrial Engineering and Industrial Management, Madrid, Spain, 5–7 September 2007; IEEE: Piscataway, NJ, USA, 2007; pp. 88–95.
14. Nasri, A.; Moghadam, M.F.; Mokhtari, H. Timetable optimization for maximum usage of regenerative energy of braking in electrical railway systems. In Proceedings of the International Symposium on Power Electronics Electrical Drives Automation and Motion, Pisa, Italy, 14–16 June 2010; IEEE: Piscataway, NJ, USA, 2010; pp. 1218–1221.
15. Kim, K.M.; Kim, K.T.; Han, M.S. A model and approaches for synchronized energy saving in timetabling. In Proceedings of the 9th World Congress on Railway Research, Lille, France, 22–26 May 2011.
16. Pe na-Alcaraz, M.; Fernández, A.; Cucala, A.P.; Ramos, A.; Pecharromán, R.R. Optimal underground timetable design based on power flow for maximizing the use of regenerative-braking energy. *Proc. Inst. Mech. Eng. Part F J. Rail Rapid Transit* **2012**, *226*, 397–408. [CrossRef]
17. Fournier, D.; Mulard, D.; Fages, F. Energy optimization of metro timetables: A hybrid approach. In Proceedings of the 18th International Conference on Principles and Practice of Constraint Programming, Quebec City, QC, Canada, 8–12 October 2012; Springer: Berlin/Heidelberg, Germany, 2012; pp. 7–12.
18. Yang, X.; Li, X.; Gao, Z.; Wang, H.; Tang, T. A cooperative scheduling model for timetable optimization in subway systems. *IEEE Trans. Intell. Transp. Syst.* **2013**, *14*, 438–447. [CrossRef]
19. Zhao, L.; Li, K.; Su, S. A multi-objective timetable optimization model for subway systems. In *Proceedings of the 2013 International Conference on Electrical and Information Technologies for Rail Transportation (EITRT2013)—Volume I*; Springer: Berlin/Heidelberg, Germany, 2014; pp. 557–565.

20. Le, Z.; Li, K.; Ye, J.; Xu, X. Optimizing the train timetable for a subway system. *Proc. Inst. Mech. Eng. Part F J. Rail Rapid Transit* **2015**, *229*, 852–862. [CrossRef]
21. Yang, X.; Chen, A.; Li, X.; Ning, B.; Tang, T. An energy-efficient scheduling approach to improve the utilization of regenerative energy for metro systems. *Transp. Res. Part C Emerg. Technol.* **2015**, *57*, 13–29. [CrossRef]
22. Li, X.; Lo, H.K. Energy minimization in dynamic train scheduling and control for metro rail operations. *Transp. Res. Part B Methodol.* **2014**, *70*, 269–284. [CrossRef]
23. Gong, C.; Zhang, S.; Zhang, F.; Jiang, J.; Wang, X. An integrated energy-efficient operation methodology for metro systems based on a real case of shanghai metro line one. *Energies* **2014**, *7*, 7305–7329. [CrossRef]
24. Xu, X.; Li, K.; Li, X. A multi-objective subway timetable optimization approach with minimum passenger time and energy consumption. *J. Adv. Transp.* **2016**, *50*, 69–95. [CrossRef]
25. Liu, H.; Zhou, M.; Guo, X.; Liu, H.; Tang, T. An abc-based subway timetable optimization model for regenerative energy utilization. In Proceedings of the International Conference on Systems, Man, and Cybernetics, Miyazaki, Japan, 7–10 October 2018; IEEE: Piscataway, NJ, USA, 2018, pp. 1412–1417.
26. Tian, Z.; Weston, P.; Zhao, N.; Hillmansen, S.; Roberts, C.; Chen, L. System energy optimisation strategies for metros with regeneration. *Transp. Res. Part C Emerg. Technol.* **2017**, *75*, 120–135. [CrossRef]
27. Liu, H.; Tang, T.; Guo, X.; Xia, X. A timetable optimization model and an improved artificial bee colony algorithm for maximizing regenerative energy utilization in a subway system. *Adv. Mech. Eng.* **2018**, *10*, 1–13. [CrossRef]
28. Liu, P.; Yang, L.; Gao, Z.; Huang, Y.; Li, S.; Gao, Y. Energy-efficient train timetable optimization in the subway system with energy storage devices. *IEEE Trans. Intell. Transp. Syst.* **2018**, *19*, 1–17. [CrossRef]
29. Gao, Z.; Fang, J.; Zhang, Y.; Sun, D. Control strategy for wayside supercapacitor energy storage system in railway transit network. *J. Mod. Power Syst. Clean Energy* **2014**, *2*, 181–190. [CrossRef]
30. Huang, Y.; Yang, L.; Tang, T.; Gao, Z.; Cao, F.; Li, K. Train speed profile optimization with on-board energy storage devices: A dynamic programming based approach. *Comput. Ind. Eng.* **2018**, *126*, 149–164. [CrossRef]
31. Ahmadi, S.; Dastfan, A.; Assili, M. Energy saving in metro systems: Simultaneous optimization of stationary energy storage systems and speed profiles. *J. Rail Transp. Plan. Manag.* **2018**, *8*, 78–90. [CrossRef]
32. Wu, P.; Che, A.; Chu, F.; Zhou, M. An improved exact *varepsilon*-constraint and cut-and-solve combined method for biobjective robust lane reservation. *IEEE Trans. Intell. Transp. Syst.* **2015**, *16*, 1479–1492. [CrossRef]
33. Chankong, V.; Haimes, Y.Y. *Multiobjective Decision Making: Theory and Methodology*; North-Holland: New York, NY, USA, 1983.
34. Carli, R.; Dotoli, M. Energy scheduling of a smart home under nonlinear pricing. In Proceedings of the 53rd IEEE Conference on Decision and Control, Los Angeles, CA, USA, 15–17 December 2014; IEEE: Piscataway, NJ, USA, 2014; pp. 5648–5653.
35. Carli, R.; Dotoli, M. Cooperative distributed control for the energy scheduling of smart homes with shared energy storage and renewable energy source. *IFAC-PapersOnLine* **2017**, *50*, 8867–8872. [CrossRef]
36. Sperstad, I.B.; Korpås, M. Energy storage scheduling in distribution systems considering wind and photovoltaic generation uncertainties. *Energies* **2019**, *12*, 1231. [CrossRef]
37. T'kindt, V.; Billaut, J.-C. Multicriteria scheduling problems: A survey. *RAIRO-Oper. Res.* **2001**, *35*, 143–163. [CrossRef]
38. Zhou, Z.; Chu, F.; Che, A.; Zhou, M. ε-constraint and fuzzy logic based optimization of hazardous material transportation via lane reservation. *IEEE Trans. Intell. Transp. Syst.* **2013**, *14*, 847–857. [CrossRef]
39. Guo, X.; Liu, S.; Zhou, M.; Tian, G. Disassembly sequence optimization for large-scale products with multiresource constraints using scatter search and petri nets. *IEEE Trans. Cybern.* **2016**, *46*, 2435–2446. [CrossRef]
40. Guo, X.; Liu, S.; Zhou, M.; Tian, G. Dual-objective program and scatter search for the optimization of disassembly sequences subject to multiresource constraints. *IEEE Trans. Autom. Sci. Eng.* **2018**, *15*, 1091–1103. [CrossRef]
41. Sun, H.; Wu, J.; Ma, H.; Yang, X.; Gao, Z. A bi-objective timetable optimization model for urban rail transit based on the time-dependent passenger volume. *IEEE Trans. Intell. Transp. Syst.* **2018**, *20*, 1–12. [CrossRef]
42. Ehrgott, M. *Multicriteria Optimization*; Springer Science & Business Media: Berlin/Heidelberg, Germany, 2005; p. 491.

43. Haimes, Y.Y.; Lasdon, L.; Wismer, D. On a bicriterion formulation of the problems of integrated system identification and system optimization. *IEEE Trans. Syst. Man Cybern.* **1971**, *1*, 296–297.
44. Berube, J.-F.; Gendreau, M.; Potvin, J.-Y. An exact ε-constraint method for bi-objective combinatorial optimization problems: Application to the traveling salesman problem with profits. *Eur. J. Oper. Res.* **2009**, *194*, 39–50. [CrossRef]
45. Feng, J.; Che, A.; Wang, N. Bi-objective cyclic scheduling in a robotic cell with processing time windows and non-euclidean travel times. *Int. J. Prod. Res.* **2014**, *52*, 2505–2518. [CrossRef]
46. Jozefowiez, N.; Semet, F.; Talbi, E.-G. The bi-objective covering tour problem. *Comput. Oper. Res.* **2007**, *34*, 1929–1942. [CrossRef]
47. Leitner, M.; Ljubic, I.; Sinnl, M. Solving the bi-objective prizecollecting steiner tree problem with the *varepsilon*-constraint method. *Electron. Notes Discret. Math.* **2013**, *41*, 181–188. [CrossRef]
48. Reiter, P.; Gutjahr, W.J. Exact hybrid algorithms for solving a bi-objective vehicle routing problem. *Cent. Eur. J. Oper. Res.* **2012**, *20*, 19–43. [CrossRef]
49. Bouziaren, S.A.; Aghezzaf, B. An improved augmented *varepsilon*-constraint and branch-and-cut method to solve the tsp with profits. *IEEE Trans. Intell. Transp. Syst.* **2018**, *20*, 1–10.
50. Marler, R.T.; Arora, J.S. Survey of multi-objective optimization methods for engineering. *Struct. Multidiscip. Optim.* **2004**, *26*, 369–395. [CrossRef]
51. Karaboga, D. *An Idea Based on Honey Bee Swarm for Numerical Optimization*; Technical Report-tr06; Erciyes University, Engineering Faculty, Computer Engineering Department: Kayseri, Turkey, 2005.
52. Deng, G.; Xu, Z.; Gu, X. A discrete artificial bee colony algorithm for minimizing the total flow time in the blocking flow shop scheduling. *Chin. J. Chem. Eng.* **2012**, *20*, 1067–1073. [CrossRef]
53. Li, J.; Pan, Q.; Gao, K. Pareto-based discrete artificial bee colony algorithm for multi-objective flexible job shop scheduling problems. *Int. J. Adv. Manuf. Technol.* **2011**, *55*, 1159–1169. [CrossRef]
54. Gu, W.; Yu, Y.; Hu, W. Artificial bee colony algorithm-based parameter estimation of fractional-order chaotic system with time delay. *IEEE/CAA J. Autom. Sin.* **2017**, *4*, 107–113. [CrossRef]
55. Singh, A. An artificial bee colony algorithm for the leaf-constrained minimum spanning tree problem. *Appl. Soft Comput.* **2009**, *9*, 625–631. [CrossRef]
56. Pan, Q.-K.; Tasgetiren, M.F.; Suganthan, P.; Chua, T. A discrete artificial bee colony algorithm for the lot-streaming flow shop scheduling problem. *Inf. Sci.* **2011**, *181*, 2455–2468. [CrossRef]
57. Karaboga, D.; Basturk, B. On the performance of artificial bee colony (abc) algorithm. *Appl. Soft Comput.* **2008**, *8*, 687–697. [CrossRef]
58. Karaboga, D.; Akay, B. A comparative study of artificial bee colony algorithm. *Appl. Math. Comput.* **2009**, *214*, 108–132. [CrossRef]
59. Tian, G.; Ren, Y.; Zhou, M. Dual-objective scheduling of rescue vehicles to distinguish forest fires via differential evolution and particle swarm optimization combined algorithm. *IEEE Trans. Intell. Transp. Syst.* **2016**, *17*, 3009–3021. [CrossRef]
60. Wang, L.; Wang, S.; Zheng, X. A hybrid estimation of distribution algorithm for unrelated parallel machine scheduling with sequence-dependent setup times. *IEEE/CAA J. Autom. Sin.* **2016**, *3*, 235–246.
61. Mareda, T.; Gaudard, L.; Romerio, F. A parametric genetic algorithm approach to assess complementary options of large scale windsolar coupling. *IEEE/CAA J. Autom. Sin.* **2017**, *4*, 260–272. [CrossRef]
62. Dong, W.; Zhou, M. A supervised learning and control method to improve particle swarm optimization algorithms. *IEEE Trans. Syst. Man Cybern. Syst.* **2017**, *47*, 1135–1148. [CrossRef]
63. Tang, Y.; Luo, C.; Yang, J.; He, H. A chance constrained optimal reserve scheduling approach for economic dispatch considering wind penetration. *IEEE/CAA J. Autom. Sin.* **2017**, *4*, 186–194. [CrossRef]
64. Kang, Q.; Feng, S.; Zhou, M.; Ammari, A.C.; Sedraoui, K. Optimal load scheduling of plug-in hybrid electric vehicles via weight-aggregation multi-objective evolutionary algorithms. *IEEE Trans. Intell. Transp. Syst.* **2017**, *18*, 2557–2568. [CrossRef]

65. Hosseini, S.M.; Carli, R.; Dotoli, M. Model predictive control for real-time residential energy scheduling under uncertainties. In Proceedings of the 2018 IEEE International Conference on Systems, Man, and Cybernetics (SMC), Miyazaki, Japan, 7–10 October 2018; IEEE: Piscataway, NJ, USA, 2018; pp. 1386–1391.
66. Touretzky, C.R.; Baldea, M. Integrating scheduling and control for economic mpc of buildings with energy storage. *J. Process.* **2014**, *24*, 1292–1300. [CrossRef]

Article

Adaptive Forgetting Factor Recursive Least Square Algorithm for Online Identification of Equivalent Circuit Model Parameters of a Lithium-Ion Battery

Xiangdong Sun *, Jingrun Ji, Biying Ren, Chenxue Xie and Dan Yan

School of Automation and Information Engineering, Xi'an University of Technology, Xi'an 710048, China; 2180320027@stu.xaut.edu.cn (J.J.); renby@126.com (B.R.); 18829028497@163.com (C.X.); 2170320042@stu.xaut.edu.cn (D.Y.)

* Correspondence: sxd1030@163.com

Received: 30 April 2019; Accepted: 6 June 2019; Published: 12 June 2019

Abstract: With the popularity of electric vehicles, lithium-ion batteries as a power source are an important part of electric vehicles, and online identification of equivalent circuit model parameters of a lithium-ion battery has gradually become a focus of research. A second-order *RC* equivalent circuit model of a lithium-ion battery cell is modeled and analyzed in this paper. An adaptive expression of the variable forgetting factor is constructed. An adaptive forgetting factor recursive least square (AFFRLS) method for online identification of equivalent circuit model parameters is proposed. The equivalent circuit model parameters are identified online on the basis of the dynamic stress testing (DST) experiment. The online voltage prediction of the lithium-ion battery is carried out by using the identified circuit parameters. Taking the measurable actual terminal voltage of a single battery cell as a reference, by comparing the predicted battery terminal voltage with the actual measured terminal voltage, it is shown that the proposed AFFRLS algorithm is superior to the existing forgetting factor recursive least square (FFRLS) and variable forgetting factor recursive least square (VFFRLS) algorithms in accuracy and rapidity, which proves the feasibility and correctness of the proposed parameter identification algorithm.

Keywords: lithium-ion battery; equivalent circuit model; recursive least square; adaptive forgetting factor; parameter identification

1. Introduction

Energy shortages and environmental pollution are becoming more and more prominent today. Therefore, electric vehicles, with many advantages such as resource conservation and environmental friendliness, have attracted more and more attention. With the rapid development of electric vehicles, industry standards of lithium-ion batteries have also been formulated. Lithium-ion batteries and their energy management have received more extensive attention [1]. An accurate state of charge (SOC) estimation of lithium-ion batteries is required in the testing and practical use of lithium-ion batteries [2]. The equivalent circuit model of lithium-ion batteries is the crucial basis for most SOC estimation algorithms, such as extended Kalman filter (EKF) [3], adaptive extended Kalman filter (AEKF) [4], etc. Although the performance of lithium-ion batteries and lead-acid batteries is very different, the reaction mechanism of the two batteries is basically the same, the conversion between chemical energy and electric energy is realized by the oxidation-reduction reaction and there is a similar response trend for the change of input current [5]. In addition, the equivalent circuit parameters are fitted to the experimental data of lithium-ion battery and lead-acid battery, and it is found that the two batteries can be characterized by a unified equivalent circuit [6]. Thus, the lithium-ion battery model can usually be established by referring to that of the lead-acid batteries. At present, the battery equivalent circuit model mainly includes R_{int} model [7], PNGV (Partnership for a New Generation of Vehicles) model [8], Thevenin model [9], and n-order RC equivalent circuit model [10]. The R_{int} model is an internal resistance model consisting of a DC source and an internal resistance. Although the model is simple, it does not take into account the internal state of the battery. Hence, the circuit structure has more defects. It is such an ideal model that it is generally only used in simple circuit simulation. The PNGV model considering capacitance characteristics accurately reflects the discharge process, but the equivalent circuit model of the charging process is not discussed. The n-order RC dynamic equivalent model can reflect the relationship between the internal parameters of the battery and the temperature or current. However, as the order increases, the complexity of the model increases, which is not conducive to real-time online calculation of the micro-controller. Therefore, the second-order RC equivalent circuit model is usually chosen, which not only has good accuracy and dynamic simulation characteristics, but also has the advantage of lower complexity [11–13].

In view of the complex chemical reaction and physical structure inside the lithium-ion battery, when the battery is actually used, the internal state of the battery will be affected by the factors such as ambient temperature, operating conditions, and battery aging degree. Some parameters in the battery equivalent model also change when the working conditions change. Therefore, it is necessary to accurately identify the parameters in the battery equivalent model in real time. The recursive least square (RLS) method is most commonly used for system parameter identification [14]. The RLS is simple and stable, but with the increase of data in the recursive process, the generation of new data will be affected by the old data, which will lead to large errors. In order to solve the above problems, reference [15] studies the forgetting factor recursive least square (FFRLS) method. The proportion of old and new data is adjusted by introducing a forgetting factor into the RLS, so that the proportion of old data is reduced when new data is available, and the algorithm can converge to the actual value more quickly. Since the forgetting factor is constant, the dynamic identification ability and accuracy of circuit parameters using FFRLS will be affected when the charging and discharging currents change frequently. Therefore, the variable forgetting factor least square (VFFRLS) method appears [16–18]. The forgetting factor is adjusted according to the square of a time-averaging estimation of the auto-correlation of a priori and a posteriori error [16]. Reference [17] analyzes the dynamic equation of the mean square error that can be used to derive a dynamic equation of the gradient of the mean square error to control the forgetting factor. Since the forgetting factor converges slowly, the tracking speed of the mutation parameter may decrease. In reference [18], the average input power estimation and exponential window size expression are introduced to update the forgetting factor. It is applied to the state regularization QR decomposition RLS method, which improves the tracking performance, steady-state mean square error, and the robustness to the input power variation. The calculation of the variable forgetting factor in the references mentioned above is rather complicated and the computational burden is heavy, which is not conducive to the real-time operation of the micro-controller. Therefore, an adaptive expression for calculating the forgetting factor relatively easily is proposed in this paper. Based on the second-order RC equivalent circuit model, it is applied to the adaptive forgetting factor recursive least square (AFFRLS) method to identify the equivalent circuit model parameters online. Experiments including the dynamic stress test (DST) are implemented to verify the real-time performance and accuracy of the AFFRLS algorithm.

2. Lithium-Ion Battery Modeling

The second-order RC equivalent circuit model of a lithium-ion battery is shown in Figure 1. It consists of an ideal voltage source U_{oc}, ohmic resistor R_0, and two RC parallel circuits. U_{oc} represents the open circuit voltage of the lithium battery. R_0 indicates the internal resistance of the battery. The two RC parallel circuits represent the electrochemical polarization and concentration polarization effects in the battery reactions. U_L is the battery terminal voltage. The following is the analysis process of the equivalent circuit model shown in Figure 1 [13–15]. According to Kirchhoff's voltage law and current law, the electrical characteristic equation of the model is expressed by (1).

$$\begin{aligned} &U_L = U_{oc}[SOC(t)] - U_1 - U_2 - I(t)\cdot R_0 \\ &C_1\cdot\frac{dU_1}{dt} = I(t) - \frac{U_1}{R_1} \\ &C_2\cdot\frac{dU_2}{dt} = I(t) - \frac{U_2}{R_2} \end{aligned} \tag{1}$$

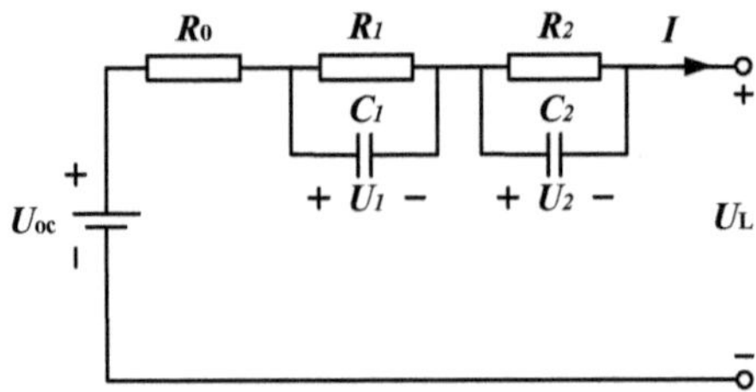

Figure 1. Second-order *RC* equivalent circuit model.

Equation (1) is written as a frequency domain expression.

$$E = U_L(s) - U_{oc}(s) = -I(s)\left(R_0 + \frac{R_1}{1+R_1C_1s} + \frac{R_2}{1+R_2C_2s}\right) \tag{2}$$

Equation (2) is rewritten to (3).

$$G(s) = \frac{E(s)}{I(s)} = -\frac{R_0s^2 + \frac{R_0R_1C_1+R_0R_2C_2+R_2R_1C_1+R_1R_2C_2}{R_1C_1R_2C_2}s + \frac{R_0+R_1+R_2}{R_1C_1R_2C_2}}{s^2 + \frac{R_1C_1+R_2C_2}{R_1C_1R_2C_2}s + \frac{1}{R_1C_1R_2C_2}} \tag{3}$$

Bilinear transformation $s = \frac{2}{T}\cdot\frac{1-z^{-1}}{1+z^{-1}}$ is brought into (3), and Equations (4) and (5) are obtained.

$$G(z^{-1}) = \frac{E(k)}{I(k)} = \frac{\theta_3 + \theta_4 z^{-1} + \theta_5 z^{-2}}{1 - \theta_1 z^{-1} - \theta_2 z^{-2}} \tag{4}$$

$$\begin{cases} \theta_1 = \frac{2T^2-8R_1C_1R_2C_2}{-T^2-2T(R_1C_1+R_2C_2)-4R_1C_1R_2C_2} \\ \theta_2 = \frac{T^2-2T(R_1C_1+R_2C_2)+4R_1C_1R_2C_2}{-T^2-2T(R_1C_1+R_2C_2)-4R_1C_1R_2C_2} \\ \theta_3 = \frac{T^2(R_0+R_1+R_2)+2T(R_0R_1C_1+R_0R_2C_2+R_1R_2C_2+R_2R_1C_1)+4R_0R_1C_1R_2C_2}{-T^2-2T(R_1C_1+R_2C_2)-4R_1C_1R_2C_2} \\ \theta_4 = \frac{2T^2(R_0+R_1+R_2)-8R_0R_1C_1R_2C_2}{-T^2-2T(R_1C_1+R_2C_2)-4R_1C_1R_2C_2} \\ \theta_5 = \frac{T^2(R_0+R_1+R_2)-2T(R_0R_1C_1+R_0R_2C_2+R_1R_2C_2+R_2R_1C_1)+4R_0R_1C_1R_2C_2}{-T^2-2T(R_1C_1+R_2C_2)-4R_1C_1R_2C_2} \end{cases} \tag{5}$$

Therefore, the recursive Equation (6) is obtained by (4).

$$E(k) = \theta_1E(k-1) + \theta_2E(k-2) + \theta_3I(k) + \theta_4I(k-1) + \theta_5I(k-2) \tag{6}$$

where $E(k-1)$ and $E(k-2)$ are the difference between the terminal voltage and the open circuit voltage at the time of $k-1$ and $k-2$. $I(k)$, $I(k-1)$, and $I(k-2)$ are input currents at the time of k, $k-1$, and $k-2$.

Suppose that a, b, c, d, f are represented by (7).

$$\begin{cases} a = R_0 \\ b = R_1C_1R_2C_2 \\ c = R_1C_1 + R_2C_2 \\ d = R_0 + R_1 + R_2 \\ f = R_0R_1C_1 + R_0R_2C_2 + R_1R_2C_2 + R_2R_1C_1 \end{cases} \tag{7}$$

Equation (7) is brought into (5) and simplified to (8).

$$\begin{cases} \theta_1 = \frac{8b-2T^2}{4b+2cT+T^2} \\ \theta_2 = \frac{4cT}{4b+2cT+T^2} - 1 \\ \theta_3 = -\frac{4ab+2eT+dT^2}{4b+2cT+T^2} \\ \theta_4 = \frac{8ab-2dT^2}{4b+2cT+T^2} \\ \theta_5 = -\frac{4ab-2eT+dT^2}{4b+2cT+T^2} \end{cases} \tag{8}$$

Therefore, Equation (9) is obtained by (8).

$$\begin{cases} a = \frac{\theta_4-\theta_3-\theta_5}{1+\theta_1-\theta_2} \\ b = \frac{T^2(1+\theta_1-\theta_2)}{4(1-\theta_1-\theta_2)} \\ c = \frac{T(1+\theta_2)}{1-\theta_1-\theta_2} \\ d = \frac{-\theta_3-\theta_4-\theta_5}{1-\theta_1-\theta_2} \\ f = \frac{T(\theta_5-\theta_3)}{1-\theta_1-\theta_2} \end{cases} \tag{9}$$

where T is the sampling time.

Suppose $\tau_1 = \frac{c+\sqrt{c^2-4b}}{2}$, $\tau_2 = \frac{c-\sqrt{c^2-4b}}{2}$. Thus, the resistance and capacitance parameters R_0, R_1, R_2, C_1, and C_2 can be obtained by (10).

$$\begin{cases} R_0 = a \\ R_1 = [\tau_1(d-a) + ac - f]/(\tau_1 - \tau_2) \\ R_2 = d - a - R_1 \\ C_1 = \tau_1/R_1 \\ C_2 = \tau_2/R_2 \end{cases} \tag{10}$$

3. Online Parameter Identification Principle

3.1. Forgetting Factor Recursive Least Square Method

The RLS method is the most commonly used method for system parameter identification [19]. This method uses the square norm of the discrete function as a metric to get the identification parameters. Equation (11) can be obtained from (6) when the system error is considered. It is a discrete expression of the system to be identified.

$$E(k) = \theta_1E(k-1) + \theta_2E(k-2) + \theta_3I(k) + \theta_4I(k-1) + \theta_5I(k-2) + e(k) \tag{11}$$

Define the parameter vector $\boldsymbol{\theta}$ and the observation data matrix $\boldsymbol{\varphi}$ as follows:

$$\boldsymbol{\theta} = [\theta_1\theta_2\theta_3\theta_4\theta_5]^T \tag{12}$$

$$\varphi = \begin{bmatrix} E(k-1) & E(k-2) & I(k) & I(k-1) & I(k-2) \\ E(k-2) & E(k-3) & I(k-1) & I(k-2) & I(k-3) \\ & & \vdots & & \\ E(k-m-1) & E(k-m-2) & I(k-m) & I(k-m-1) & I(k-m-2) \end{bmatrix} \tag{13}$$

where k denotes the current moment. m is the observation times. φ is the known observation data matrix. $\boldsymbol{\theta}$ is the parameter vector to be estimated. The matrix form of (11) can be expressed from (12) and (13).

$$\boldsymbol{E} = \varphi\boldsymbol{\theta} + \boldsymbol{e} \tag{14}$$

where $\boldsymbol{e}$ is the systematic error vector $\boldsymbol{e} = [e(k)e(k-1)\cdots e(k-m)]^T$. $\boldsymbol{E}$ is the system output vector, and its data is the observation value of system output $\boldsymbol{E} = [E(k)E(k-1)\cdots E(k-m)]^T$. The evaluation function of the RLS method is given by (15).

$$J = \sum_{t=0}^{m} [e(k-t)]^2 = \boldsymbol{e}^T\boldsymbol{e} \tag{15}$$

If the derivative of J is zero, the parameter vector $\boldsymbol{\theta}'$ can be obtained in the smallest case of (14).

$$\begin{gathered} \left.\frac{\partial J}{\partial \theta}\right|_{\theta=\theta'} = \frac{\partial J}{\partial \theta}\left[(\boldsymbol{E}-\varphi\boldsymbol{\theta})^T(\boldsymbol{E}-\varphi\boldsymbol{\theta})\right] = 0 \\ \varphi^T\boldsymbol{E} = \varphi^T\varphi\boldsymbol{\theta}' \end{gathered} \tag{16}$$

When $\varphi^T\varphi$ is a full rank matrix, the parameter estimation of the RLS method is expressed by (17).

$$\boldsymbol{\theta}' = (\varphi^T\varphi)^{-1}\varphi^T\boldsymbol{E} \tag{17}$$

On the basis of the RLS method, the FFRLS method is to add the forgetting factor λ as a coefficient in the observed data matrix φ and the system output vector $\boldsymbol{E}$, they are expressed by (18) and (19). When each observation obtains new data, the proportion of new and old data is adjusted by exponential weighting, and then the last obtained identification parameter is corrected. Thus, when the input variables change, the FFRLS method can respond quickly and obtain better identification parameters as the system observation data increase.

$$\boldsymbol{E} = [E(k)\lambda E(k-1)\cdots\lambda^m E(k-m)]^T \tag{18}$$

$$\varphi = \begin{bmatrix} E(k-1) & E(k-2) & I(k) & I(k-1) & I(k-2) \\ \lambda E(k-2) & \lambda E(k-3) & \lambda I(k-1) & \lambda I(k-2) & \lambda I(k-3) \\ & & \vdots & & \\ \lambda^m E(k-m-1) & \lambda^m E(k-m-2) & \lambda^m I(k-m) & \lambda^m I(k-m-1) & \lambda^m I(k-m-2) \end{bmatrix} \tag{19}$$

3.2. *Adaptive Forgetting Factor Analysis*

λ allocates the weights of old and new data, and usually takes a constant of 0.98. When $\lambda = 1$, the FFRLS method degenerates into the RLS method. Since the forgetting factor is constant, when the online identification parameter error is very small, the introduction of the forgetting factor may increase the online identification parameter error. When the online identification parameter error is very large, it is desirable to optimize the forgetting factor to make the online identification have faster convergence speed and reduce the identification error. Therefore, it is expected that the forgetting factor can vary adaptively with the identification parameter error.

The most critical part of the variable forgetting factor least squares algorithm (VFFRLS) is how to make the forgetting factor adaptively change. In the steady state, the forgetting factor λ is close to or equal to 1. On the contrary, the forgetting factor λ tends to be a suitable value, which only affects the error of the nearby moment, so that the online identification parameter can be quickly tracked to the actual value, and λ is gradually increased to the optimum value at steady state. An equation for calculating the adaptive forgetting factor is proposed to achieve the above purpose, it is expressed by (20).

$$\begin{aligned} \lambda(k) &= \lambda_{\min} + (1-\lambda_{\min}) \cdot h^{\varepsilon(k)} \\ \varepsilon(k) &= round((\tfrac{e(k)}{e_{base}})^2) \end{aligned} \tag{20}$$

where $\lambda_{\min}$ is the minimum value of the forgetting factor. Usually, the range of forgetting factor is 0.95~1 [15,20] and it is found by the experimental data that the equivalent circuit model parameters identified by the AFFRLS algorithm are accurate and fast as the range of forgetting factor is selected as 0.98~1, and, therefore, $\lambda_{\min}$ is 0.98. h is the sensitivity coefficient. h may be selected as any value between 0 and 1, which indicates the sensitivity of forgetting factor to the errors. When h is close to 1 (e.g., 0.99), the forgetting factor changes slowly from 1 to 0.98, which leads to the slow response speed of parameter identification. Conversely, when h is close to 0 (e.g., 0.01), the forgetting factor quickly changes from 1 to 0.98, which results in the response speed of parameter identification too fast and reduces the accuracy. Therefore, h is generally chosen to be 0.9, which takes into account the balance between the rapidity and accuracy of identification parameters. $e(k)$ is the error at k time and e_{base} is the allowable error reference. Equation (20) shows that the forgetting factor λ decreases rapidly when the kth error $e(k)$ exceeds e_{base}; hence, e_{base} is usually chosen according to the magnitude of the expected error. When the identification parameter error is less than e_{base}, the identification parameters are considered stable and λ changes to a larger value. When the error of identification parameters is greater than e_{base}, the identification parameters are considered unstable and the change of e_{base} is smaller. The function $round(n)$ represents the integer closest to n. It can be seen from (20) that the larger the error value, the smaller the forgetting factor e_{base}, and its variation range is between 0.98 and 1; thus, the forgetting factor can be adaptively changed with the error of identification parameters.

3.3. Implementation of Online Parameter Identification Algorithm Based on AFFRLS

It is seen from the above analysis that each parameter in the second-order RC equivalent circuit model of the lithium-ion battery can be calculated by (10) as long as θ_1, θ_2, θ_3, θ_4, and θ_5 in (4) are estimated. Therefore, it is necessary to identify θ_1, θ_2, θ_3, θ_4, and θ_5 by using the online parameter identification algorithm based on AFFRLS. The overall block diagram is shown in Figure 2. The specific implementation flow chart of AFFRLS is shown in Figure 3.

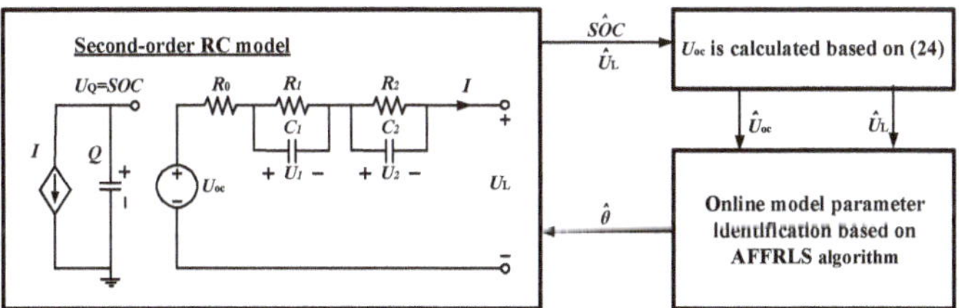

Figure 2. Overall block diagram of the online identification parameters.

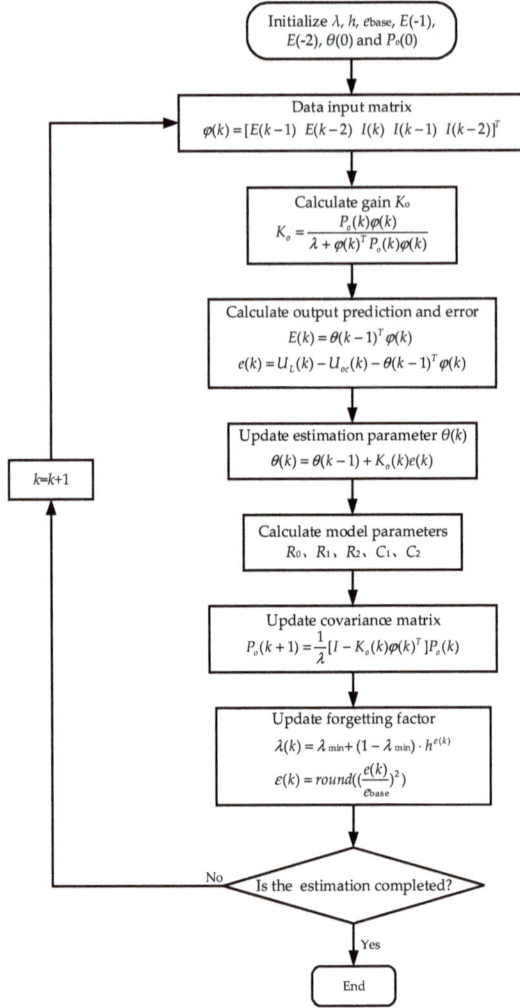

Figure 3. Flow chart of the adaptive forgetting factor recursive least square (AFFRLS) algorithm.

The online parameter identification algorithm is performed by the AFFRLS. It is known from (12) and (13) that $\theta(k) = [\theta_1\theta_2\theta_3\theta_4\theta_5]^T$ and $\varphi(k) = [E(k-1)E(k-2)I(k)I(k-1)I(k-2)]^T$. Where $\varphi(k)$ is the known data at time k, and $\theta(k)$ is the parameter to be estimated at time k.

The given initial value $\theta(0)$ generally is a sufficiently small real matrix. At the two moments before the start of the algorithm, the input current is zero, and the open circuit voltage U_{oc} is equal to the terminal voltage U_L, so $E(-1) = E(-2) = 0$, and the initial $\phi(0)$ value is [0 0 0 0 0]. The gain matrix K_o is calculated by (21) [14].

$$K_o = \frac{P_o(k-1)\varphi(k)}{\lambda + \varphi(k)^T P_o(k-1)\varphi(k)} \tag{21}$$

where $P_o(k)$ is the covariance matrix at time k and its initial value is an identity matrix.

Hence, the estimated parameter $\theta(k)$ is updated by (22).

$$\theta(k) = \theta(k-1) + K_o(k)[U_L(k) - U_{oc}(k) - \theta(k-1)^T\varphi(k)] \tag{22}$$

where the open circuit voltage $U_{oc}(k)$ at time k is given by the polynomial between the open circuit voltage (OCV) and the SOC.

λ is obtained from (20) and taken into (21) to obtain the gain matrix K_o. Bringing K_o into (22) and θ_1, θ_2, θ_3, θ_4, and θ_5 are obtained. The estimated values of R_0, R_1, R_2, C_1, and C_2 at time k can then be obtained by (9) and (10).

According to (23), the covariance matrix $P_o(k)$ is updated by the obtained gain matrix K_o. The parameter identification at the next moment is performed again. Where I is an identity matrix.

$$P_o(k) = \frac{1}{\lambda}[I - K_o(k)\varphi(k)^T]P_o(k-1) \tag{23}$$

4. Experimental Verification and Analysis

The special power supply is used to charge and discharge the 3.2 V/36 Ah lithium iron phosphate battery produced by Shandong Wina Green Power Co., Ltd in Weifang, China. The sampling time is T = 10 s and the environment temperature is 25 °C. The experimental platform and the specification of the battery are shown in Figure 4 and Table 1, respectively.

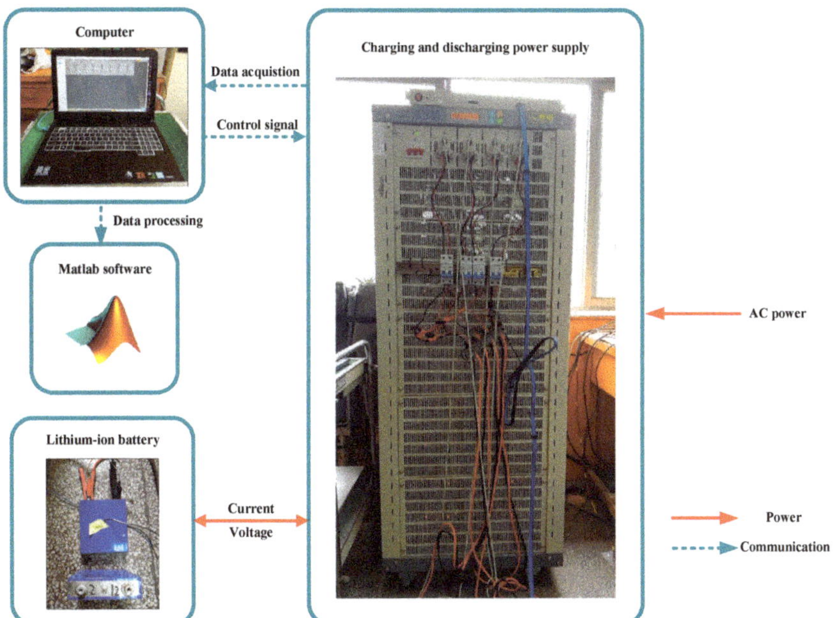

Figure 4. Experimental platform.

Table 1. The battery specifications.

Parameter	Value
Rated capacity (Ah)	36
Nominal voltage (V)	3.2
Standard charging/discharging current (A)	12
Charging cut-off voltage (V)	3.7
Discharging cut-off voltage (V)	2.5
Maximum continuous discharging current (A)	108

4.1. OCV-SOC Curve

Intermittent constant current charging and discharging experiments with 0.33 C standard rate current recommended by a company are carried out. The charging and discharging experimental curves are shown in Figure 5, and the obtained SOC-OCV data are given by Table 2. Polynomial fitting of the experimental data is performed by using Matlab software, Equation (24) is obtained, which provides an open circuit voltage U_{oc} for the FFRLS or AFFRLS algorithm. The OCV-SOC relationship curve under this condition is shown in Figure 6. It is seen from Figure 6 that the OCV-SOC curve of the entire charging and discharging process is approximately a hysteresis curve. Therefore, the influence of the charging and discharging current direction on the open circuit voltage needs to be considered during the online parameter identification.

$$\begin{aligned} &U_{ocDis} = 1813.4*SOC^9 - 8629.9*SOC^8 + 17470*SOC^7 - 19595*SOC^6 \\ &+13285*SOC^5 - 5570.7*SOC^4 + 1419.9*SOC^3 - 208.1*SOC^2 \\ &+15.953*SOC + 2.7228 \\ &U_{ocCha} = 3060.5*SOC^9 - 13713*SOC^8 + 25909*SOC^7 - 26862*SOC^6 \\ &+16655*SOC^5 - 6310.9*SOC^4 + 1434.4*SOC^3 - 185.1*SOC^2 \\ &+12.471*SOC + 2.9002 \end{aligned} \quad (24)$$

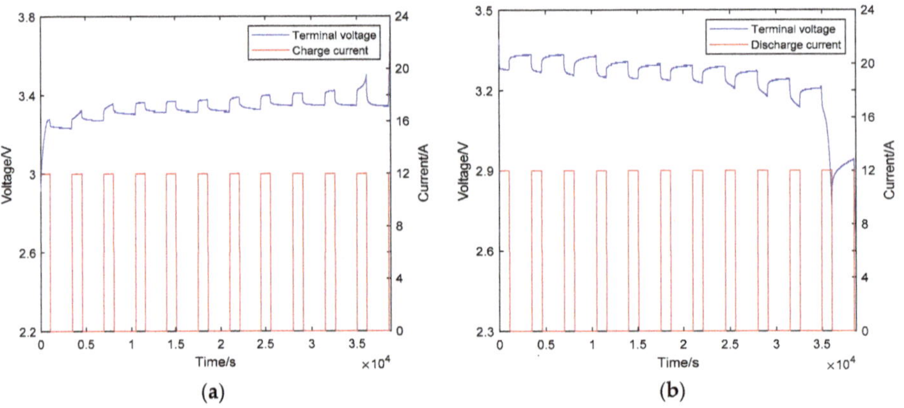

Figure 5. Voltage and current curves of the intermittent constant current charging and discharging experiments: (**a**) Charging process; (**b**) Discharging process.

Table 2. The state of charge (SOC)-open circuit voltage (OCV) data.

Item	1	2	3	4	5	6	7	8	9	10	11	12	13
Intermittent constant current charging experiments with 0.33 C standard current													
SOC/%	0	8.96	17.92	26.88	35.84	44.80	53.76	62.72	71.68	80.64	89.60	98.56	100
OCV/V	2.902	3.233	3.270	3.304	3.311	3.311	3.311	3.326	3.344	3.348	3.344	3.341	3.615
Intermittent constant current discharging experiments with 0.33 C standard current													
SOC/%	0	1.49	10.45	19.40	28.36	37.31	46.27	55.22	64.18	73.13	82.09	91.04	100
OCV/V	2.683	2.939	3.214	3.244	3.270	3.289	3.292	3.300	3.307	3.330	3.333	3.333	3.393

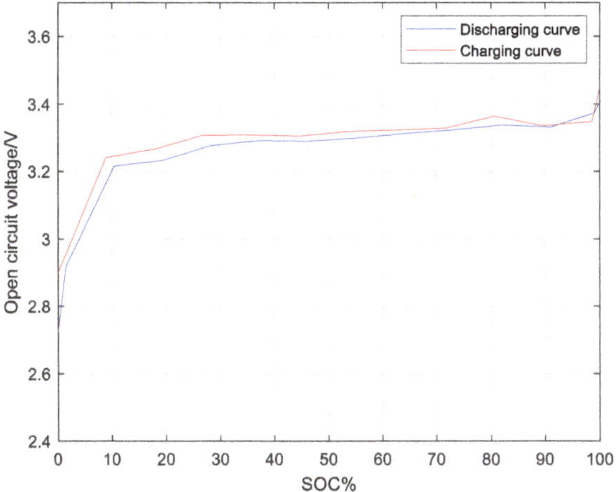

Figure 6. OCV-SOC relationship curve.

4.2. Online Parameter Identification of Lithium-Ion Battery Equivalent Model

The dynamic stress test (DST) experiment has a strict charge and discharge process, it is shown in Figure 7. The cyclic charge current rates are 0.22 C, 0.33 C, and 0.5 C, and the cyclic discharge current rates are 0.22 C and 0.33 C, respectively. Under these conditions, the validity of the online parameter identification algorithm can be more rigorously verified.

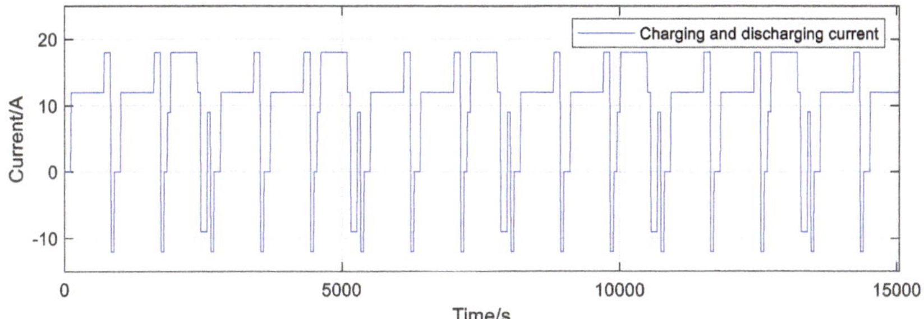

Figure 7. Charge and discharge current waveform of dynamic stress testing (DST) experiment.

Figures 8 and 9 are the identified parameter curves of the FFRLS algorithm and the AFFRLS algorithm under the DST conditions, respectively. Comparing Figures 8 and 9, it can be seen that the parameters identified by the FFRLS algorithm are relatively stable, but the identification ability of dynamic parameter change is insufficient. The parameters identified by the AFFRLS algorithm have obvious fluctuations, which more accurately reflect the complex characteristics of real-time variation of each parameter with the change of charging and discharging current. The dynamic parameters also have more spikes, which fully highlights the identification ability when charging and discharging currents are frequently switched. Figure 10 shows the adaptive forgetting factor λ. It can be seen that the forgetting factor λ has many spikes. And it is adaptively varied with the change of charging and discharging current, which is beneficial to enhance the dynamic parameter identification ability of the AFFRLS algorithm.

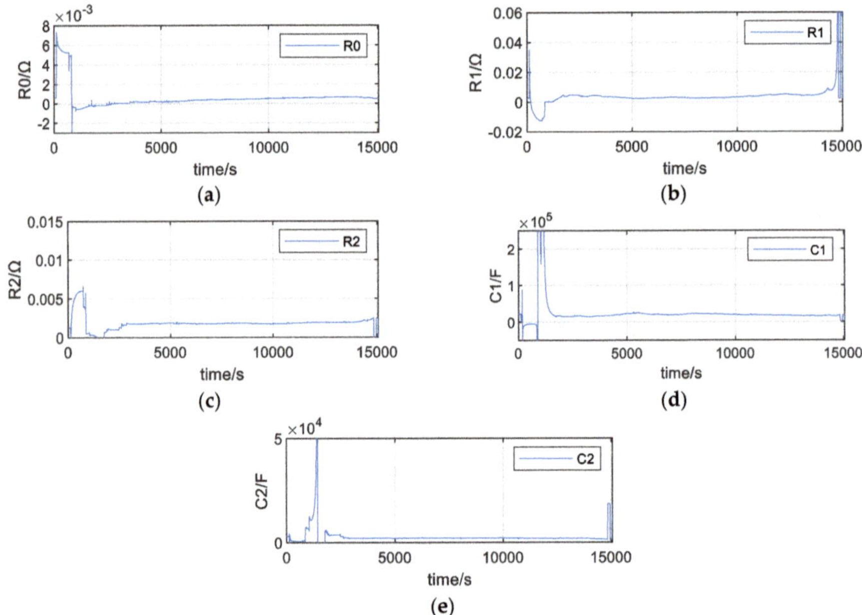

Figure 8. Parameter identification results of the forgetting factor recursive least square (FFRLS) algorithm: (**a**) Identification curve of R_0; (**b**) Identification curve of R_1; (**c**) Identification curve of R_2; (**d**) Identification curve of C_1; (**e**) Identification curve of C_2.

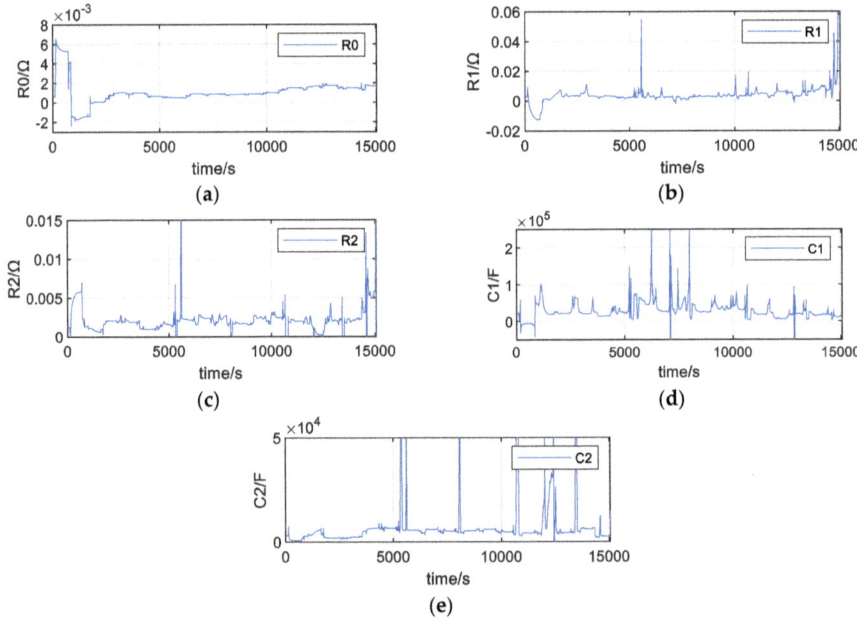

Figure 9. Parameter identification results of the AFFRLS algorithm: (**a**) Identification curve of R_0; (**b**) Identification curve of R_1; (**c**) Identification curve of R_2; (**d**) Identification curve of C_1; (**e**) Identification curve of C_2.

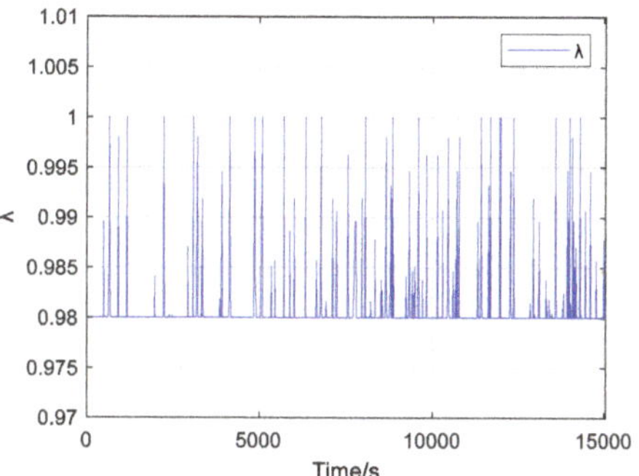

Figure 10. Adaptive forgetting factor λ.

4.3. Comparative Analysis of the Prediction Effect of the Lithium-Ion Battery Terminal Voltage

The experimental conditions in Figures 11 and 12 are the same as those in Figure 7. The FFRLS algorithm and the AFFRLS algorithm are used to predict the lithium-ion battery terminal voltage respectively on the basis of the identification parameters shown in Figures 8 and 9, and Figure 11 is a comparison of the measured terminal voltage and the terminal voltage identified by the FFRLS and AFFRLS algorithms. Figure 12 is a comparison of the measured terminal voltage and the terminal voltage identified by the VFFRLS algorithm of literature [16] with a certain weight coefficient. It may be determined from Figures 11 and 12 which algorithm identifies the circuit parameters more accurately and responds faster.

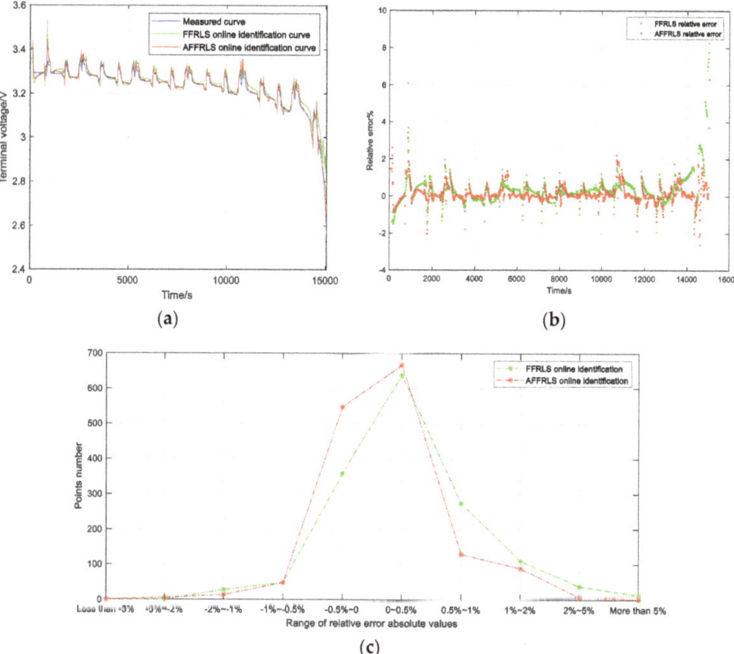

Figure 11. Comparison of the measured terminal voltage and the terminal voltage identified by the FFRLS and AFFRLS algorithms: (**a**) Terminal voltage comparison curve; (**b**) Scatter plots of relative errors; (**c**) Distribution statistics of relative errors.

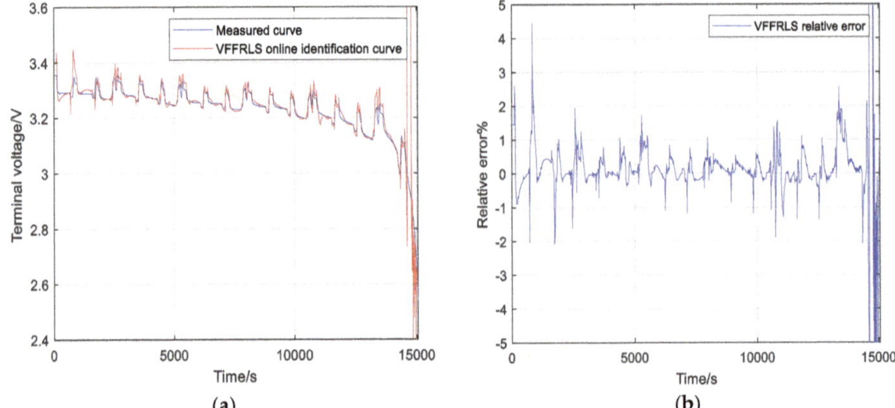

Figure 12. Comparison of the measured terminal voltage and the terminal voltage identified by the VFFRLS algorithm of the literature [16] with a certain weight coefficient: (**a**) Terminal voltage comparison curve; (**b**) Relative error of terminal voltages.

Figure 11a is the measured terminal voltage and the terminal voltage identified by the FFRLS and AFFRLS algorithms. Figure 11b shows the scatter plot of relative error for FFRLS and AFFRLS algorithms. The range of absolute value of relative errors in Figure 11b is divided into 10 intervals, and the points of relative errors falling into each interval are counted, and Figure 11c is obtained. As can be seen from Figure 11c, the relative error distribution of AFFRLS in the range of (±0.5%) is significantly higher than that of FFRLS, while the relative error distribution in other ranges is mostly lower than that of FFRLS.

From the aspect of the sample average and standard deviation, sample average value of the relative errors is 0.372% and sample standard deviation of the relative errors is 0.947 for FFRLS. The AFFRLS sample average value is 0.136%, and the sample standard deviation is 0.526. The sample average and standard deviation of AFFRLS algorithm are smaller than those of FFRLS algorithm.

The FFRLS algorithm and AFFRLS algorithm are tested by *F*-test. Assume H_0: There is no significant difference in the total variance between the two algorithms. H_1: There is a significant difference in the total variance between the two algorithms. Significance level is set to 0.05 and tail type is bilateral, $H = 1$, $p = 3.9233 \times 10^{-109}$ is obtained. The confidence interval of mean difference is [2.9293, 3.5856], and we can see from $H = 1$, $p = 3.9233 \times 10^{-109} < 0.05$ that the original hypothesis is not accepted, i.e., there is a significant difference in variance between the two algorithms.

The FFRLS algorithm and AFFRLS algorithm are tested by *t*-test. Assume H_0: There is no significant difference in the average value of the two algorithms. H_1: There is a significant difference in the average value of the two algorithms. The significance level is 0.05, the tail type is bilateral, and the variance type is unequal. $H = 1$, $p = 3.9716 \times 10^{-17}$ is obtained, the confidence interval of mean difference is [0.18189, 0.29135]. We can see from $H = 1$, $p = 3.9716 \times 10^{-17} < 0.05$ that the original hypothesis is not accepted, i.e., there is a significant difference in the average value between the two algorithms.

In summary, the average value of AFFRLS algorithm is closer to zero than that of FFRLS algorithm, and the variance is smaller, which shows that the parameter identification result of AFFRLS algorithm is more accurate than that of FFRLS algorithm.

Figure 13 is a real-time variation curve of the forgetting factor obtained by the algorithm of literature [16] under a certain weight coefficient. It can be seen from Figure 12 that this algorithm also has a good terminal voltage prediction capability, but it is more demanding on the weight coefficient. When the weight coefficient is not appropriate, the forgetting factor will be too small and the parameter changes drastically, which may lead to the divergence of the algorithm. While the AFFRLS algorithm in this paper limits the variation range of the forgetting factor, it has better stability and the range of correlation coefficients is more relaxed.

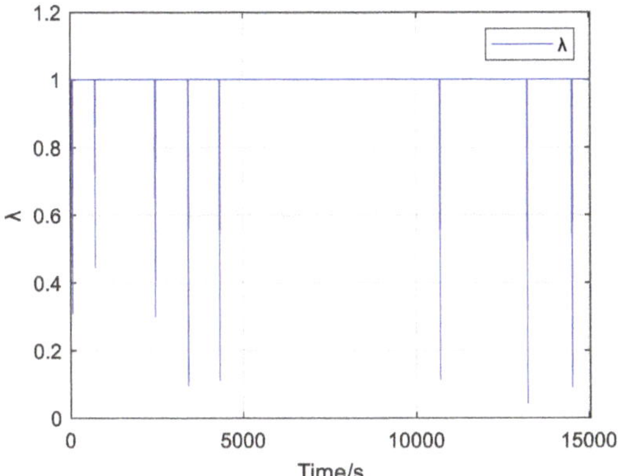

Figure 13. Real-time variation curve of the forgetting factor λ in the literature [16].

The VFFRLS algorithm of the literature [16] to be compared is written into the same program together with the AFFRLS algorithm proposed in this paper. The two algorithms run in parallel and measure the computing time corresponding to each experimental data respectively. Last, the total computing time of the two algorithms is obtained by accumulating the computing time, respectively. The average calculation time of the AFFRLS algorithm in this paper is 17.65 ms, and that of the literature [16] is 26.05 ms. The average calculation time is saved by 32.25%. This indicates that the adaptive algorithm of this paper is simpler, the operation time is shorter, and the real-time performance is better, which is beneficial to the practical application of the algorithm in the micro-controller such as digital signal processor (DSP).

5. Conclusions

In this paper, the second-order *RC* equivalent circuit model of the lithium-ion battery is analyzed, and the online identification algorithm of the equivalent circuit model parameters based on the AFFRLS is studied. The correctness of the equivalent circuit model parameter identification in the case of charging and discharging is verified by the DST experiment, and the prediction terminal voltage obtained by the model parameters are compared with the actual terminal voltage. The experimental results show that the proposed AFFRLS algorithm has a more accurate parameter identification ability than the original FFRLS algorithm. Compared with other VFFRLS algorithms, it has better stability of parameter identification and shorter operation time.

Author Contributions: Conceptualization, X.S. and J.J.; methodology, X.S.; software, J.J.; validation, X.S., J.J., B.R., C.X., and D.Y.; formal analysis, J.J.; investigation, X.S.; resources, B.R.; data curation, J.J., C.X., and D.Y.; writing—original draft preparation, X.S., J.J.; writing—review and editing, X.S.; visualization, J.J.; supervision, X.S.; project administration, X.S.; funding acquisition, X.S.

Funding: This research was funded by the National Natural Science Foundation of China grant number 51577155 and the Natural Science Foundation of Shaanxi Province grant number 2018 JZ5006.

Conflicts of Interest: The authors declare no conflict of interest.

References

1. Hannan, M.A.; Hoque, M.M.; Hussain, A.; Yusof, Y.; Ker, P.J. State-of-the-art and energy management system of lithium-ion batteries in electric vehicle applications: Issues and recommendations. *IEEE Access* **2018**, *6*, 19362–19378. [CrossRef]

2. Xu, J.; Mi, C.C.; Cao, B.; Deng, J.; Chen, Z.; Li, S. The state of charge estimation of lithium-ion batteries based on a proportional-integral observer. *IEEE Trans. Veh. Technol.* **2014**, *63*, 1614–1621.
3. Paschero, M.; Storti, G.L.; Rizzi, A.; Mascioli, F.M.F.; Rizzoni, G. A novel mechanical analogy based battery model for SoC estimation using a multi-cell EKF. *IEEE Trans. Sustain. Energy* **2016**, *7*, 1695–1702. [CrossRef]
4. Zhang, Z.L.; Cheng, X.; Lu, Z.Y.; Gu, D.J. SOC estimation of lithium-ion batteries with AEKF and wavelet transform matrix. *IEEE Trans. Power Electron.* **2017**, *32*, 7626–7634. [CrossRef]
5. Hardik, K.; Jesse, T.; Taha, S.U. Comparison of lead-acid and lithium ion batteries for stationary storage in off-grid energy systems. In Proceedings of the 4th IET Clean Energy and Technology Conference (CEAT 2016), Kuala Lumpur, Malaysia, 14–15 November 2016; pp. 1–7.
6. Daniel, R.; Alfredo, P.; Marco, R.; Michele, O. 12V battery modeling: Model development, simulation and validation. In Proceedings of the 2017 International Conference of Electrical and Electronic Technologies for Automotive, Torino, Italy, 15–16 June 2017; pp. 1–5.
7. Yang, J.; Wei, X.; Dai, H.; Zhu, J.; Xu, X. Lithium-ion battery internal resistance model based on the porous electrode theory. In Proceedings of the 2014 IEEE Vehicle Power and Propulsion Conference (VPPC), Coimbra, Portugal, 27–30 October 2014; pp. 1–6.
8. Liu, X.; Li, W.; Zhou, A. PNGV equivalent circuit model and SOC estimation algorithm for lithium battery pack adopted in AGV vehicle. *IEEE Access* **2018**, *6*, 23639–23647. [CrossRef]
9. Putra, W.S.; Dewangga, B.R.; Cahyadi, A.; Wahyunggoro, O. Current estimation using Thevenin battery model. In Proceedings of the Joint International Conference on Electric Vehicular Technology and Industrial, Mechanical, Electrical and Chemical Engineering (ICEVT & IMECE), Surakarta, Indonesia, 4–5 November 2015; pp. 5–9.
10. Ceraolo, M. New dynamical models of lead-acid batteries. *IEEE Trans. Power Syst.* **2000**, *15*, 1184–1190. [CrossRef]
11. He, H.; Xiong, R.; Zhang, X.; Sun, F.; Fan, J.X. State-of-charge estimation of the lithium-ion battery using an adaptive extended Kalman filter based on an improved Thevenin model. *IEEE Trans. Veh. Technol.* **2011**, *60*, 1461–1469.
12. Chen, S.; Fu, Y.; Mi, C.C. State of charge estimation of lithium ion batteries in electric drive vehicles using extended Kalman filtering. *IEEE Trans. Veh. Technol.* **2013**, *62*, 1020–1030. [CrossRef]
13. Zhang, Q.Z.; Wang, X.Y.; Yuan, H.M. Estimation for SOC of Li-ion battery based on two-order RC temperature model. In Proceedings of the 13th IEEE Conference on Industrial Electronics and Applications (ICIEA), Wuhan, China, 31 May–2 June 2018; pp. 2158–2297.
14. Diniz, P.S.R. *Fundamentals of adaptive filtering. Adaptive Filtering*; Springer: Boston, MA, USA, 2013; pp. 13–78.
15. Tian, Y.; Zeng, Z.; Tian, J.; Zhou, S.; Hu, C. Joint estimation of model parameters and SOC for lithium-ion batteries in wireless charging systems. In Proceedings of the 2017 IEEE PELS Workshop on Emerging Technologies: Wireless Power Transfer (WoW), Chongqing, China, 20–22 May 2017; pp. 263–267.
16. Albu, F. Improved variable forgetting factor recursive least square algorithm. In Proceedings of the 2012 12th International Conference on Control Automation Robotics & Vision (ICARCV), Guangzhou, China, 5–7 December 2012; pp. 1789–1793.
17. Leung, S.H.; So, C.F. Gradient-based variable forgetting factor RLS algorithm in time-varying environments. *IEEE Trans. Signal Process.* **2005**, *53*, 3141–3150. [CrossRef]
18. Chan, S.C.; Chu, Y.J. A new state-regularized QRRLS algorithm with a variable forgetting factor. *IEEE Trans. Cir. Syst. II Express Briefs* **2012**, *59*, 183–187. [CrossRef]
19. Chen, Q.; Gu, Y.; Ding, F. Data filtering based recursive least squares estimation algorithm for a class of Wiener nonlinear systems. In Proceedings of the 11th World Congress on Intelligent Control and Automation, Shenyang, China, 29 June–4 July 2014; pp. 1848–1852.
20. Chen, X.P.; Shen, W.X.; Dai, M.X.; Cao, Z.W.; Jin, J.; Ajay, K. Robust adaptive sliding-mode observer using RBF neural network for lithium-ion battery state of charge estimation in electric vehicles. *IEEE Trans. Veh. Technol.* **2016**, *65*, 1936–1947. [CrossRef]

Article

The Effects of Lithium Sulfur Battery Ageing on Second-Life Possibilities and Environmental Life Cycle Assessment Studies

Deidre Wolff [1,*], Lluc Canals Casals [1,2], Gabriela Benveniste [2], Cristina Corchero [1,2] and Lluís Trilla [1]

1 Catalonia Institute for Energy Research (IREC), Sant Adrià de Besòs 08930, Spain; lcanals@irec.cat (L.C.C.); ccorchero@irec.cat (C.C.); lltrilla@irec.cat (L.T.)
2 Universitat Politècnica de Catalunya (UPC), Barcelona 08034, Spain; gaby.benve@gmail.com
* Correspondence: dwolff@irec.cat; Tel.: +34-933-562-615

Received: 29 May 2019; Accepted: 24 June 2019; Published: 25 June 2019

Abstract: The development of Li-ion batteries has enabled the re-entry of electric vehicles into the market. As car manufacturers strive to reach higher practical specific energies (550 Wh/kg) than what is achievable for Li-ion batteries, new alternatives for battery chemistry are being considered. Li-Sulfur batteries are of interest due to their ability to achieve the desired practical specific energy. The research presented in this paper focuses on the development of the Li-Sulfur technology for use in electric vehicles. The paper presents the methodology and results for endurance tests conducted on in-house manufactured Li-S cells under various accelerated ageing conditions. The Li-S cells were found to reach 80% state of health after 300–500 cycles. The results of these tests were used as the basis for discussing the second life options for Li-S batteries, as well as environmental Life Cycle Assessment results of a 50 kWh Li-S battery.

Keywords: energy storage ageing and degradation; life cycle assessment; second-life energy storage applications; Li-Sulfur batteries

1. Introduction

The history of the electric vehicle (EV) is full of back and forth. It was born in the 19th century before the first internal combustion engine vehicle (ICEV), but was soon abandoned. It resurged in the 1890s by the hand of General Motors but was also soon abandoned. It was not until the arrival of lithium ion batteries, with their clearly higher performance in comparison to other energy storage systems, that the EV again entered the market in 2010. This time, though, apart from technical and economic issues, the development and implementation of country and region-specific environmental policies and directives was crucial for market penetration [1].

With the market share of electric vehicles (EVs) increasing and EV adoption being widely debated [2], research related to EV energy consumption, environmental impact and economic impact has increased on a yearly basis [3]. As part of this, and due to the increasing interest of adopting Circular Economy principles, Life Cycle Assessment studies have been conducted to quantify the environmental impact of EVs with the goal of reducing the pressure on ecosystems and natural resources [4].

Life Cycle Assessment (LCA) is the current state-of-the-art for quantifying the environmental life cycle impact and is thought to be valuable for assessing the potential impact of moving towards an electrified transportation infrastructure [5]. LCA is defined as the "compilation and evaluation of the inputs, outputs and the potential environmental impacts of a product system throughout its life cycle" [6]. LCA studies of EVs have focused on impact categories such as climate change and energy demand. This is due to the fact that variation in the electricity grid mix has a large influence on the overall result, and thus decarbonization of the grid mix will lead to further improvements in the environmental impact of the EV [7]. Along the same line, improvements in driving range and efficiency of the battery will also lead to a lower environmental impact. Therefore, many efforts have focused on the environmental analysis of the use phase of the EV. However, components such as batteries, that generally use scarce and precious materials, also present environmental concerns that need to be addressed, such as resource depletion. An LCA approach is necessary to give a more complete picture of the environmental burdens caused by EVs, from raw material extraction through to final disposal.

In order to normalize the results from one LCA study to the next, most studies have assumed a total lifetime driving distance above 100,000 km [8,9], a consumption of between 0.12 [10] and 0.2 kWh/km [11,12], and a battery lifespan that is equivalent to that of the EV. Other studies have conducted scenario analysis on the driving distance, the consumption, and the battery lifespan (including one or more battery replacements [13]). In terms of End-of-Life (EoL) of the EV battery, degradation to 80% of the initial battery capacity is considered the appropriate lifespan for mobility purposes, after which the battery should be replaced [14]. However, this 80% limit has been debated in relation to the real needs of the EV owner [15], as trips are often well below 100 km [16] and may still be supported with a battery below the suggested 80% State of Health (SoH). Once the battery reaches its EoL, there is an opportunity to reuse the battery in stationary applications [17], referred to as the second life of the battery [18].

There are three main strategies to consider regarding second life batteries, each having positive and negative aspects. The first suggests that the best option from an economical perspective is to use the batteries exactly as they are when extracted from the vehicle, without any further manipulation. The battery pack is installed as one unit in a portable container [19,20], or a tertiary building knowing that the battery might not be the most suitable for the stationary application. The second strategy is based on the concept that the battery re-use should concentrate on modules, which are relatively easy to dismantle from the battery pack and will allow for the battery to be sized according to the second life application. In this case, the repurposed battery can use modules from different car manufacturers [21]. Finally, the third strategy suggests that the dismantling of the EV battery should be at cell level in order to select the cells that have similar degradation. This selection allows perfectly homogeneous batteries to be built [22]. However, besides the choice of the strategy and stationary application to use, there are still other issues to consider before a positive revenue is generated from the defined business case, such as battery ownership and battery collection, among others [23,24].

Battery performance is another aspect being considered for increased deployment of EVs. Not all Li-ion batteries are equal, differing in the chemical composition of the anode, cathode and doping elements to provide various performance characteristics, such as higher energy density, higher power density, longer lifespan or improved safety. Currently, nickel manganese cobalt oxide (NMC) batteries are preferred by the automotive sector [25] due to their relatively high energy density, acceptable lifespan and safety level. Another chemical composition used by Chinese car manufacturers is Iron phosphate (LFP) that has a lower cost and good lifespan but provides lower energy and power densities compared to NMC. Finally, nickel cobalt aluminum oxide batteries (NCA) provide higher energy and power densities than NMC, but have lower lifespans and safety inconveniences [26]. Due to the different options available for Li-ion batteries, research has been done to analyze the environmental impact of battery manufacturing. Studies have indicated that the preferred NMC batteries perform worse than the other types from an environmental perspective [13,27].

Despite the quite good technical performance of Li-ion batteries that allowed the return of EVs, the cost of the batteries is still too expensive for a massive deployment [28]. Moreover, car manufacturers aim to reach specific energies of approximately 550 Wh/kg to increase the battery capacity and reduce the overall weight of EVs, and in turn eliminate range anxiety concerns of EV owners. Since Li-ion batteries are thought to have achieved their practical specific energy limit [29], which ranges between 100 and 250 Wh/kg, new alternatives for battery chemistry are being considered that have higher practical specific energy limits, such as Li-Sulfur (Li-S) [30], lithium air, and all-solid-state batteries [31].

From the aforementioned alternatives to Li-ion, the research presented in this study focuses on Li-S technology as part of the work conducted under the framework of the HELIS H2020 project [32] that aims to develop Li-S batteries for automotive purposes. This paper first presents the analysis of the Li-S battery ageing tests that were conducted on in-house manufactured Li-S cells (achieving around 300–500 cycles at 80% SoH). From these results, the second life options and possibilities are discussed together with the results for an environmental Life Cycle Assessment (LCA) study. The LCA is conducted on a 50 kWh Li-S battery and uses the results from the ageing tests to define the lifespan of the battery. The work presented here is focused on the analysis of the evolution of the capacity related to the SoH and the efficiency of the cells. The study of the internal mechanisms that lead to degradation, material activation and self-discharge will not be addressed in this paper, but will be included in future work.

2. Materials and Methods

This section is divided into two subsections. The description of the cell ageing tests and how the results will be treated is presented in the first section. The second section presents the environmental LCA methodology.

2.1. Ageing Tests

The ageing tests of the Li-S technology were performed by exposing in-house manufactured coin cells to endurance tests under laboratory conditions in a thermal chamber. Note that these coin cells were part of the second generation of cells resulting from the HELIS project. The cell composition was based on a sulfur-carbon composite cathode and Li-metal anode. The cathode was fabricated using a conventional doctor blade approach, consisting of 80% sulfur-carbon composite, 10% conductive carbon and 10% Polyvinylidene fluoride (PVDF) binder. Electrodes were punched in a disc and dried at 80 °C prior to the manual coin cell assembly. The cells contained 2 mgS/cm^2/side with a theoretical capacity of 1675 mAh/gS yielding the final capacity in the range of 1 mAh, although some variability was observed due to the manual manufacturing process. An optimized amount of 1M lithium bistrifluoromethanesulfonimidate (LiTFSI) in 1:1 dimethyl glycol (DME) and dioxolane (DOL) electrolyte [33] was used, resulting in a ratio of about 35 µL/mgS. It is noted here that this ratio was used in the coin cells for project purposes to ensure cell performance, but was reduced when scaling to larger cell sizes.

The testing platform included a Bio-Logic BCS series potentiostat with 24 channels for multiple simultaneous testing, an Angelantoni FM600BT climatic chamber for low temperature testing and a DRY-line VWR oven to regulate high temperatures (Figure 1). At the end of the tests, EC-lab software was used to extract all data for further analysis.

Figure 1. Image of the testing equipment.

A total of six cells were tested combining different temperatures (−10 °C, room temperature and 45 °C) and current (C-rate) conditions (from C/5 to 2C) to determine which (if any) of these factors can be considered the principal ageing factors that accelerate the ageing phenomena that occurs in all types of batteries:

- Room Temperature (RT), C-rate: 2C (J26)
- Room Temperature (RT), C-rate: C/10 (provided by SAFT Battery Manufacturer)
- Temperature −10 °C, C-rate: C/2 (J4)
- Temperature −10 °C, C-rate: C/5 (J1)
- Temperature 45 °C, C-rate: 2C (J33)
- Temperature 45 °C, C-rate: C/5 (J32)

Note that there was only one cell per test due to the channel limitations of the equipment and the duration of the experiments. Although one cell might not be enough to ensure the absolute validity of results, it was preferred to test different scenarios rather than just a few but with a redundancy in the number of cells following the same profiles. It should also be noted that the second cell, cycled at room temperature (RT) and following a 10 h charge/discharge cycle profile, had the particularity to be the only cell manufactured by SAFT, the battery manufacturer in the HELIS project. All the other cells (J1, J4, J26, J32 and J33) were built in the Catalonia Institute for Energy Research (IREC) facilities following manual processes. Furthermore, at −10 °C, the operative capacity of the cells submitted to relatively high rates was residual (less than 10% of the capacity identified at room temperature), being impossible to retrieve reliable information from these tests, which is the reason that the maximum cycling rate at −10 °C was done at C/2 instead of 2C.

All the endurance tests at low C-rates followed non-stop symmetric constant current capacity cycles. That is, charges and discharges, which have the same C-rate (no matter if it is a charge or discharge process) occurred consecutively without any pause between cycles and without having a constant voltage period to achieve a full charge. All charges stopped at a maximum voltage of 2.6 V while discharges stopped when the minimum of 1.9 V was reached. Moreover, the capacity fade presented in the results section are directly extracted from these continuous cycling and not from specific "control cycles." Note that the continuous constant current cycling allows the batteries to age relatively quickly compared to using constant current-constant voltage strategies, but it goes in detriment of reliability, as the resulting data might present higher dispersion.

On the other hand, due to the particularities of the charge/discharge voltage profile of Li-S, the effective or functional capacity of a cell might dramatically change depending on the C-rate but independently of the ageing of the cell. Consequently, the instant performance of Li-S should be clearly differentiated. Figure 2 (left) shows that the behavior of the discharge of Li-S batteries clearly has three phases, an initial small voltage drop (an abrupt step just after the beginning of a charge or discharge) of about 0.2 V (from 2.6 to 2.4 V), followed by a continuous voltage decrease and, finally, a plateau that has a slight voltage recovery prior to the final descent of voltage until reaching the minimum limit of 1.9 V. This is the common behavior of a Li-S battery as the kinetics of the polysulfides inside the cell are related to voltage [34]. However, when exposed to higher currents (Figure 2 right), the initial voltage drop caused by the internal resistance is much higher (around 0.4 V) forcing the minimum voltage of 1.9 V to be reached during the continuous voltage decrease of the second phase and before entering in the last plateau [35]. In consequence, the functional capacity of the cell is divided by almost 2, and thus should be considered during the ageing analysis.

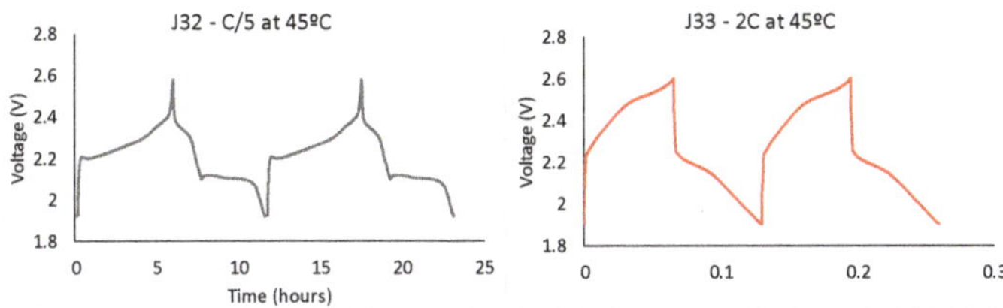

Figure 2. Charge/discharge voltage profile of a cell cycling at C/5 (left) and at 2C (right).

Additionally, as five of the cells were manufactured in-house following a non-industrialized process, their capacity is substantially lower than the cell built by SAFT (Table 1) and they also have quite a large dispersion between themselves. It is worth to remark that all cells were manufactured using the same active materials and electrolyte and contain the same sulfur loading, the differences observed here are due to the manufacturing process (manual and industrial) and the inherent imperfections linked to the manual processing of the components.

Table 1. Initial capacity of cells.

Cell Number	Initial Capacity (mAh)
J1	1.97
J4	1.18
J26	0.83
J32	0.92
J33	0.68
SAFT	2.62

Due to the low capacity and large dispersion, the battery degradation was evaluated by analyzing the evolution of SoH through the endurance cycling tests. In this study, the SoH is calculated as the ratio between the capacity at the current cycle discharge divided by the capacity of the first discharge done by the cell (Equation (1)).

$$SoH = Cap_i/Cap, \quad (1)$$

where Cap_i is the initial Capacity and Cap is the capacity at the current cycle.

Using this process, the degradation of the battery can be easily compared between the different endurance tests to be able to extract the functional effects of temperature and current intensity to the available capacity. Note that for cells having lower capacity, the small dispersion caused by the measurement equipment is amplified when relating it to SoH. To ease the interpretations of the evolution of SoH results and the trends derived from them, one data point from every 100 cycles is presented in the graphics in the results section (to have fewer overlapping data points in the same graph). Note that the presentation of results as SoH versus cycles instead of SoH versus capacity throughput (Ah) is also clearer due to the relatively important data dispersion of the initial capacity of the in-house manufactured cells. In addition, the study also analyzes the degradation of the battery in terms of efficiency, which is related to the internal resistance increase of the cells [36]. To do so, the study considers the ratio between the total capacity (Ah) charged to the cell divided by the capacity discharged from the cell for each cycle (Equation (2)).

$$Eff = Ah\ Charge/Ah\ Discharge. \quad (2)$$

To understand the exact evolution of the resistance, pulse tests [37,38], or even more precise methods, such as Electrochemical Impedance Spectroscopy (EIS) [39–41], could have been used. However, as the main scope of the study was to evaluate the functional characteristics of the cells and their relation to the End-of-Life, Second Life applications and LCA, it was decided that the SoH and efficiency were enough for this analysis and thus these results were not included in this study. The analysis of the internal mechanisms that explains the exact behavior of cells at every instant will be performed in future work.

2.2. LCA Methodology

LCA is divided into four stages including, Goal and Scope definition, Life Cycle Inventory (LCI), Life Cycle Impact Assessment (LCIA) and Interpretation. The Goal and Scope definition states the overall goal of the study and defines the system boundary, functional unit, and all other methodological choices required to meet the goal. The functional unit describes the function of the product system being assessed and is the unit for which the data is collected. Often the functional unit is scaled to a more appropriate unit for quantifying the outputs of processes within the system boundary that fulfill the function, referred to as the reference flow [6]. The LCI is the data collection step, and the LCIA categorizes the LCI data into impact categories defined in the scope, applies the associated characterization model and quantifies the overall environmental impact for each category assessed. The interpretation stage checks that the LCI and LCIA have met the requirements defined in the goal and scope.

For this study, an environmental attributional LCA was conducted for the production, use and disposal of a 50 kWh Li-S battery in accordance to ISO 14044 [6]. The Li-S battery is based on the composition of the Li-S coin cells manufactured in-house and considers the ageing tests as described in Section 2.1. The goal of the study was to quantify the environmental impact of a Li-S battery for use in an electric vehicle from cradle-to-grave, which includes raw material extraction, materials production, battery manufacturing, use and final End-of-Life disposal (Figure 3). It should be noted that the system boundary of the study does not include transportation, the production of the Battery Management System (BMS), or the production of the electric vehicle.

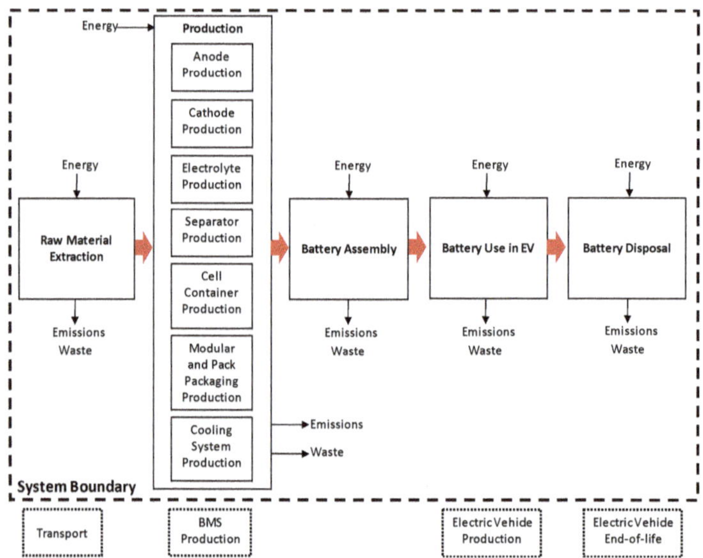

Figure 3. System boundary for the Life Cycle Assessment (LCA) of the 50 kWh Li-Sulfur (Li-S) Battery.

The functional unit of the LCA is defined as 1 km of driving based on an average of 0.17 kWh/km for EVs [42] and an 85% efficiency [43] that takes into account both the charge/discharge efficiency confirmed in the results section as well as the efficiency of the charger. The reference flow that is commonly applied in LCA studies of EVs is 150,000 km, particularly for comparisons between EVs with Li-ion batteries to ICEVs [44]. Historically, LCA studies conducted on Li-S batteries assumed that the Li-S chemistry would be able to achieve the 150,000 km, however, the degradation curves from laboratory performance tests on Li-S batteries were not included in this assumption. Therefore, another goal of the study was to use the ageing data to determine the total kilometers reached during the lifespan of the battery.

In order to include the actual performance of the Li-S battery in the LCA, an alternative reference flow was thus defined as total km for one Li-S battery. The total km was quantified using ageing test data from laboratory tests on Li-S coin cells to calculate the SoH of the battery per cycle. It should be noted that calendar ageing tests were also conducted on the coin cells, however, these results were not considered in the ageing tests for this study as further investigation is required to determine the relationship, if any, between cycle and calendar ageing for Li-S cells. The end-of-life of the battery was defined as 60% SoH for several reasons further discussed in the results section regarding the ageing test performance.

The LCA was modeled using GaBi Professional software, a tool designed for LCA studies assessing a variety of impact categories. Both GaBi 8 Professional and EcoInvent 3.5 datasets were used in the study. The electricity mix shown in Figure 4. was used for this study and corresponds to the EU-28 grid mix.

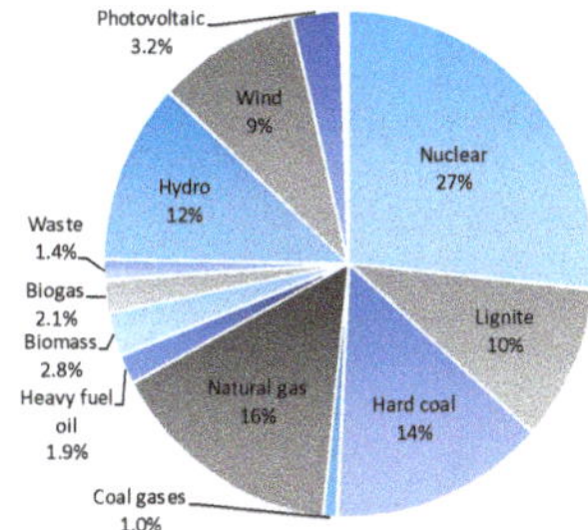

Figure 4. Electricity grid mix used in the LCA (EU-28 mix). Source: Adapted from GaBi Professional Database.

A key limitation of this study includes the scaling up from the composition of a coin cell to a battery. However, including laboratory data from Li-S ageing tests in an LCA case study of Li-S batteries is a step forward for the environmental assessment of this technology.

2.2.1. Li-S Battery Manufacturing Inventory Data

The mass of the active material (electrolyte, anode and cathode) in the Li-S coin cells was scaled to the mass of active material required for a 50 kWh Li-S battery based on the gravimetric energy density (GED) and the ratio of total mass to active mass. The GED and active mass were calculated as given in Equations (3) and (4), respectively. As indicated previously, the electrolyte ratio used in the coin cells is 35 µL/mgS for project purposes to ensure proper cell performance. However, since this amount of electrolyte is not optimal and it is assumed that larger cells can reach at least 6 µL/mgS, if not lower, 11.5 µL of electrolyte was estimated as the amount per coin cell for the LCA. It is worth to remark that the electrolyte ratio assumed is closer to the actual for EV-size batteries and provides a more realistic approach for LCA studies. This is further confirmed in [45], where a ratio of 10 µL/mgS is recommended when scaling up from coin cells. Similarly, the quantity of lithium anode in the coin cell is in excess, therefore, the diameter of the Lithium ribbon was assumed to be the same diameter as the cathode. For the active material in the cathode, a 1:1 ratio of carbon to sulfur was used. Table 2 gives the specifications of the coin cells used in scaling up to the 50 kWh battery. The mass of each active material in the battery was calculated using Equations (5) and (6).

$$GED = \frac{C_{CC}}{M_{AM}} \times V_{CC} \quad (3)$$

$$M_{AM} = M_{El_{CC}} + M_{A_{CC}} + M_{C_{CC}}, \quad (4)$$

$$Scaling\ Factor = \frac{1x10^6}{M_{AM}} \times \frac{E_B}{GED}, \quad (5)$$

$$M_{X_B} = M_{X_{CC}} \times \frac{Scaling\ Factor}{1000}, \quad (6)$$

where C_{CC} is the capacity (mAh) of the coin cell, V_{CC} is the voltage of the coin cell, M_{AM} is the mass (g) of the active material in the coin cell, $M_{El_{CC}}$, $M_{A_{CC}}$ and $M_{C_{CC}}$ are the masses (g) of electrolyte, anode and cathode in the coin cell, E_B is the energy (kWh) of the battery, and GED is the gravitational energy density (Wh/kg). M_{X_B} (kg) and $M_{X_{CC}}$ (g) are the masses in kilograms and grams for X (electrolyte, anode or cathode) in the battery (B) and coin cell (CC), respectively.

Table 2. Inventory data to calculate Scaling Factor.

Specification	Quantity	Unit
Anode ($M_{A_{CC}}$)	0.0064	g
Cathode ($M_{C_{CC}}$)	0.0078	g
Electrolyte ($M_{El_{CC}}$)	0.015	g
Mass active material (M_{AM})	0.026	g
Mass coin cell (TM_{CC})	3.59	g
Capacity (C_{CC})	3.3	mAh
Voltage (V_{CC})	2.3	V
Capacity Density ($\frac{C_{CC}}{M_{AM}}$)	128.4	Ah/kg
Gravimetric Energy Density (GED)	295.4	Wh/kg
Energy Li-S Battery (E_B)	50	kWh

The mass of the other battery components, including the cell container, separator, module and pack packaging, and cooling system were taken from a previous LCA study conducted on Li-S batteries [43], which used the BatPac software for sizing Li-ion batteries and adapted it to a Li-S system. Similarly, data estimated in [43] for industry manufacturing of Li-S batteries was used for the energy consumption. Table 3 gives the quantities used for the 50 kWh Li-S battery.

Table 3. Bill of materials for the 50 kWh Li-S battery.

Li-S Battery Composition	Quantity	Unit	Data Source
Cathode (M_{C_B})	51.4	kg	Equation (6)
Anode (M_{A_B})	42.0	kg	Equation (6)
Electrolyte (M_{El_B})	75.9	kg	Equation (6)
Separator	6.9	kg	[43]
Cell container	19.6	kg	[43]
Module packaging	22.6	kg	[43]
Cooling system	27	kg	[43]
Pack packaging	41.8	kg	[43]
Assembly Energy consumption [1]	12,016	MJ	[43]

[1] average value was used.

2.2.2. Li-S Battery Use and End-of-Life Inventory Data

The EU-28 grid mix was used for the use phase. Tests for the ageing of coin cells as described in Section 2.1 were used to compute the total amount of kilometers and, thus, extract the energy used during the use phase. To do so, the linear relationship between SoH and number of cycles from the results of the ageing tests served to obtain Equation (7), which was then used to quantify the total use phase energy requirement for each scenario. Three scenarios were defined, being a minimum, average and maximum number of cycles achievable by the battery to reach 60% SoH according to the ageing results for the various cells tested. It should be noted that these scenarios were defined from the cells that cycled for more than 800 cycles during the ageing tests.

$$E_T = E_B \sum_{i=1}^{n} (mn + b) = E_B \left(m \frac{n(n+1)}{2} + nb \right), \quad (7)$$

where E_T is total accumulated energy (kWh), E_B is the energy of the battery (kWh), m and b are the slope and intercept of the fit linear curve for SoH versus cycle number, respectively, and n is the cycle number.

Data for the recycling of Li-S batteries was provided by ACCUREC (project partner) who developed a recycling process for Li-S cells that is in compliance with EU-directive 66/2006. This directive sets a minimum recycling efficiency requirement for batteries of 50% of the mass of the battery.

2.2.3. Impact Categories Assessed

The LCA data was aggregated into impact categories and summed to give a total result. The impact categories assessed include resource depletion, acidification, eutrophication, climate change, photochemical ozone formation and energy demand. The characterization model and characterization factors used are defined in Table 4.

Table 4. Description of impact categories assessed.

Impact Category	Characterization Factor	Unit	Model	Description
Resource Depletion	Abiotic Depletion Potential (ADP elements)	kg Sb-eq.	CML 2001- Jan 2016	The depletion of reserves due to the unsustainable extraction of non-renewable minerals
Acidification	Acidification Potential (AP)	kg SO_2-eq.	CML 2001- Jan 2016	The emission of substances that lead to the change in soil acidity and ecosystem damage
Eutrophication	Eutrophication Potential (EP)	kg PO_4^{-3}-eq.	CML 2001- Jan 2016	The release of nutrients that lead to growth of algae and cyanobacteria and a relative loss in species diversity
Climate Change	Global Warming Potential (GWP)	kg CO_2-eq.	CML 2001- Jan 2016	The emission of greenhouse gases that lead to increased radiative forcing and raise in mean global temperature
Photochemical Ozone Formation	Photochem. Ozone Creation Potential (POCP)	kg C_2H_4-eq.	CML 2001- Jan 2016	The emission of substances that undergo photochemical reactions to form ozone at ground level which leads to human health impacts and ecosystem damage
Energy Demand	Primary energy demand (ren. and non-ren. resources, PED)	MJ	PED, gross calorific value	The consumption of both renewable and non-renewable primary energy sources measured prior to processing

3. Results and Discussion

Following the same structure as in Section 2, this section first discusses the results of the ageing endurance tests and then the LCA results.

3.1. Ageing Results

The results obtained from the data analysis are presented together in Figure 5 for all the endurance tests to highlight the similarity of the degradation trends (SoH reduction) that all cells follow independently of the cell.

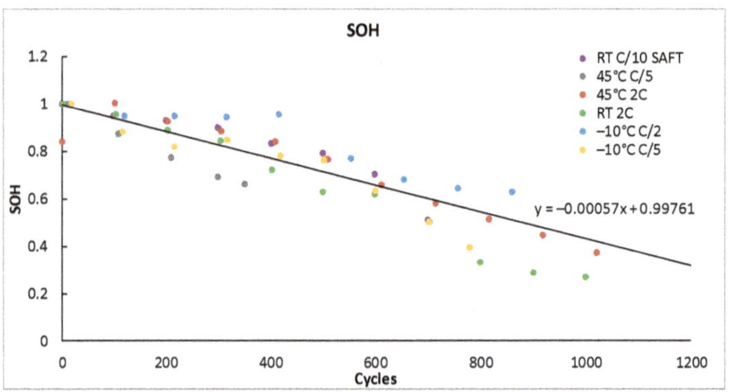

Figure 5. State of Health (SoH) evolution through cycling.

Although there is a large dispersion in the measurements (Figure 5), cells seem to follow a linear degradation of about −0.00057 SoH per cycle on average. This value is close to the values obtained by Shang et al. [46] where, after a first rapid decrease, cells continued with a constant (0.000625 SoH per cycle) capacity decrease for over 200 cycles. Additional relevant information can be extracted from Figure 5, for instance, there seems to be no clear relationship between the C-rate and the acceleration of cell degradation. This is demonstrated by the fact that the cell from SAFT (purple dots), which cycled at C/10 (the lowest current under test) and J33 (red dots), which cycled at 2C (20 times faster) follow almost the same trend. Similarly, temperature also does not appear to be a relevant factor that accelerates the ageing of Li-S cells. This can be observed in the figure by comparing the cells J4 (blue dots), SAFT (purple dots) and J33 (red dots) that cycled at −10 °C, room temperature (RT) and 45 °C respectively, and for which similar degradation trends are obtained.

To clearly state this first impression, Table 5 presents the slope of the linear curve that best fits each cell submitted to endurance tests and the corresponding R square value. Effectively, these three cells (J4, J33 and SAFT) have a slope that is close to the average, reinforcing the idea that temperature and current have no relevant effect on ageing. Notice that the most rapid degradation occurs at C/5 at 45 °C (J32), however, there are too few points to ensure that this trend is going to be sustained after more cycles. In fact, it is worth mentioning that the behavior of the cell cycling at C/5 at −10°C (J4) initially had a similar degradation pattern but then stabilized for some cycles before again decreasing more rapidly. This behavior might indicate that the initial aggressive slope would be softened if cycled for a longer period. Similarly, after more than 600 cycles, cell 26 (cycling at 2C and RT) reaches the 60% SoH and then seems to suffer a substantial drop but in fact it re-stabilizes at 40% SoH.

The differences in capacity fading observed with Li-S cell cycling have also been noted in previous research. The trend that shows an initial linear capacity fade transitioning into a stabilization period and then continuing with a capacity loss was indicated in [47] for cells cycled at C/10 and C/5, where stabilization periods of 150 to 350 cycles were observed. Furthermore, an increase of the capacity during the initial cycles, as observed for cell J33, has been highlighted in [48] for Li-S cells cycled at C/10 and C/5 and temperatures of 20, 30 and 40 °C.

Table 5. Initial capacity of cells.

Cell Identifier	Endurance Test	Share of Capacity Loss per Cycle	R^2
J1	C/5 at −10 °C	0.00069	90.45%
J4	C/2 at −10 °C	0.00049	88.36%
J26	2C at RT [1]	0.00080	97.96%
J32	C/5 at 45 °C	0.00100	99.60%
J33	2C at 45 °C	0.00058	89.61%
SAFT	C/10 at RT [1]	0.00061	88.62%
Average		0.00057	74.29%

[1] RT = Room Temperature.

Thus, it seems reasonable to state that Li-S battery ageing does not behave like the Lithium ion in response to changes in temperature and C-rate. In fact, Lithium ion battery ageing is strongly affected by several factors that accelerate the ageing in different ways [49,50]. These factors are Temperature, State of Charge (SoC), C-rate and Depth of Discharge (DoD) [51]. Typically, temperature has an exponential effect on ageing, meaning that the battery lifespan shortens as temperature increases [52,53]. The SoC and C-rate, on the other hand, follow first and second polynomial relations being more severe when the battery remains fully charged or suffers from higher intensive discharges [54,55]. The DoD follows a logarithmic behavior, which reverts in almost no ageing effect during small ripples or cycles that increases rapidly as the DoD increases, becoming relatively stable after 40% DoD [56].

In comparison to the Li-ion ageing behavior, the endurance tests presented in this section indicate that Li-S ageing seems independent of changes in temperature and C-rate confirmed by the linear degradation trends with similar slopes under varying test conditions. However, this statement does not mean that the C-rate and temperature have no effect on Li-S battery performance. In fact, the tested coin cells displayed higher stability when working at higher temperatures and low C-rates but also presented poorer efficiency, as described later in this section. In addition, these results show that the sudden death or ageing knee that typically occurs in Lithium ion batteries [57,58] is not appreciable in Li-S batteries (some cells achieved 40% SoH and continued working).

There is another important aspect to look at related to battery ageing performance for traction purposes, which is the efficiency and loss of power. These two aspects are closely related to the internal resistance of the battery by the Ohm law, the higher the internal resistance, the higher the losses. Lithium ion batteries generally suffer an exponential internal resistance increase as SoH decreases, that is, the internal resistance increase is quite low at the beginning but is more and more noticeable as the battery ages. For instance, a study regarding the battery ageing of real electric vehicles using the internal resistance shows how at 88% SoH, the internal resistance of all the cells in the battery was slightly higher than at the beginning, but at 82% SoH their internal resistance was already 20% higher [59] and it may rise even higher if the SoH goes beyond this point [57], up to 200% at 60% SoH [60].

To analyze what occurs with Li-S, the evolution of the efficiency measured for all cells during the endurance tests versus the SoH (Equation 1) was plotted (Figure 6). Note that the dispersion is relevant due to the constant current cycling method, the particularities of the entrance into the second plateau, and the fact that the efficiency versus SoH is presented in the figure. However, it is difficult to identify any correlation between the evolution of efficiency versus SoH and the temperature or C-rate. The shuttle effect has an important impact on efficiency [61], making Li-S cells less efficient at higher temperatures and at lower C-rates [62]. It should be noted, however, that the cell with a lower efficiency and a quicker efficiency loss is the SAFT cell, which was manufactured following an industrial process. This cell was cycled at a lower C-rate where the self-discharge of the battery might noticeably interfere [63]. Yet, the loss of efficiency through ageing is relatively low decreasing from 99% to 96% in all the cases except for the SAFT cell and cell J33. This is in accordance with the quite stable efficiency values presented in [64], where three cells containing different separators to inhibit the shuttle effect were tested for more than 500 cycles. It further demonstrates that the internal resistance does not seem to increase in an exponential way as occurs with Li-ion batteries.

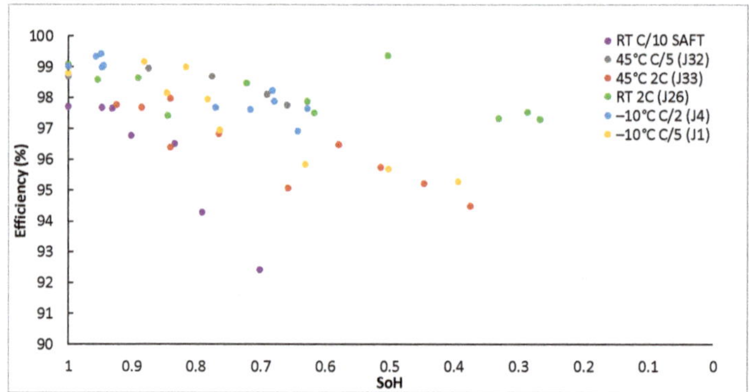

Figure 6. Evolution of the efficiency of the cells in relation to SoH.

All these aforementioned issues indicate that, effectively, aged Li-S batteries perform similarly to new batteries but differ in initial capacity. Therefore, from a strictly ageing perspective, it seems that there should not be much difference in using brand new Li-S cells or aged cells apart from the bigger volume of the battery built from re-used cells and an uncertainty of a sudden malfunction of the cell corresponding to a shorter lifespan. However, some of the cells continued working well below a 50% SoH and, in one case (J26), 30% SoH was reached before the test was finally stopped. The stabilization of the capacity fade at below 50% SoH was also observed in previous research on Li-S pouch cells [65], however the number of cycles achieved by these cells was reported to be significantly lower, reaching a 20% capacity fade before 50 cycles. The observed stabilization resulted from the inhibition of polysulfide diffusion caused by similar concentrations of sulfur/polysulfide in the electrolyte and carbon interface being reached [65].

It should be noted that most applications will fail before such a low SoH is reached, and thus, a limit of 60% SoH at the end of the first and second life is acceptable, which is the value used for the LCA discussed in the following section. Furthermore, the preliminary ageing results presented for the few cells studied should be confirmed by future research.

The following recommendations should be considered when using Li-S batteries in stationary applications for both new or re-used batteries:

- A loss of capacity occurs when cycled at high current rates (caused by the impossibility to enter the second plateau shown in Figure 2).
- Low temperatures result in a sudden decrease in performance of Li-S batteries (at temperatures below 0 °C)
- High temperatures result in a loss in efficiency (due to an increase of the shuttle effect)
- Very low C-rates or long durations without use result in a loss in efficiency (due to the shuttle effect)

3.2. *Life Cycle Impact Assessment Results and Interpretation*

The Life Cycle Impact Assessment results are presented and discussed in this section. For the system boundary defined in Figure 3, the ageing test results from Section 3.1 were used to define three scenarios for the analysis. From Figure 5, the number of cycles achieved by the battery to reach 60% SoH can range from 552 to 912 cycles. The scenarios defined for the use phase are summarized in

Table 6. From this table, Scenario 2 is very close to and Scenario 3 exceeds the 150,000 km defined for most LCA studies on EVs, however, Scenario 1 does not quite meet this distance. Particularly for Scenario 1, the options for battery replacement may be considered depending on the End-of-Life of the EV.

Table 6. Battery cycling scenarios from ageing test results

Scenario	Number of Cycles	Cell Identifier	Ageing Test	E_T (kWh)	Distance (km)
S1	552	J26	2C at RT	22,660	113,290
S2	722	-	Average [1]	28,790	143,955
S3	912	J4	C/2 at −10 °C	35,350	176,750

[1] average of all tests.

Figure 7 shows the impact per total kilometers achieved for each scenario defined in Table 6 and for all impact categories assessed as described in Table 4. The figure also shows the contribution of the production, use and disposal life cycle stages to the overall result. As can be seen, the production and disposal stages are the same for all scenarios. This is due to the fact that only one battery was considered for each scenario. For the use phase, however, the impact for each scenario differs depending on the quantified energy required as calculated with Equation (7). The energy required is dependent on the total number of cycles the battery is able to achieve before reaching its defined End-of-Life of 60% SoH. Therefore, the life cycle environmental impact of the Li-S battery changes based on the number of cycles (and hence the total kilometers) the battery is able to achieve. However, the amount of this change differs depending on the impact category being assessed. For example, the contribution of the use phase to the ADP elements (referred to as ADP from here forward) is insignificant compared to the production stage and therefore changes in the use phase will not significantly change the overall result.

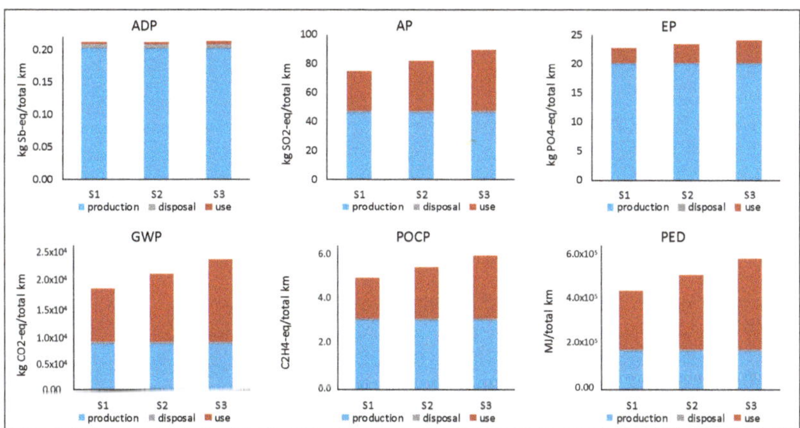

Figure 7. Impact per total kilometers achieved for each scenario defined in Table 6 with a breakdown of the contribution from production, disposal and use phases of the life cycle.

For all the other impact categories assessed (AP, EP, GWP, POCP, PED), however, the use phase has a more significant contribution to the overall result than seen for ADP. Thus, the result for these impact categories increases with an increase in the number of cycles achieved, as defined in each scenario. It can further be seen for these impact categories that the production phase also has a contribution to the overall result, and thus both the production and use life cycle stages are important for quantification of the overall environmental impact for these impact categories.

In order to see the environmental benefits of the extended lifespan of the battery due to more cycles being achieved, it is necessary to look at the results per functional unit of 1 km (Figure 8). In Figure 8, the trend clearly shows that as the number of cycles achievable by the battery improves from 552 towards 912 cycles, the impact per km also improves. Table 7 further summarizes the results per functional unit (per km) and per reference flow (per total km for one battery).

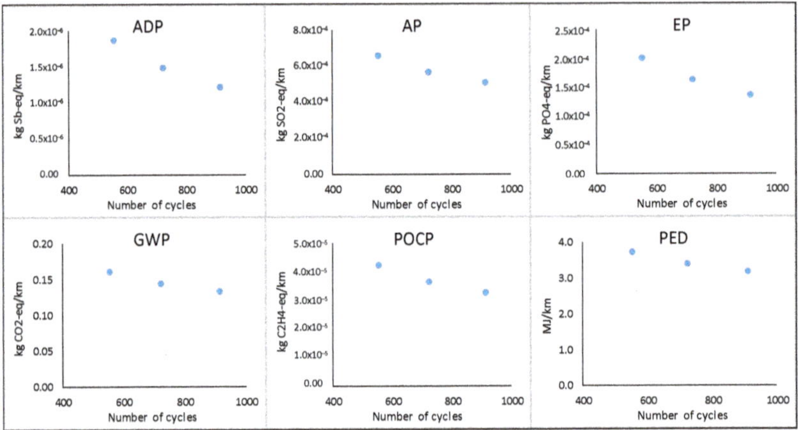

Figure 8. Impact per km for each impact category assessed versus the number of cycles achieved during the lifespan of the battery.

Table 7. Life Cycle Impact Assessment (LCIA) results per functional unit and per reference flow.

Impact Category	Unit	Total per Functional Unit [1]			Total per Reference Flow [2]		
		S1	S2	S3	S1	S2	S3
ADP	kg Sb-eq.	1.9×10^{-6}	1.5×10^{-6}	1.2×10^{-6}	0.2	0.2	0.2
AP	kg SO_2-eq.	6.6×10^{-4}	5.7×10^{-4}	5.1×10^{-4}	74.5	81.8	89.5
EP	kg PO_4^{-3}-eq.	2.0×10^{-4}	1.6×10^{-4}	1.4×10^{-4}	22.8	23.4	24.2
GWP	kg CO_2-eq.	0.16	0.14	0.13	18,198	20,765	23,504
POCP	kg C_2H_4-eq.	4.2×10^{-5}	3.6×10^{-5}	3.2×10^{-5}	4.8	5.3	5.7
PED	MJ	3.7	3.4	3.2	423,641	490,775	562,595

[1] Functional unit is 1 km. [2] Reference flow for S1, S2 and S3 is 113,290, 143,955 and 176,750 km, respectively.

Therefore, for the use phase, improvements in the environmental impact will come from extending the lifespan of the battery, as well as from the improvement in efficiencies and decarbonization of the electricity grid mix, as discussed in the introduction. Furthermore, it is shown that one 50 kWh Li-S battery, taking into consideration the effects from cycle ageing, has the potential to reach the 150,000 km reference that is often used in comparative LCA studies for EVs and ICEVs.

Since it was further found that the production phase also contributes to the results (Figure 7), the contribution of the battery components to the production phase was further investigated. Figure 9 shows the percent contribution of each component to the total result due to production of the battery. The material components are defined in the bill of materials in Table 3.

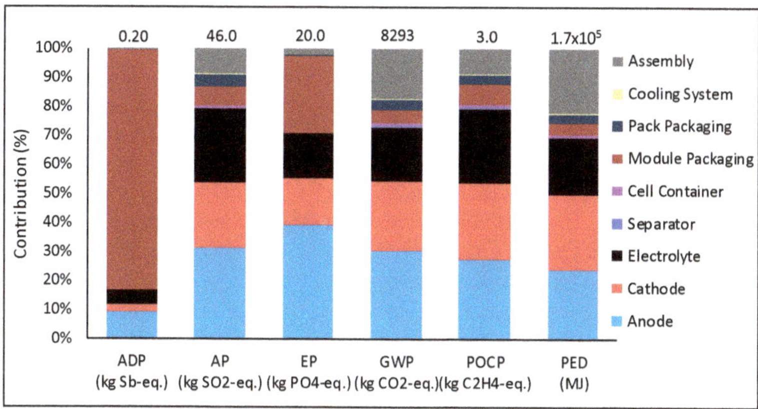

Figure 9. Percent contribution of each component to each impact category for the production of the Li-S battery. The numbers at the top of the stacked columns indicate the total value.

From Figure 9., the module packaging contributes significantly to the ADP. The module packaging consists of electronic components, the contribution of which was found to range from 80% to 99% of the total impact of the module packaging production depending on the impact category (Figure 10). A complete list of the components of the module packaging can be found in [43]. Therefore, the main contributing component to the ADP, which was found to come from the production stage (Figure 7), is due to the electronic components in the modular packaging (Figures 9 and 10).

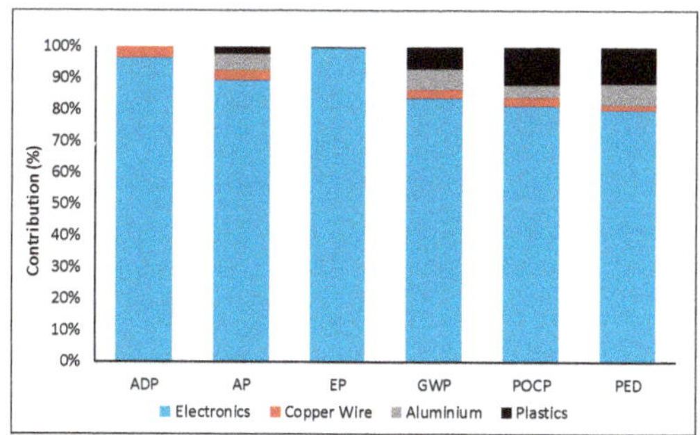

Figure 10. Percent contribution of the components of the module packaging to the result for each impact category.

For all other impact categories, the contribution of the active material (cathode, anode and electrolyte) is greater than 70% to the total impact of battery production (Figure 9). This material has been scaled from that in a coin cell with the use of Equations (3)–(6). Therefore, this data should be updated when more tests are conducted on Li-S batteries and more data is available for the composition and performance of larger batteries. This is important as the production of the battery contributes to the overall result for all impact categories, as was seen in Figure 7. Therefore, improvements in this data will influence the overall result for each of the environmental impact categories assessed.

4. Conclusions

This work presented the results for ageing tests conducted on Li-S coin cells. These results were used to discuss the second-life battery applications and were further applied in an environmental attributional LCA case study for the use of a 50 kWh Li-S battery in an electric vehicle.

After analyzing the results from the endurance ageing tests, it seems that Li-S batteries do not follow the same patterns of Li-ion batteries. It was found that the temperature and C-rate seem to produce no acceleration of ageing, and that there is no dramatic change in the ageing tendency as is seen in Li-ion batteries when they reach the "ageing knee" or "sudden death." For Li-S batteries, the cells either continue to cycle or crash instantaneously, meaning they can no longer absorb or deliver energy. Furthermore, the efficiency of Li-S batteries decreases constantly, that is, an exponential loss of efficiency is not visible which is in contrast to the behavior of Li-ion batteries. From an ageing perspective, these factors make Li-S batteries preferable to Li-ion batteries, as their behavior appears "more predictable" and they seem to be unaltered due to external factors related to the application environment.

In terms of battery second life applications, this may not be foreseen for Li-S batteries. Future batteries are expected to have larger capacities of up to the 50 kWh as described in this study, and thus the End-of-Life of the battery in an EV (first life) could be lower than the targeted 80% SoH (60% or even lower) and still be capable of satisfying all the driving needs up until the car is recycled. Therefore, it is not only the inherent complexities of their normal operability that limits the second life applications, but also both the low SoH at the beginning of the second life corresponding to a lower capacity and the fact that the batteries will be quite old (possibly 15 years old) at the end of the first life. In this duration, newer and more interesting batteries will likely be on the market at reasonable prices.

In terms of the LCA case study conducted for a 50 kWh Li-S battery, it was found that both the production and use stages of the life cycle contribute to the overall environmental impact for all impact categories assessed, except for resource depletion (ADP) where the production stage is the key contributor. For the use phase, three scenarios for the cycle life of the battery were defined based on the ageing test results conducted on Li-S cells. It was found here that the 50 kWh Li-S battery has the potential to achieve the 150,000km usually defined in LCA case studies of EVs. For the production stage of the life cycle, the active material in the battery (anode, cathode and electrolyte) contributes greater than 70% to all impact categories assessed except for resource depletion (ADP) where the electronics in the module packaging is the largest contributor. It is noted here that the data quality for scaling the active material from a coin cell to that in a 50 kWh battery will be improved and should be updated in the LCA as better data from larger batteries is available, along with the efficiencies and driving ranges defined for the Li-S technology. However, in this study it was shown that with improvements in the Li-S technology, the environmental impact per kilometer will improve as the number of cycles the battery achieves during its first life improves. Furthermore, the use of laboratory data for the ageing of Li-S cells in an LCA study is a step forward for the assessment of the environmental impact of this technology.

Author Contributions: Conceptualization, C.C.; Data curation, L.C.C.; Formal analysis, D.W. and L.C.C.; Funding acquisition, C.C.; Investigation, D.W., L.C.C., G.B. and L.T.; Methodology, D.W. and G.B.; Project administration, G.B., C.C .and L.T.; Resources, G.B. and L.T.; Software, G.B.; Supervision, C.C.; Validation, C.C.; Visualization, D.W. and L.C.C.; Writing—original draft, D.W. and L.C.C.; Writing—review and editing, D.W., L.C.C., G.B., C.C. and L.T.

Funding: This research received funding from the European Union's Horizon 2020 research and innovation programme under Grant Agreement No 666221 (www.helis-project.eu). C. Corchero's work is supported by the grant IJCI-2015-26650 (MICINN).

Acknowledgments: We would like to acknowledge ACCUREC for their contribution of data for the recycling process of the Li-S cells.

Conflicts of Interest: The authors declare no conflict of interest.

References

1. Sierzchula, W.; Bakker, S.; Maat, K.; Van Wee, B. The influence of financial incentives and other socio-economic factors on electric vehicle adoption. *Energy Policy* **2014**, *68*, 183–194. [CrossRef]
2. Ortar, N.; Ryghaug, M. Should All Cars Be Electric by 2025? The Electric Car Debate in Europe. *Sustainability* **2019**, *11*, 1868. [CrossRef]

3. Uribe-Toril, J.; Ruiz-Real, J.; Milán-García, J.; de Pablo Valenciano, J. Energy, Economy, and Environment: A Worldwide Research Update. *Energies* **2019**, *12*, 1120. [CrossRef]
4. European Commission Implementation of the Circular Economy Action Plan. Available online: http://ec.europa.eu/environment/circular-economy/index_en.htm (accessed on 28 May 2019).
5. Nordelöf, A.; Messagie, M.; Tillman, A.-M.; Ljunggren Söderman, M.; Van Mierlo, J. Environmental impacts of hybrid, plug-in hybrid, and battery electric vehicles—What can we learn from life cycle assessment? *Int. J. Life Cycle Assess.* **2014**, *19*, 1866–1890. [CrossRef]
6. *ISO 14044:2006 Environmental management—Life Cycle Assessment—Requirements and Guidelines*; ISO: Geneva, Switzerland, 2006.
7. Rangaraju, S.; De Vroey, L.; Messagie, M.; Mertens, J.; Van Mierlo, J. Impacts of electricity mix, charging profile, and driving behavior on the emissions performance of battery electric vehicles: A Belgian case study. *Appl. Energy* **2015**, *148*, 496–505. [CrossRef]
8. Bradley, T.H.; Frank, A.A. Design, demonstrations and sustainability impact assessments for plug-in hybrid electric vehicles. *Renew. Sustain. Energy Rev.* **2009**, *13*, 115–128. [CrossRef]
9. Egede, P.; Dettmer, T.; Herrmann, C.; Kara, S. Life Cycle Assessment of Electric Vehicles—A Framework to Consider Influencing Factors. *Procedia CIRP* **2015**, *29*, 233–238. [CrossRef]
10. Faria, R.; Marques, P.; Moura, P.; Freire, F.; Delgado, J.; De Almeida, A.T. Impact of the electricity mix and use profile in the life-cycle assessment of electric vehicles. *Renew. Sustain. Energy Rev.* **2013**, *24*, 271–287. [CrossRef]
11. Girardi, P.; Gargiulo, A.; Brambilla, P.C. A comparative LCA of an electric vehicle and an internal combustion engine vehicle using the appropriate power mix: The Italian case study. *Int. J. Life Cycle Assess.* **2015**, 1127–1142. [CrossRef]
12. Jochem, P.; Babrowski, S.; Fichtner, W. Assessing CO_2 emissions of electric vehicles in Germany in 2030. *Transp. Res. Part A Policy Pract.* **2015**, *78*, 68–83. [CrossRef]
13. Almeida, A.; Sousa, N. Quest for Sustainability: Life-Cycle Emissions Assessment of Electric Vehicles Considering Newer Li-Ion Batteries. *Sustainability* **2019**, *11*, 2366. [CrossRef]
14. Cusenza, M.A.; Guarino, F.; Longo, S.; Mistretta, M.; Cellura, M. Reuse of electric vehicle batteries in buildings: An integrated load match analysis and life cycle assessment approach. *Energy Build.* **2019**, *186*, 339–354. [CrossRef]
15. Saxena, S.; Le Floch, C.; Macdonald, J.; Moura, S. Quantifying EV battery end-of-life through analysis of travel needs with vehicle powertrain models. *J. Power Sources* **2015**, *282*, 265–276. [CrossRef]
16. Shi, X.; Pan, J.; Wang, H.; Cai, H. Battery electric vehicles: What is the minimum range required? *Energy* **2019**, *166*, 352–358. [CrossRef]
17. Reinhardt, R.; Christodoulou, I.; Gassó-Domingo, S.; Amante García, B. Towards sustainable business models for electric vehicle battery second use: A critical review. *J. Environ. Manag.* **2019**, *126*, 432–446. [CrossRef] [PubMed]
18. Canals Casals, L.; Amante García, B.; Cremades, L. V Electric Vehicle Battery Reuse: Preparing for a Second Life. *J. Ind. Eng. Manag.* **2017**, *10*, 266–285. [CrossRef]
19. Gohla-Neudecker, B.; Bowler, M.; Mohr, S. Battery 2nd life: Leveraging the sustainability potential of EVs and renewable energy grid integration. In Proceedings of the 5th International Conference on Clean Electrical Power: Renewable Energy Resources Impact (ICCEP), Taormina, Italy, 16–18 June 2015.
20. Canals Casals, L.; Amante García, B.; Canal, C. Second life batteries lifespan: Rest of useful life and environmental analysis. *J. Environ. Manag.* **2019**, *232*, 354–363. [CrossRef] [PubMed]
21. Zhang, C.; Marco, J.; Fai Yu, T. Hardware Platform Design of Small Energy Storage System Using Second Life Batteries. In Proceedings of the 2018 UKACC 12th International Conference on Control (CONTROL), Sheffield, UK, 5–7 September 2018.
22. Lee, K.; Kum, D. Development of cell selection framework for second-life cells with homogeneous properties. *Int. J. Electr. Power Energy Syst.* **2019**. [CrossRef]
23. Pistoia, G.; Liaw, B. *Behaviour of Lithium-Ion Batteries in Electric Vehicles*; Green Energy Action Alliance: Toronto, ON, Canada, 2018; ISBN 9783319699493.
24. Hossain, E.; Murtaugh, D.; Mody, J.; Faruque, H.M.R.; Haque Sunny, M.S.; Mohammad, N. A Comprehensive Review on Second-Life Batteries: Current State, Manufacturing Considerations, Applications, Impacts, Barriers & Potential Solutions, Business Strategies, and Policies. *IEEE Access* **2019**, *7*, 73215–73252.

25. Anderman, M. Assessing the Future of Hybrid and Electric Vehicles: The xEV Industry Insider Report. *Adv. Automot. Batter.* **2014**, 1–161.
26. Olivetti, E.A.; Ceder, G.; Gaustad, G.G.; Fu, X. Lithium-Ion Battery Supply Chain Considerations: Analysis of Potential Bottlenecks in Critical Metals. *Joule* **2017**, *1*, 229–243. [CrossRef]
27. Raugei, M.; Winfield, P. Prospective LCA of the production and EoL recycling of a novel type of Li-ion battery for electric vehicles. *J. Clean. Prod.* **2019**, *213*, 926–932. [CrossRef]
28. Fang, X.; Peng, H. A Revolution in Electrodes: Recent Progress in Rechargeable Lithium-Sulfur Batteries. *Small* **2015**, *11*, 1488–1511. [CrossRef] [PubMed]
29. Barchasz, C.; Molton, F.; Duboc, C.; Leprêtre, J.-C.; Patoux, S.; Alloin, F. Lithium/Sulfur Cell Discharge Mechanism: An Original Approach for Intermediate Species Identification. *Anal. Chem.* **2012**, *84*, 3973–3980. [CrossRef] [PubMed]
30. Benveniste, G.; Rallo, H.; Canals Casals, L.; Merino, A.; Amante, B. Comparison of the state of Lithium-Sulphur and lithium-ion batteries applied to electromobility. *J. Environ. Manag.* **2018**, *226*, 1–12. [CrossRef] [PubMed]
31. Ding, Y.; Cano, Z.P.; Yu, A.; Lu, J.; Chen, Z. Automotive Li-Ion Batteries: Current Status and Future Perspectives. *Electrochem. Energy Rev.* **2019**, *2*, 1–28. [CrossRef]
32. H2020 Helis Project. Available online: https://www.helis-project.eu/ (accessed on 28 May 2019).
33. Tang, Q.; Shan, Z.; Wang, L.; Qin, X.; Zhu, K.; Tian, J.; Liu, X. Nafion coated sulfur-carbon electrode for high performance lithium-sulfur batteries. *J. Power Sources* **2014**, *246*, 253–259. [CrossRef]
34. Propp, K.; Marinescu, M.; Auger, D.J.; Neill, L.O.; Fotouhi, A.; Somasundaram, K.; Offer, G.J.; Minton, G.; Longo, S.; Wild, M.; et al. Multi-temperature state-dependent equivalent circuit discharge model for lithium-sulfur batteries. *J. Power Sources* **2016**, *328*, 289–299. [CrossRef]
35. Fotouhi, A.; Auger, D.J.; Member, S.; Propp, K.; Member, S.; Purkayastha, R.; Neill, L.O.; Walus, S. Lithium-Sulfur Cell Equivalent Circuit Network Model Parameterization and Sensitivity Analysis. *IEEE Trans. Veh. Technol.* **2017**, *66*, 7711–7721. [CrossRef]
36. Huria, T.; Ceraolo, M.; Gazzarri, J.; Jackey, R. High Fidelity Electrical Model with Thermal Dependence for Characterization and Simula- tion of High Power Lithium Battery Cells. In Proceedings of the Electric Vehicle Conference (IEVC), Greenville, SC, USA, 4–8 March 2012.
37. Schweiger, H.-G.; Obeidi, O.; Komesker, O.; Raschke, A.; Schiemann, M.; Zehner, C.; Gehnen, M.; Keller, M.; Birke, P. Comparison of several methods for determining the internal resistance of lithium ion cells. *Sensors (Basel)* **2010**, *10*, 5604–5625. [CrossRef]
38. Kolosnitsyn, V.S.; Kuzmina, E.V.; Mochalov, S.E. Determination of lithium sulphur batteries internal resistance by the pulsed method during galvanostatic cycling. *J. Power Sources* **2014**, *252*, 28–34. [CrossRef]
39. Zhu, X.; Wang, Z.; Wang, C.; Huang, L. Overcharge Investigation of Large Format Lithium-Ion Pouch Cells with $Li(Ni_{0.6}Co_{0.2}Mn_{0.2})O_2$ Cathode for Electric Vehicles: Degradation and Failure Mechanisms. *J. Electrochem. Soc.* **2018**, *165*, A3613–A3629. [CrossRef]
40. Wagner, N.; Eneli, H.; Ballauff, M.; Friedrich, K.A. Correlation of capacity fading processes and electrochemical impedance spectra in lithium / sulfur cells. *J. Power Sources* **2016**, *323*, 107–114.
41. Stroe, D.I.; Knap, V.; Swierczynski, M.; Schaltz, E. Electrochemical impedance spectroscopy-based electric circuit modeling of lithium-sulfur batteries during a discharging state. *IEEE Trans. Ind. Appl.* **2019**, *55*, 631–637. [CrossRef]
42. Canals Casals, L.; Martinez-Laserna, E.; Amante García, B.; Nieto, N. Sustainability analysis of the electric vehicle use in Europe for CO_2 emissions reduction. *J. Clean. Prod.* **2016**, *127*, 425–437. [CrossRef]
43. Deng, Y.; Li, J.; Li, T.; Gao, X.; Yuan, C. Life cycle assessment of lithium sulfur battery for electric vehicles. *J. Power Sources* **2017**, *343*, 284–295. [CrossRef]
44. Hawkins, T.R.; Singh, B.; Majeau-Bettez, G.; Strømman, A.H. Comparative Environmental Life Cycle Assessment of Conventional and Electric Vehicles. *J. Ind. Ecol.* **2013**, *17*, 53–64. [CrossRef]
45. Zhang, S.S. Improved cyclability of liquid electrolyte lithium/sulfur batteries by optimizing electrolyte/sulfur ratio. *Energies* **2012**, *5*, 5190–5197. [CrossRef]
46. Shang, X.; Guo, P.; Qin, T.; Liu, M.; Lv, M.; Liu, D.; He, D. Sulfur Immobilizer by Nanoscale TiO2 Trapper Deposited on Hierarchical Porous Carbon and Graphene for Cathodes of Lithium–Sulfur Batteries. *Adv. Mater. Interfaces* **2018**, *5*, 1–8. [CrossRef]

47. Knap, V.; Stroe, D.-I.; Purkayastha, R.; Walus, S.; Auger, D.J.; Fotouhi, A.; Propp, K. Reference Performance Test Methodology for Degradation Assessment of Lithium-Sulfur Batteries. *J. Electrochem. Soc.* **2018**, *165*, A1601–A1609. [CrossRef]
48. Hunt, I.; Zhang, T.; Patel, Y.; Marinescu, M.; Purkayastha, R.; Kovacik, P.; Walus, S.; Swiatek, A.; Offer, G.J. The Effect of Current Inhomogeneity on the Performance and Degradation of Li-S Batteries. *J. Electrochem. Soc.* **2018**, *165*, A6073–A6080. [CrossRef]
49. Barré, A.; Deguilhem, B.; Grolleau, S.; Gérard, M.; Suard, F.; Riu, D. A review on lithium-ion battery ageing mechanisms and estimations for automotive applications. *J. Power Sources* **2013**, *241*, 680–689. [CrossRef]
50. Vetter, J.; Nov, P.; Wagner, M.R.R.; Veit, C.; Novák, P.; Möller, K.-C.; Besenhard, J.O.; Winter, M.; Wohlfahrt-Mehrens, M.; Vogler, C.; et al. Ageing mechanisms in lithium-ion batteries. *J. Power Sources* **2005**, *147*, 269–281. [CrossRef]
51. Baghdadi, I.; Briat, O.; Delétage, J.Y.; Gyan, P.; Vinassa, J.M. Lithium battery aging model based on Dakin's degradation approach. *J. Power Sources* **2016**, *325*, 273–285. [CrossRef]
52. Waldmann, T.; Wilka, M.; Kasper, M.; Fleischhammer, M.; Wohlfahrt-Mehrens, M. Temperature dependent ageing mechanisms in Lithium-ion batteries – A Post-Mortem study. *J. Power Sources* **2014**, *262*, 129–135. [CrossRef]
53. Li, J.; Zhang, J.; Zhang, X.; Yang, C.; Xu, N.; Xia, B. Study of the storage performance of a Li-ion cell at elevated temperature. *Electrochim. Acta* **2010**, *55*, 927–934. [CrossRef]
54. Ecker, M.; Gerschler, J.B.; Vogel, J.; Käbitz, S.; Hust, F.; Dechent, P.; Sauer, D.U. Development of a lifetime prediction model for lithium-ion batteries based on extended accelerated aging test data. *J. Power Sources* **2012**, *215*, 248–257. [CrossRef]
55. Warnecke, A. Ageing effects of Lithium-ion batteries. In Proceedings of the 17th Conference on Power Electronics and Applications, Geneva, Switzerland, 8–10 September 2015.
56. Wikner, E.; Thiringer, T. Extending Battery Lifetime by Avoiding High SOC. *Appl. Sci.* **2018**, *8*, 1825. [CrossRef]
57. Martinez-Laserna, E.; Sarasketa-Zabala, E.; Villarreal, I.; Stroe, D.I.; Swierczynski, M.; Warnecke, A.; Timmermans, J.-M.; Goutam, S.; Omar, N.; Rodriguez, P. Technical Viability of Battery Second Life: A Study from the Ageing Perspective. *IEEE Trans. Ind. Appl.* **2018**, *54*, 2703–2713. [CrossRef]
58. Ecker, M.; Nieto, N.; Käbitz, S.; Schmalstieg, J.; Blanke, H.; Warnecke, A.; Sauer, D.U. Calendar and cycle life study of Li(NiMnCo)O2-based 18650 lithium-ion batteries. *J. Power Sources* **2014**, *248*, 839–851. [CrossRef]
59. Canals Casals, L.; Amante García, B.; Castellà Dagà, S. The electric vehicle battery ageing and how it is perceived by its driver | El envejecimiento de las baterías de un vehículo eléctrico y cómo lo percibe el conductor. *Dyna* **2016**, *91*, 188–195. [CrossRef]
60. Canals Casals, L.; Amante García, B.; González Benítez, M. Aging Model for Re-used Electric Vehicle Batteries in Second Life Stationary Applications. In *Lecture Notes in Management and Industrial Engineering*; Springer International Publishing: Cham, Switzerland, 2017; pp. 139–151, ISBN 978-3-319-51858-9.
61. Moy, D.; Manivannan, A.; Narayanan, S.R. Direct Measurement of Polysulfide Shuttle Current: A Window into Understanding the Performance of Lithium-Sulfur Cells. *J. Electrochem. Soc.* **2015**, *162*, A1–A7. [CrossRef]
62. Hofmann, A.F.; Fronczek, D.N.; Bessler, W.G. Mechanistic modeling of polysulfide shuttle and capacity loss in lithium–sulfur batteries. *J. Power Sources* **2014**, *259*, 300–310. [CrossRef]
63. Knap, V.; Stroe, D.-I.; Swierczynski, M.; Teodorescu, R.; Schaltz, E. Investigation of the Self-Discharge Behavior of Lithium-Sulfur Batteries. *J. Electrochem. Soc.* **2016**, *163*, A911–A916. [CrossRef]
64. Yang, Y.; Zhang, J. Highly Stable Lithium–Sulfur Batteries Based on Laponite Nanosheet-Coated Celgard Separators. *Adv. Energy Mater.* **2018**, *8*, 1–9. [CrossRef]
65. Sedlakova, V.; Sikula, J.; Sedlak, P.; Cech, O.; Urrutia, L. A Simple Analytical Model of Capacity Fading for Lithium-Sulfur Cells. *IEEE Trans. Power Electron.* **2019**, *34*, 5779–5786. [CrossRef]

Article

Using Self Organizing Maps to Achieve Lithium-Ion Battery Cells Multi-Parameter Sorting Based on Principle Components Analysis

Bizhong Xia [1], Yadi Yang [1,*], Jie Zhou [1], Guanghao Chen [1], Yifan Liu [1], Huawen Wang [2], Mingwang Wang [2] and Yongzhi Lai [2]

[1] Graduate School at Shenzhen, Tsinghua University, Shenzhen 518055, China
[2] Sunwoda Electronic Co. Ltd., Shenzhen 518108, China
* Correspondence: yangyd16@mails.tsinghua.edu.cn; Tel.: +86-1324-386-7516

Received: 28 June 2019; Accepted: 29 July 2019; Published: 1 August 2019

Abstract: Battery sorting is an important process in the production of lithium battery module and battery pack for electric vehicles (EVs). Accurate battery sorting can ensure good consistency of batteries for grouping. This study investigates the mechanism of inconsistency of battery packs and process of battery sorting on the lithium-ion battery module production line. Combined with the static and dynamic characteristics of lithium-ion batteries, the battery parameters on the production line that can be used as a sorting basis are analyzed, and the parameters of battery mass, volume, resistance, voltage, charge/discharge capacity and impedance characteristics are measured. The data of batteries are processed by the principal component analysis (PCA) method in statistics, and after analysis, the parameters of batteries are obtained. Principal components are used as sorting variables, and the self-organizing map (SOM) neural network is carried out to cluster the batteries. Group experiments are carried out on the separated batteries, and state of charge (SOC) consistency of the batteries is achieved to verify that the sorting algorithm and sorting result is accurate.

Keywords: lithium-ion battery; cell sorting; multi-parameters sorting; principal component analysis; self-organizing maps clustering

1. Introduction

The initial differences among batteries which lead to inconsistency after charging and discharging are the main reasons for the shortened life and low safety in the use of battery packs. It is not feasible to completely eliminate the inconsistency of battery packs, but there are some ways to reduce inconsistency's negative impact [1,2]. Cell sorting in lithium-ion battery industry is an indispensable process to assure the reliability and safety of cells that are assembled into strings, blocks, modules and packs [3]. In the current lithium-ion power battery pack production line, cell sorting refers to the selection of qualified cells from raw ones according to quantitative criterions in terms of accessible descriptors such as battery resistance, open circuit voltage (OCV), charging/discharging capacity, etc. Correspondingly, resistance sorting, voltage sorting and capacity sorting are the main single parameter sorting methods used in battery pack production lines at present [4–7]. The single-parameter sorting method can quickly screen batteries whose parameters are in the qualified range from raw ones [8], but this method ignores the relationship among battery parameters. In contrast, the multi-parameter sorting method which combines static and dynamic characteristics of batteries is a more accurate and comprehensive solution. In order to realize multi-parameter sorting of lithium batteries in the battery pack production line, it is necessary to test battery parameters, such as the resistance, voltage and capacity, using the existing equipment.

We note that a certain number of open publications focusing on sorting methods can be found, and clustering algorithms [9,10], including the fuzzy C-means algorithm (FCM) [11,12], k-means algorithm [13], and self-organizing maps (SOM) neural networks [14–16], are the main research direction. Nevertheless, the sorting parameters and their relationships and constraints adopted in the sorting process, which are of fundamental significance, have not been investigated. Therefore, in this study, a series of parameters as sorting criterions which can characterize the performance and difference of batteries are investigated, and they are divided into three categories: general parameters, static performance parameters and dynamic characteristics.

General parameters refer to the appearance parameters such as mass (m) and volume (v) of batteries and the derived parameters such as mass specific energy and volume specific energy caused by them. Static performance parameters include battery resistance, voltage, charge/discharge capacity, self-discharge rate, charge/discharge efficiency and charge/discharge time. They describe the performance of batteries under non-working conditions, but their acquisition needs certain conditions. For instance, charging and discharging [17] experiments are needed to obtain the capacity data of batteries. As for static characteristics of a battery, charge-discharge voltage curve and electrochemical impedance spectroscopy (EIS) [18] are typical methods to show the dynamic performance. Some previous studies utilize an overall voltage-current or voltage-time curve comparing the differences among batteries to complete sorting. However, drawing an overall voltage-time curve or EIS takes a long time to complete, so it is unrealistic to use it in production line.

Under such circumstances, this study compares several parameters as descriptors for battery sorting and finally selects the following sorting indicators: battery mass (m), volume (v), voltage (V), capacity (C) and surface temperature (T) [19] as static indicators. As aforementioned, an overall voltage-time curve or EIS spectrum takes a long time to obtain, therefore, in this study the impedance characteristics of batteries at several key frequencies in EIS spectrum [20] are extracted, and the AC (Alternative Current) impedance (R_A) and reactance (X) of batteries are taken as the evaluation criteria of dynamic characteristics. These parameters can be obtained by testing equipment now available on the production line. The principle components analysis (PCA) [21] method is used to pre-process the data of battery parameters (clarifying the relationship between parameters), and obtain the principal components that can reflect the characteristics of the battery. The clustering and sorting process of batteries is accomplished by a self-organizing map (SOM) neural network.

The rest of this paper is organized as follows: The acquisition of battery parameters and data analysis are completed in Section 2, followed by the results of data processing using PCA. Section 3 introduces the principle of SOM algorithm and its application in battery clustering. Section 4 carries out the battery sorting experiments and operation tests to obtain the state of charge (SOC) curves of the batteries to be sorted under special conditions. Finally, in Section 5 the performance of batteries after grouping is studied, and the results of battery sorting are verified. Section 6 concludes the major findings.

2. Parameter Acquisition and Data Processing

This section describes the process to obtain the parameters of the batteries to be sorted, so as to get a sufficient number of parameter inputs to start a follow-up clustering algorithm.

A set of 58 lithium iron phosphate ($LiFePO_4$) batteries are considered in this work. The batteries to be sorted are INR 18650-33G cylindrical batteries produced by Samsung SDI Company (Suwon, South Korea). The cathode is made of lithium iron phosphate and the anode is made of graphite. Tests were performed in laboratories in Tsinghua University Shenzhen Graduate School and Sunwoda electronics Co., Ltd. in Shenzhen to obtain experimental data such as mass, volume, voltage and currents under charge/discharge and capacity, etc. and to identify internal resistance. These 58 Samsung 18650 cells with a nominal capacity of 2700 mAh and nominal voltage of 3.6V were tested under certain circumstances. Table 1 shows the cell specification.

Table 1. Cell Specification.

Item	Value
Manufacturer	Samsung INR18650-33G
Nominal capacity	2700 mAh
Charging cut-off voltage	4.2 V
Discharging cut-off voltage	2.6 V
Nominal voltage	3.6 V
Standard weight	47 ± 2.0 g
Dimensions	Height: 65.35 ± 0.15 mm Diameter: φ18.33 + 0.1/−0.13 mm
Working temperature	Charge: −10~45 °C Discharge: −20~60 °C

2.1. General Parameters

The general parameters of batteries can be directly reflected in their physical attributes, such as the mass, volume and surface temperature of battery cells, as well as the derived mass specific energy and volume specific energy. Because all the batteries selected in this paper are of the same type, the battery energy is the same by default. Therefore, the mass (m), volume (v) and surface temperature (T) of the battery cells are selected as the general parameters.

Parameters m and v of batteries were tested with Keyence pressure sensor and vision measurement sensor. The quality measurement with 10 μg accuracy can be achieved by pressure conversion measured by pressure gauge. The volume of the battery is calculated by measuring the geometric size. The surface temperature (T) of battery is measured by thermocouple.

2.2. Static Parameters

The voltage (V) of the batteries to be sorted can be measured with a high precision voltmeter. The total capacity (C) of batteries is a unique characteristic, which is different between different batteries. It can be obtained from the charge/discharge test, by means of the capacity data provided from the Coulomb counting method, shown as Equation (1), directly implemented in the testing equipment. The charge/discharge test is carried out using Arbin BT-5HC battery testing equipment, with battery cells (to be sorted) in a Sanwood constant temperature and humidity box. Figure 1 indicates the curve of voltage and current during the charge/discharge test.

$$C = It. \tag{1}$$

C denotes charging/discharging capacity, I and t represent charging/discharging current and time. In the process of capacity testing, the lithium battery that has been stationary for 2 h is discharged to the discharge cut-off voltage (2.6 V) at a rate of 1 C. Then the battery is charged by means of CC-CV (Constant Current-Constant Voltage) at a rate of 0.5 C to the charging cut-off voltage (4.2 V), and then is discharged at a rate of 1 C until the discharge cut-off voltage. The capacity test curve as shown in Figure 1 is obtained after several cycles, and the capacity is calculated using Equation (1).

There are several ways to measure DC (Direct Current) internal resistance of batteries. In this study, according to Equation (2), the DC internal resistance is derived by dividing the voltage drop during the transition to rest by the constant discharge current:

$$R_D = \frac{\Delta V}{I} = \frac{V_2 - V_1}{I}. \tag{2}$$

Due to the polarization phenomenon, the voltage of lithium batteries presents a dynamic process of rapid rise, slow rise, rapid fall and slow fall during pulse charging. The change of voltage ΔV during discharging is contrary to that during charging. In this process, the sharp voltage drop is caused by the ohmic polarization inside the battery, so the ohmic internal resistance of the battery can be identified by using the voltage and current data.

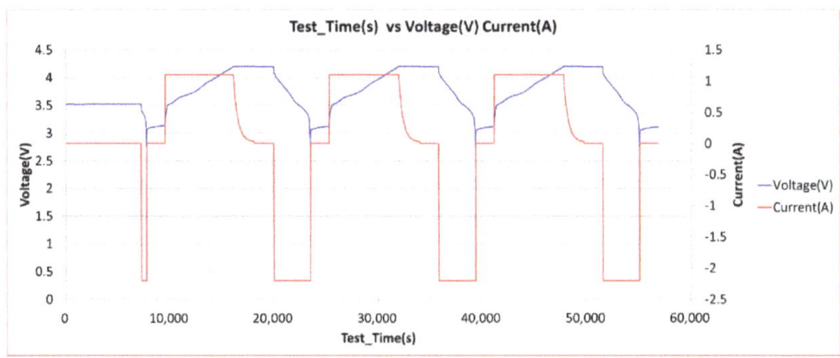

Figure 1. Capacity testing process.

2.3. *Impedance Characteristics*

The electrochemical impedance spectroscopy (EIS) method is used to measure the change of the impedance or phase angle of the battery system with frequency, by applying sinusoidal alternating currents of different frequencies to the battery. The EIS of a battery is an inherent characteristic. It is a feasible method to use the similarity of EIS drawings to sort batteries. The EIS method is a quasi-dynamic process, and small sinusoidal alternating current will not cause great disturbance to the battery, so the EIS sorting method is more accurate than the static method.

In the EIS of lithium-ion batteries, the different parts corresponding to different frequencies can reflect the process of lithium ion removal and embedding in the embedded electrode during the charging and discharging process of lithium-ion batteries.

(1) Ultra high frequency (UHF) part: The section where the impedance curve intersects the transverse axis in EIS, reflecting the ohmic impedance (R_b) of lithium ion batteries.
(2) High frequency part: A semicircle related to the diffusion and migration of lithium ions through the insulating layer on the surface of active material particles. This process can be represented by a R_{sei}/C_{sei} parallel circuit.
(3) Intermediate frequency (IF) part: A semicircle reflecting the charge transfer process. The process is represented by a R_{ct}/C_{dl} parallel circuit. R_{ct} is the charge transfer impedance, or the polarization impedance of the battery, and C_{dl} is double-layer capacitor.
(4) Low frequency part: A line with a slope of 45 degrees, which is related to the solid diffusion process of lithium ions in active material particles. In the equivalent circuit model, a Warburg impedance Z_W describing diffusion can be used, also known as concentration polarization impedance.
(5) Ultra low frequency: A semicircle and a vertical line, reflecting the change in crystal structure of active material particles and the accumulation and consumption of lithium ions in active material respectively.

Figure 2a indicates the equivalent circle model for EIS used for curve fitting. In this study, using the HIOKI BT4560 Impedance Tester (HIOKI, Shanghai, China), the impedance experiments are carried out on one of the lithium battery cells using 0.1–1050 Hz AC current, and the impedance spectra are plotted, as shown in Figure 2b, which reflects the intermediate frequency part mentioned above. BT4560 can realize simultaneous measurement of battery voltage and impedance with high accuracy. The measurement accuracy of voltage and impedance can reach 10 μV and 1 μΩ respectively. It can also measure the surface temperature of batteries by connecting thermocouples, and the accuracy can reach 0.1 °C.

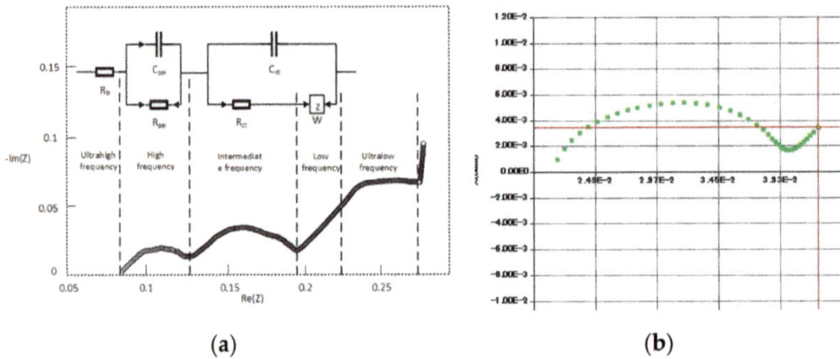

Figure 2. Electrochemical impedance spectroscopy (EIS) of lithium-ion battery: (**a**) EIS and equivalent circle model of Li-ion battery; (**b**) EIS in intermediate frequency (IF) part of one 18,650 cell using BT4560.

As described in Section 1, we collect the impedance characteristics of batteries at several key frequencies in EIS spectrum, so that the measurement process can be rapid. In this study, two characteristic frequencies, 0.1 Hz and 1000 Hz, are selected to obtain the impedance (R_A) and reactance (X) of the battery as the sorting parameters. These two frequencies are between high, medium and low frequencies. The results of impedance can reflect the polarization impedance and Warburg impedance of batteries, respectively.

2.4. Principle Components Analysis

The parameters of each battery cell to be sorted are tested using the test equipment suitable for the production line. The results are shown in Table 2; two cells are selected to show the input parameters.

Table 2. Examples of parameters for two batteries.

No.	1	2
Manufacturer	Samsung INR18650-33G	Samsung INR18650-33G
Mass (g)	46.92499999	47.60953785
Volume (mm^3)	16,540	16,540
Capacity (mAh)	2600	2650
Voltage (V)	3.524017	3.568177
Impedance R1 (1000 Hz) (Ω)	0.0219427	0.0223717
Reactance X1 (1000 Hz) (Ω)	−0.0002701	−0.0006503
Impedance R2 (0.1 Hz) (Ω)	0.0355982	0.0401869
Reactance X2 (0.1 Hz) (Ω)	−0.0022978	−0.0024677
Surface temperature (°C)	26.2	26.5

The input data matrix of 58 × 9 is obtained after the test of 58 battery cells. Principle component analysis (PCA) is used to analyze data. PCA is a statistical method, which uses linear transformation to achieve dimension reduction. Its basic principle is to use the idea of dimension reduction and linear transformation to transform multiple indicators into several unrelated comprehensive indicators without losing much information, which are called the principle components. In this study, the data of 58 batteries are analyzed by principal component analysis in the following steps.

1. Arrange the raw data in rows to form a matrix X.
2. Standardize the data of matrix X, change its mean to zero.
3. Compute Covariance Matrix C.
4. Calculate eigenvalues according to the covariance matrix C. The eigenvectors are arranged from large to small eigenvalues, and the first k matrices are composed of rows P.

5. Calculate by Equation (3) to get the reduced dimension data matrix Y.

$$Y = PX. \tag{3}$$

6. Compute the contribution rate V_i of each eigenvalue as shown in Equation (4)

$$V_i = \frac{x_i}{x_1 + x_2 + \cdots x_n}. \tag{4}$$

7. Sort the principal components according to their contribution rate and explain the physical significance of principal components based on eigen roots and their eigenvectors.

The result of PCA shows that three principle components can cover more than 90% of data variability, as shown in Table 3. By analyzing the proportion of principal components of each parameter, the proportion results are obtained.

$$\begin{array}{l} t_1 = -0.09011m - 0.09044v - 0.08985C - 0.34997V \\ +0.461975R_1 + 0.458163X_1 + 0.465659R_2 + 0.456563X_2 + 0.06728T. \end{array} \tag{5}$$

$$\begin{array}{l} t_2 = 0.564575m + 0.564293v + 0.564558C - 0.14541V \\ +0.046857R_1 + 0.079282X_1 - 0.00029R_2 + 0.083406X_2 + 0.086721T. \end{array} \tag{6}$$

$$\begin{array}{l} t_3 = -0.05338m - 0.0557v - 0.05278C - 0.17591V \\ -0.08143R_1 - 0.08759X_1 - 0.04988R_2 - 0.08818X_2 + 0.967342T. \end{array} \tag{7}$$

Table 3. (**a**) Principal component analysis of 58 batteries with nine parameters. (**b**) Proportional relationship between the original parameters and principal components.

(**a**)

Principle Component	Eigenvalue	D-Value	Contribution Rate	Cumulative Contribution Rate
1	4.513363	1.502216	50.14848	50.14848
2	3.011147	2.013297	33.45719	83.60566
3	0.99785	0.546226	11.08722	94.69288
4	0.451624	0.427816	5.018046	99.71093
5	0.023808	0.022719	0.264529	99.97546
6	0.001089	0.000355	0.012096	99.98755
7	0.000733	0.000383	0.008149	99.9957
8	0.000351	0.000315	0.003898	99.9996
9	3.6×10^{-5}	0	0.000399	100

(**b**)

Standardized Variables	Feature Vector (t_1)	Feature Vector (t_2)	Feature Vector (t_3)
Mass (g)	−0.09011	0.564575	−0.05338
Volume (mm^3)	−0.09044	0.564293	−0.0557
Capacity (mAh)	−0.08985	0.564558	−0.05278
Voltage (V)	−0.34997	−0.14541	−0.17591
Impedance R1 (1000 Hz) (Ω)	0.461975	0.046857	−0.08143
Reactance X1 (1000 Hz) (Ω)	0.458163	0.079282	−0.08759
Impedance R2 (0.1 Hz) (Ω)	0.465659	−0.00029	−0.04988
Reactance X2 (0.1 Hz) (Ω)	0.456563	0.083406	−0.08818
Surface temperature (°C)	0.06728	0.086721	0.967342

Equations (5)–(7) use eigenvectors to show the relationship with nine initial characteristic parameters. It can be seen that these nine characteristics are confirmed and can be divided into three categories: appearance parameters, impedance characteristics and safety parameters. The characteristic vector t_1 is mainly positively correlated with the impedance characteristics of batteries, and negatively correlated with mass, volume and capacity, so t_1 can be regarded as the principal component reflecting the impedance characteristics of batteries; similarly, the vector t_2 can be regarded as the principal component reflecting the appearance parameters of batteries. The characteristic vector t_3 mainly reflects the temperature parameters of the battery, which is related to the safety and thermal management of the battery.

3. SOM Neutral Networks and SOM Clustering

3.1. Introduction of SOM Neural Networks

The self-organizing map (SOM) neural network is an unsupervised learning clustering algorithm which realizes high-dimension visualization. It is an artificial neural network developed by simulating the characteristics of human brain to signal. The model was first proposed by Teuvo Kohonen [22], a professor at Helsinki University in Finland in 1981, so it is also called the Kohonen network.

The application of SOM neural networks in clustering is mainly based on the following advantages:

1. Dimension reduction can be achieved, and clustering results have good visibility. High dimensional input space can be mapped to low dimensional output space maintain the original topological relationship.
2. With self-organizing and unsupervised learning, it can be applied to situations where the characteristics of input data are not fully understood.
3. The algorithm is clear and the calculation is simple.

These features enable SOMs to be widely used in clustering, helping to recognize homogeneous groups of generous and complex inputs. Reference [14–16] indicates research and applications where SOMs are used in clustering processes in the last decades. Compared with traditional clustering algorithms, such as the k-means algorithm and FCM algorithm, SOM is more accurate and not affected by the selection of initial clustering centers, although its operation process is relatively complex.

3.2. Principle and Learning Algorithm of SOM

The structure of a SOM neural network is shown in Figure 3. It consists of an input layer and output layer (competition layer). The number of neurons in the input layer is determined by the number of vectors in the input network, and the output layer is arranged into a two-dimensional node matrix by neurons. The neurons in the input layer and the neurons in the output layer are fully connected by weights W. When the network receives the external input signal, the neurons in the output layer excite and distribute the neurons in the region with the highest input spatial density through competition.

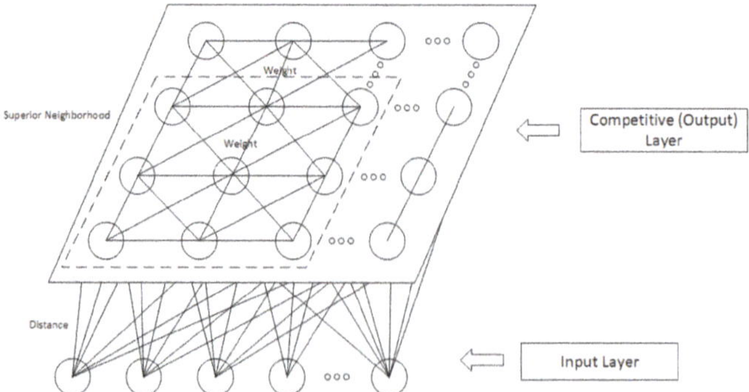

Figure 3. The structure of a self-organizing map (SOM) neural network.

A SOM is a competitive neural network, which follows the rules of competitive learning (referring to lateral inhibition of neurons in human brain). For all neurons in the output layer (competitive layer), the winner neurons are found according to the principle of minimum distance, then only the winner neurons and the neurons in the winning neighborhood are adjusted by the weight vector. Based on the WTA (winner-takes-all) rule, the specific algorithm steps of the SOM are as follows:

Step 1: Data initialization. The weight vectors of the output layer are given smaller random numbers and normalized, which are recorded as $\hat{W}_j(j = 1, 2, \ldots, m)$, m is number of neurons in the output layer. The initial optimal neighborhood is established as $N_j^*(0)$, and the initial value of learning efficiency α and the number of iterations T are set.

Step 2: Input data and normalize. Normalize the input vector $X^p(p = 1, 2, \ldots, n)$ and the weight vector W_j in the competition layer as shown in Equation (8), where n is the dimension of the input data.

$$\hat{X} = \frac{X}{||X||}, \hat{W}_j = \frac{W_j}{||W_j||} \tag{8}$$

The normalized input vector is denoted as $\hat{X}^p(p = 1, 2, \ldots, n)$, where n is the dimension of the input data.

Step 3: Find the winning node. Calculate the distance between input $\hat{X}^p$ and all output neurons, and select the winning neuron with the smallest distance. Euclidean distance is chosen for distance formula, as shown in Equation (9). Set the winning node as the center and determine the weight adjustment region at t-time. Setting the neighborhood distance d, the region within the D range of the distance from the winning node is regarded as the winning neighborhood.

$$d_j = ||\hat{X} - \hat{W}_j|| = \sqrt{\sum_{j=1}^{m} [X - W_j]^2}. \tag{9}$$

Step 4: Definition of superior neighborhood. Taking the winning node j^* as the center, the weight adjustment region of t time is determined. Setting the neighborhood distance d, the region within the d range of the distance from the winning node is regarded as the winning neighborhood.

$$N_j(t) = \left\{j, d_{j,j*} \leq d\right\} \tag{10}$$

Step 5: Weight adjustment. Based on the gradient descent method, the connection weights between the winning node and all other nodes in the winning neighborhood are adjusted. Equation (11) indicates the adjustment process as follows:

$$W_{j*}(t+1) = \hat{W}_{j*}(t) + \Delta W_{j*} = \begin{cases} \hat{W}_{j*}(t) + \alpha(\hat{X}^p - \hat{W}_{j*}(t)) & j = j^* \\ \hat{W}_{j*}(t) + \alpha N(t)(\hat{X}^p - \hat{W}_{j*}(t)), & j \in N_j(t) \quad i = 1, 2, \ldots, n, j \in N_j^*(t) \\ \hat{W}_{j*}(t) & j \notin N_j(t) \end{cases} \tag{11}$$

where, $w_{ij}(t)$ denotes the weight of neuron i to neuron j at t time; $0 < \alpha \leq 1$ denotes the learning rate, which affects the convergence and stability of the algorithm and decreases with time; $N(j, t)$ denotes the topological distance between the jth neuron and the winning neuron j^* in the neighborhood at training time t.

Step 6: End Judgment: When the learning rate $\alpha(t) \leq \alpha_{\min}$, or iterations $t \geq T$, the training process is terminated, otherwise return back to Step 2 and continue the iteration.

The advantages of SOM unsupervised learning and visualization of results enable it to be well applied in clustering analysis, as well as in lithium cell sorting.

3.3. SOM Clustering in Battery Sorting

The SOM is able to map any high-dimensional inputs to low-dimensional outputs, such as one-dimensional linear array or two-dimensional grid. Therefore, this feature of the algorithm provides the possibility of sorting battery cells from raw cell groups according to single or multiple parameters or characteristics. As aforementioned in Section 2, multiple parameters tested by testing equipment in a production line are obtained, and they can be regarded as input data. The output layer of the SOM map shows the sorting types each cell belongs to.

Nine parameters are tested including mass, voltage, impedance and parameters data. These are pre-processed using PCA and transformed to three components in Section 2. These parameters are used as inputs of the SOM neural network for battery cell sorting and the results of classification are the output.

This study uses a SOM neural network to sort battery cells. The data of m battery cells with n parameters are input in the form of matrix of $m \times n$, and finally the cells are classified into k classes. The learning rate α and neighborhood radius r of the network are updated in the way shown in Equations (12) and (13), respectively.

$$\alpha = \alpha_0 \times (1 - \frac{t}{T_{\max}}), \tag{12}$$

$$r = r_0 \times (1 - \frac{t}{T_{\max}}) \tag{13}$$

where, $T_{\max}$ means the maximum numbers of iteration. $\alpha_0 = 0.99$ and $r_0 = 1$ are the initial learning rate and winning neighborhood radius respectively, and these two parameters decrease with time.

Figure 4 demonstrates the process of battery sorting using SOM in this study and the specific operation will be introduced in next section.

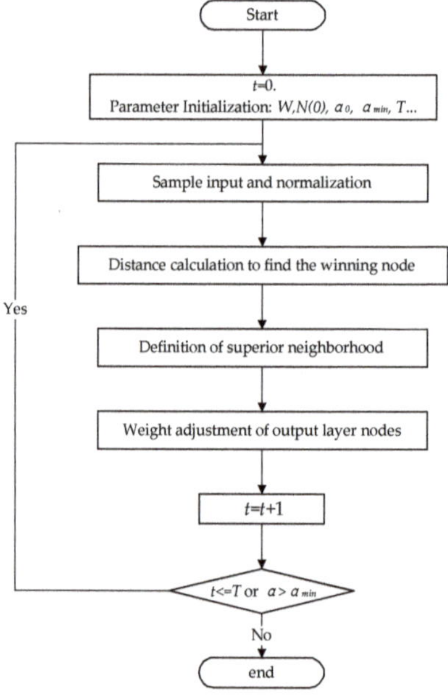

Figure 4. Procedure of battery sorting using a SOM.

4. Cell Sorting Using a SOM

As described in Section 2, the vector of nine parameters is considered as input for SOM neural network clustering. We tested 58 battery cells and recorded nine parameters of each cell. The input of the SOM is the parameter matrix with 58 × 9. After PCA, in order to reduce the data dimension, the matrix can be transformed to 58 × 3 as the nine parameters are replaced with the three principle components. Therefore, the first group of input data including nine parameters is marked as Group 1, and the second group of input data including three principle components is marked as Group 2. Table 4 denotes part of the input data (eight cells).

Table 4. Examples of principal component scores for eight batteries (Group 2).

Battery Number	Principal Component Score First (t_1)	Principal Component Score Second (t_2)	Principal Component Score Third (t_3)
01	−1.42814	7.525299	−0.02753
02	−1.38087	7.602607	0.678699
03	−0.97269	2.441534	−0.89297
04	−0.96143	2.449494	−0.78799
05	−0.96094	2.466339	−0.71301
06	−0.95142	2.465672	−0.68406
07	−0.90951	−1.02728	−1.6356
08	−0.88487	2.546314	0.351155

This section compiles the algorithm of a SOM neural network and sets its initial algorithm parameters by using the Neural Network Toolbox in MATLAB. The initial number of neurons in the output layer, is set to four, as shown in Figure 5.

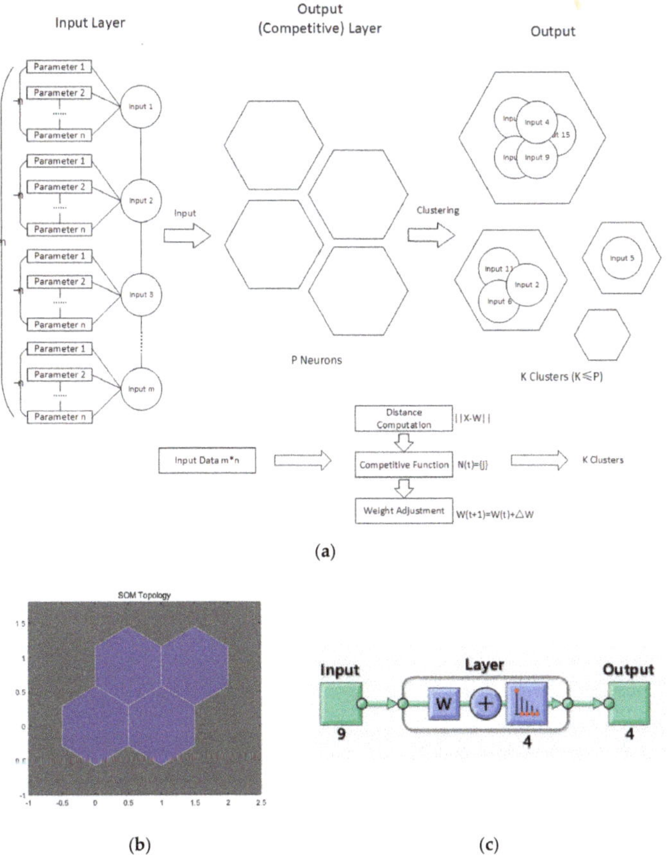

Figure 5. Example of SOM mapping and clustering process: (**a**) SOM mapping and learning process in clustering; (**b**) SOM topology in MATLAB; (**c**) SOM structure in view in MATLAB (e.g., nine inputs in the input layer and four neurons in the output layer).

Figure 6 shows the SOM clustering results of Group 1 and Group 2. In both sets of cells, eight battery cells are clearly separated from the whole ones, which means that these eight cells are different from the others.

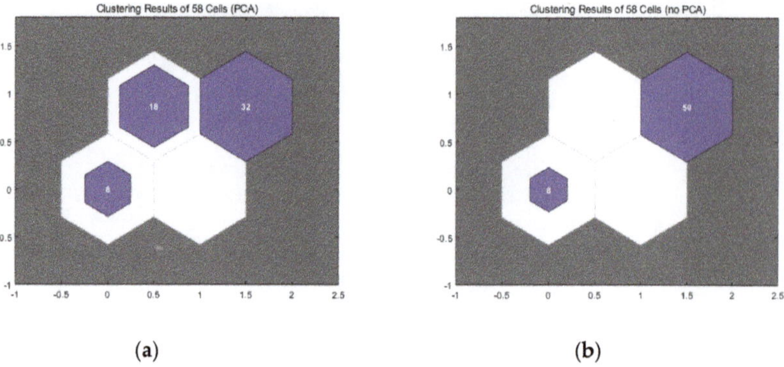

(a) (b)

Figure 6. SOM clustering results (four neurons), (**a**) Group 1, (**b**) Group 2.

The same procedure using a SOM is carried out to obtain the sorting results from the remaining 50 battery cells and the number of sorting is adjusted to six, as shown in Figure 7. Six groups are obtained after being classified. Table 5 shows the categories of 50 numbered batteries based on the SOM clustering results, both in Group 1 and Group 2.

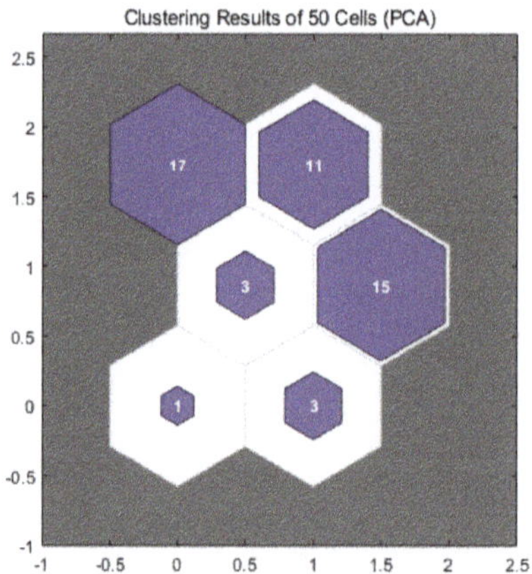

Figure 7. SOM clustering results of Group 1 (six neurons in the output layer).

Table 5. Categories of 50 battery cells.

Group No.	Battery Cell Number in Group 1	Battery Cell Number in Group 2
Category 1	1 2 7 15 21 22 24 27 32 34 35 36 38 40 45	2 7 15 21 22 34 35 36 38 40 45
Category 2	3 4 5 6 9 13 16 23 28 33 41 42 43 46 47 48 49	3 4 5 6 9 13 28 33 41 42 43 46 47 49
Category 3	8 10 14 19 20 29 30 31 37 39 44	8 10 14 19 20 29 30 31 37 39 44
Category 4	11 17 25	11 17 25 24 27 32
Category 5	12 18 26	12 16 18 23 26 48
Category 6	50	1 50

The contents mentioned above in this section illustrates a qualitative result of SOM battery cell sorting, therefore a validation experiment of the sorting results must be operated.

It can be noted that the sorting result of Group 2 is basically coincident with Group 1, which means SOM sorting, using the three principle components obtained from nine parameters after PCA, is feasible. Consequently, pretreatment of PCA is also a workable method when the sorting batteries have multiple dimension parameters. It also effectively reduces the workload of the subsequent scoring algorithm such as SOM. According to the classification results, eight batteries with good consistency (i.e., classified into the same category) can be selected from each group to form a module. This result will be validated by experiments in the next section.

5. Verification for Sorting Results

In response to different operating conditions, the SOC of lithium-ion batteries will vary with the voltage and current. Despite of their different dynamic responses, battery cells with good consistency should have less SOC variation differences. Therefore, two typical power load profiles of electric vehicles, the New European Driving Cycle (NEDC) and the Urban Dynamo-meter Driving Schedule (UDDS), shown in Figure 8, are chosen to test the battery modules sorted and formed from Section 4. The changes of the SOC curve and differences in the module under working conditions are recorded and calculated in the module under working conditions to test the clustering algorithm.

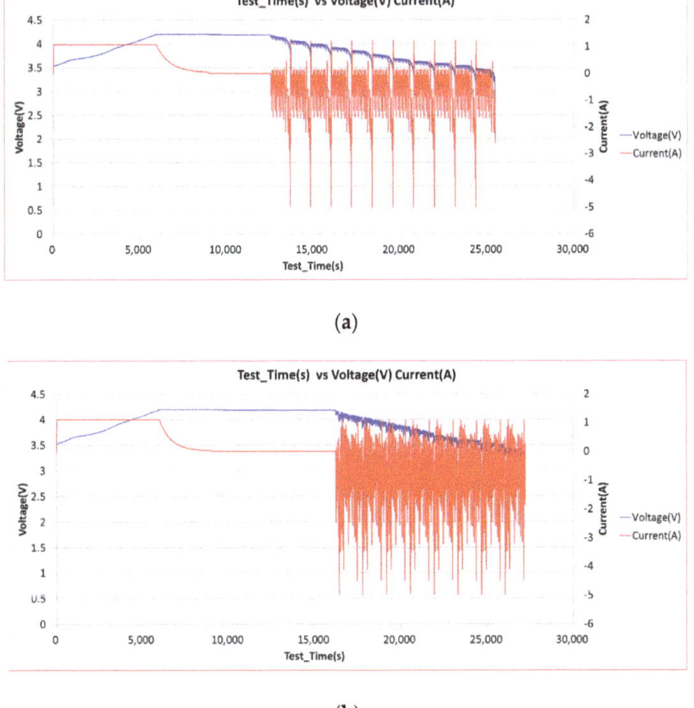

Figure 8. Power load profile: (**a**) New European Driving Cycle (NEDC) working conditions; (**b**) Urban Dynamo-meter Driving Schedule (UDDS) working conditions.

Eight battery cells of Module 1 and Module 2 are chosen from the same categories of Group 1 and Group 2 in Section 4, respectively. Eight cells (No. 3, 4, 5, 6, 9, 13, 16, 23), which are all classified to category 2 in Group 1, are connected in series to form Module 1. The same procedure is carried out to category 1 in Group 2, so that eight cells (No. 2, 7, 15, 21, 22, 34, 35, 36) are connected in series to form Module 2. For validation of consistency, another eight cells are chosen randomly from the whole cell stack to form Module 3. To ensure the randomness of the selection of the eight batteries, one cell is selected from each seven cells according to number order, which means the first cell is selected from No.1 to No.7 randomly, the second cell is selected from No. 8 to No. 14 and so on. The final number of these eight cells in Module 3 are: No. 1, 8, 17, 26, 32, 37, 45, 54.

The battery module is charged and discharged using battery charging and discharging equipment under the working conditions shown in Figure 8, and the current and voltage changes of each battery are recorded.

The main result of battery module inconsistency is that the state of charge (SOC) of each cell is not uniform, but differs from one another, because every cell is different in available capacity, dynamics, and imbalance of the individual cells in a series or parallel chain. So, the SOC curve and variability can be used to determine whether the battery module has good consistency or not. Currently, a lot of research has been done on the estimation methods of SOC. Among them, the most typical methods are the time integration method, open-circuit voltage method and Kalman filter method. In this section, extended Kalman filtering (EKF) [23,24] is used to estimate SOC change under two working conditions.

The Kalman filter is a common algorithm in SOC estimation. It can realize the optimal estimation of the state of a discrete-time linear system based on the minimum mean square error. The Kalman filter method mainly includes two parts: prediction and correction. The prediction process refers to updating the state estimates at the last time based on the state equation obtained from the battery model. The correction process is to update the state predicted value, according to the observed value obtained from the test. The extended Kalman filter (EKF) is based on the Kalman filter, which expands the non-linear function of the system into a Taylor expansion of the first order, and obtains the linearized system equation to complete the filtering and estimation of the state parameters.

In this study, EKF is one of the SOC estimation methods used for example, so the specific process of EKF is not discussed here.

In order to save time, a section of NEDC working conditions is selected as shown in Figure 9. Figure 10 shows the SOC of these three modules subject to the power profile of the NEDC working conditions shown above. The SOC in the battery module appears as a consistent trend over time, but the SOC between each cell at the same time is clearly different, especially in different modules. The curves of cells in Module 1 and Module 2 have relatively better consistency, while in Module 3 the difference between the SOC of batteries increases obviously, showing that cells after clustering and sorting have better homogeneity than those chosen randomly.

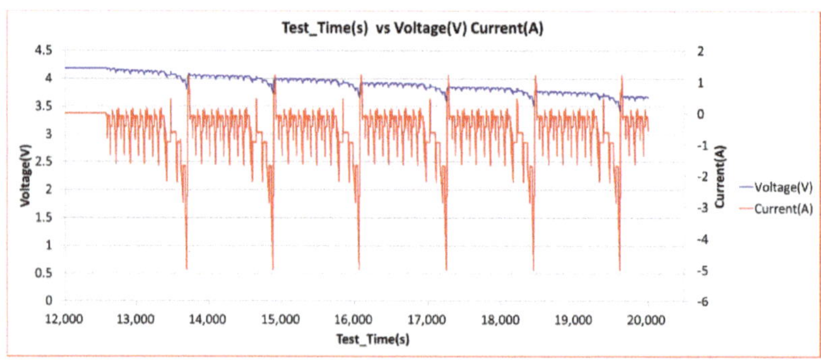

Figure 9. Local display of NEDC working conditions.

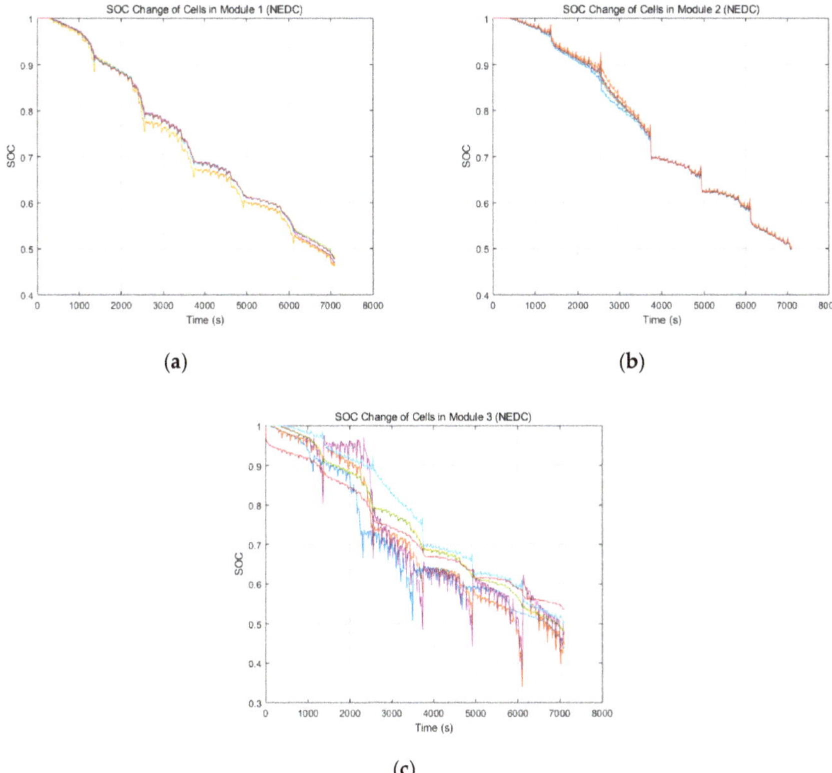

Figure 10. State of charge (SOC) curve of cells in modules: (**a**) Module 1, (**b**) Module 2, (**c**) Module 3.

The same validation procedure is performed under UDDS working conditions to verify the SOC changes in battery modules. The standard deviation of the SOC after load profile of Figure 11 is chosen as another testing criterion. Figure 11, and Tables 6 and 7 illustrate the standard deviation of state of charge of these cells in respective modules.

Table 6. SOC standard deviation of cells in modules under NEDC working conditions.

Module No.	Mean Standard Deviation	Maximum Standard Deviation
Module 1	0.0058	0.0208
Module 2	0.0026	0.0142
Module 3	0.0356	0.1274

Table 7. SOC standard deviation of cells in modules under UDDS working conditions.

Module No.	Mean Standard Deviation	Maximum Standard Deviation
Module 1	0.0069	0.0413
Module 2	0.0034	0.0184
Module 3	0.0472	0.1591

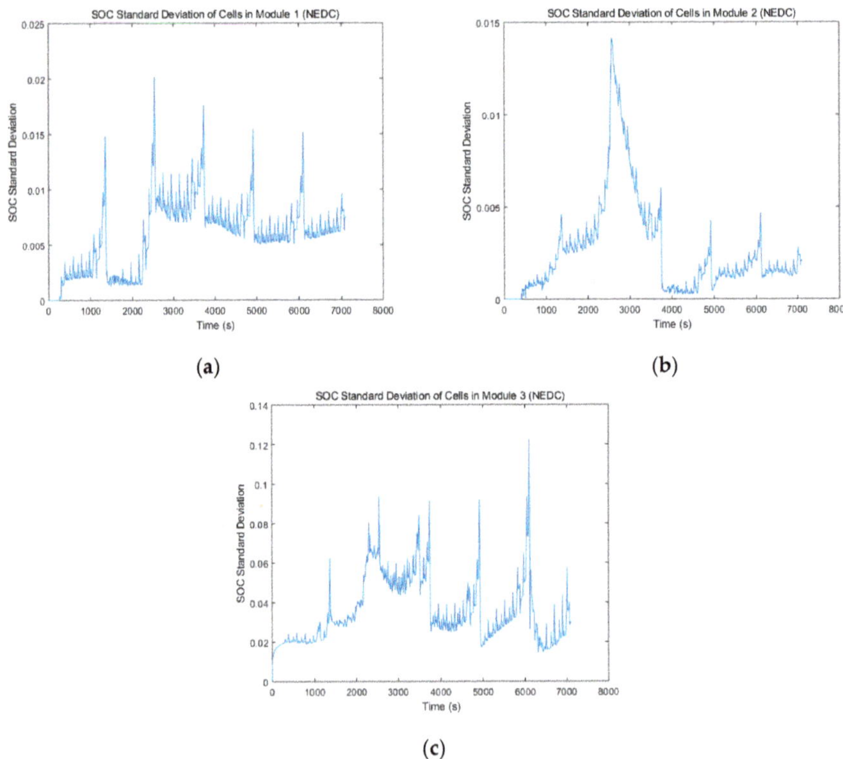

Figure 11. SOC standard deviation curve of cells in modules: (**a**) Module 1, (**b**) Module 2, (**c**) Module 3.

The results clearly show a reduction in the SOC variability after power profile charge and discharge. The SOC standard deviation of Module 1 or Module 2 is notably lighter than that of Module 3, which means there is a significant improvement in the consistency of the battery module.

At the same time, it can be seen that the standard deviation of Module 2 is slightly less than Module 1, indicating the clustering based on PCA gives the best results overall.

6. Conclusions

Effective sorting of lithium batteries is a means to eliminate the inconsistency of battery modules and battery modules. Selecting appropriate sorting parameters and using appropriate sorting algorithms can effectively improve the accuracy and efficiency of battery sorting. This work analyzes the static and dynamic performance of 18650-cylinder lithium battery cell and selects appropriate parameters to form feature characteristics. Based on the unsupervised learning and self-organization characteristics of SOM neural networks and the dimensionality reduction characteristics of PCA, the multi-parameter sorting of lithium battery combining dynamic and static characteristics was carried out. The results show that sorting significantly improves the SOC consistency of batteries in the module after grouping. The consistency after grouping can be further validated by other methods, and the elimination of inconsistency during the using process is also the author's future research direction. Whether different numbers of batteries will affect the accuracy of sorting results, and whether the sorting method can ultimately achieve the purpose that batteries are more accurately sorted, also remains to be further explored by the author and his team.

Author Contributions: B.X. and Y.Y. proposed the SOM sorting algorithm based on PCA analyzed the data; Y.Y. designed the testing experiment and performed the data processing; J.Z. performed testing and simulation; G.C.

and Y.L. (Yifan Liu) provided experiment guidance; H.W., M.W. and Y.L. (Yongzhi Lai) prepared necessary data and equipment; Y.Y. wrote the paper.

Funding: This research was funded by the Shenzhen Economic, Trade and Information Commission of Shenzhen Municipality Strategic Emerging Industries and Future Industrial Development "Innovation Chain + Industrial Chain" Project (2017) and National Natural Science Foundation of China (Grant No. 51877120).

Conflicts of Interest: The authors declare no conflict of interest.

References

1. Paul, S.; Diegelmann, C.; Kabza, H.; Tillmetz, W. Analysis of ageing inhomogeneities in lithium-ion battery systems. *J. Power Sources* **2013**, *239*, 642–650. [CrossRef]
2. Moore, S.W.; Schneider, P.J. A review of cell equalization methods for lithium ion and lithium Polymer battery systems. In Proceedings of the SAE 2001 World Congress, Detroit, MI, USA, 5–8 March 2001; SAE International: Detroit, MI, USA, 2001.
3. Wei, X.; Zhu, B. The research of vehicle power Li-ion battery pack balancing method. In Proceedings of the International Conference on Electronic Measurement and Instruments (ICEMI), Beijing, China, 16–19 August 2009; pp. 2–498.
4. Li, J.; Wang, Y.; Tan, X. Research on the Classification Method for the Secondary Uses of Retired Lithium-ion Traction Batteries. *Energy Procedia* **2017**, *105*, 2843–2849. [CrossRef]
5. Kim, J.; Cho, B.H. Screening process-based modeling of the multi-cell battery string in series and parallel connections for high accuracy state-of-charge estimation. *Energy* **2013**, *57*, 581–599. [CrossRef]
6. An, F.; Huang, J.; Wang, C.; Li, Z.; Zhang, J.; Wang, S.; Li, P. Cell sorting for parallel lithium-ion battery systems: Evaluation based on an electric circuit model. *J. Energy Storage* **2016**, *6*, 195–203. [CrossRef]
7. Li, X.; Wang, T.; Pei, L.; Zhu, C.; Xu, B. A comparative study of sorting methods for Lithium-ion batteries. In Proceedings of the 2014 IEEE Transportation Electrification Conference and Expo, ITEC Asia-Pacific 2014, Beijing, China, 31 August–3 September 2014; Institute of Electrical and Electronics Engineers Inc.: Beijing, China, 2014.
8. Jong-Hoon, K.; Jong-Won, S.; Chang-Yoon, J.; Bo-Hyung, C. Screening process of Li-Ion series battery pack for improved voltage/SOC balancing. In Proceedings of the 2010 International Power Electronics Conference (IPEC), Sapporo, Japan, 21–24 June 2010; pp. 1174–1179.
9. Hartigan, A.J. *Clustering Algorithms*; John Wiley & Sons: Hoboken, NJ, USA, 1975; Volume 31, p. 793.
10. Miyamoto, S. Data clustering algorithms for information systems. In Proceedings of the 11th International Conference on Rough Sets, Fuzzy Sets, Data Mining, and Granular Computer, RSFDGrC 2007, Toronto, ON, Canada, 14–17 May 2007; Springer: Toronto, ON, Canada, 2007; pp. 13–24.
11. Guo, L.; Liu, G.W. Research of Lithium-ion Battery Sorting Method Based on Fuzzy C-Means Algorithm. In *Progress in Power and Electrical Engineering, Pts 1 and 2*; Zhang, H., Fu, Y., Tang, Z., Eds.; Trans Tech Publications Ltd.: Stafa-Zurich, Switzerland, 2012; Volume 354–355, pp. 983–988.
12. Zhang, R.; Zhou, Y.; Li, R. A Battery Sorting Scheme Based on Fuzzy C-mean Clustering, Taking Advantage of the Flatness of Discharge Voltage Curve. *Qiche Gongcheng Automot. Eng.* **2017**, *39*, 864.
13. Kumar, U.A.; Dhamija, Y. Comparative Analysis of SOM Neural Network with K-means Clustering Algorithm. In Proceedings of the 2010 IEEE International Conference on Management of Innovation & Technology (ICMIT 2010), Singapore, 2–5 June 2010; pp. 55–59.
14. He, F.X.; Shen, W.X.; Song, Q.; Kapoor, A.; Honnery, D.; Dayawansa, D. Self-organising map based classification of LiFePO4 cells for battery pack in EVs. *Int. J. Veh. Des.* **2015**, *69*, 151–167. [CrossRef]
15. Fang, K.Z.; Chen, S.; Mu, D.B.; Wu, B.R.; Wu, F. Investigation of nickel-metal hydride battery sorting based on charging thermal behavior. *J. Power Sources* **2013**, *224*, 120–124. [CrossRef]
16. Raspa, P.; Frinconi, L.; Mancini, A.; Cavalletti, M.; Longhi, S.; Fulimeni, L., Bellesi, P., Isidori, R. Selection of Lithium cells for EV battery pack using Self–Organizing maps. In Proceedings of the 25th World Battery, Hybrid and Fuel Cell Electric Vehicle Symposium and Exhibition: Sustainable Mobility Revolution, EVS 2010, Shenzhen, China, 5–9 November 2010; Electric Drive Transportation Association: Shenzhen, China, 2010.
17. Miyatake, S.; Susuki, Y.; Hikihara, T.; Itoh, S.; Tanaka, K. Discharge characteristics of multicell lithium-ion battery with nonuniform cells. *J. Power Sources* **2013**, *241*, 736–743. [CrossRef]

18. Huang, J.; Li, Z.; Zhang, J.B. Dynamic electrochemical impedance spectroscopy reconstructed from continuous impedance measurement of single frequency during charging/discharging. *J. Power Sources* **2015**, *273*, 1098–1102. [CrossRef]
19. Fang, K.Z.; Mu, D.B.; Chen, S.; Wu, B.R.; Wu, F. A prediction model based on artificial neural network for surface temperature simulation of nickel-metal hydride battery during charging. *J. Power Sources* **2012**, *208*, 378–382. [CrossRef]
20. Wang, H.; Dong, K.; Li, G.; Wu, F. Research on the consistency of the power battery based on multi-points impedance spectrum. In Proceedings of the 2010 International Forum on Strategic Technology, IFOST 2010, Ulsan, Korea, 13–15 October 2010; IEEE Computer Society: Washington, DC, USA, 2010; pp. 1–4.
21. Jolliffe, I. *Principal Component Analysis and Factor Analysis*; Springer: Berlin, Germany, 1986; pp. 115–128.
22. Kohonen, T. *Self-Organizing Maps*, 2nd ed.; Springer: Berlin, Germany, 1995.
23. He, H.W.; Xiong, R.; Zhang, X.W.; Sun, F.C.; Fan, J.X. State-of-Charge Estimation of the Lithium-Ion Battery Using an Adaptive Extended Kalman Filter Based on an Improved Thevenin Model. *IEEE Trans. Veh. Technol.* **2011**, *60*, 1461–1469.
24. Xia, B.Z.; Guo, S.K.; Wang, W.; Lai, Y.Z.; Wang, H.W.; Wang, M.W.; Zheng, W.W. A State of Charge Estimation Method Based on Adaptive Extended Kalman-Particle Filtering for Lithium-ion Batteries. *Energies* **2018**, *11*, 2755. [CrossRef]

Article

Analysis of the Effect of the Variable Charging Current Control Method on Cycle Life of Li-ion Batteries

In-Ho Cho [1,*], Pyeong-Yeon Lee [2] and Jong-Hoon Kim [2,*]

[1] Smart Electrical & Signaling Division, Korea Railroad Research Institute, Uiwang-si 16105, Korea
[2] Department of Electrical Engineering, Chungnam National University, Daejeon 34134, Korea
* Correspondence: inhocho@krri.re.kr (I.-H.C.); whdgns0422@cnu.ac.kr (J.-H.K.); Tel.: +82-31-460-5469 (I.-H.C.); +82-42-821-5657 (J.-H.K.)

Received: 28 May 2019; Accepted: 1 August 2019; Published: 6 August 2019

Abstract: Applications of rechargeable batteries have recently expanded from small information technology (IT) devices to a wide range of other industrial sectors, including vehicles, rolling stocks, and energy storage system (ESS), as a part of efforts to reduce greenhouse gas emissions and enhance convenience. The capacity of rechargeable batteries adopted in individual products is meanwhile increasing and the price of the batteries in such products has become an important factor in determining the product price. In the case of electric vehicles, the price of batteries has increased to more than 40% of the total product cost. In response, various battery management technologies are being studied to increase the service life of products with large-capacity batteries and reduce maintenance costs. In this paper, a charging algorithm to increase the service life of batteries is proposed. The proposed charging algorithm controls charging current in anticipation of heating inside the battery while the battery is being charged. The validity of the proposed charging algorithm is verified through an experiment to compare charging cycles using high-capacity type lithium-ion cells and high-power type lithium-ion cells.

Keywords: battery charging; cycle-life; state-of-health (SOH); battery cycle-life extension

1. Introduction

The trend of wider diffusion of mobile electronics, such as mobile phones and laptops, and the increasing demand for high performance devices have led to growth of both the battery industry and the information technology (IT) industry. Consumers need mobile devices that can be used for long periods of time on a single charge, even with short battery charging time. This has motivated research on rapid charging technology and improving the energy density of batteries. Furthermore, the growth of the electric vehicle (EV) market has given rise to the new issue in the battery industry of managing the battery's state-of-health (SOH). Unlike small mobile devices, which are often replaced within two to three years, EVs are relatively expensive products and their batteries, which generally have a life expectancy of more than eight years, account for the largest cost among all parts. As a result, batteries in the automotive industry must be capable of maintaining battery capacity for a considerable period of time, and many studies have been conducted to manage the battery's SOH.

The cycle life of batteries that generate electricity through chemical reactions is determined by various external factors [1–8]. First, battery charging operation in low temperature conditions reduces the chemical reaction rate of lithium ions. This causes the accumulation of lithium-ion metals in the anode layer, which reduces the capacity and in turn is directly linked to the battery's lifetime [1]. Second, high C-rate charging and high depth of discharge (DOD) range result in loss of active material and the formation of a thick layer of solid electrolyte interphase (SEI) at the surface of electrodes.

This increases internal impedance and reduces the capacity of the battery [2,3]. Third, overcharge battery operation causes unwanted heating inside the battery. This results in cracking of the SEI and loss of active area inside the battery, which reduces the total capacity of the battery [4,5]. Fourth, over-discharge of batteries also reduces the number of ions participating in electrochemical reaction, which causes capacity loss [4]. Fifth, battery cycle life can be shortened by charging/discharging operation at high temperatures. Electrolytes and the binders of the battery are destroyed during the charging/discharging operation at high temperatures and this reduces the capacity of the batteries [6,7]. In addition to the aforementioned reasons, the battery SOH is affected by various factors and many of these factors are related to the charging conditions. Thus, the aging of batteries can be mitigated by developing the charging algorithm.

Recently, various battery charging algorithms have been investigated to extend the battery cycle life and reduce restrictions on battery use [7–19]. Most of these studies focused on developing charging algorithms and profiles to reduce the battery charging time by adopting a high C-rate current. One of the most common fast charging algorithms is the CC-CV, constant current-constant voltage, charging method [8,9]. Batteries are charged with constant current until the maximum acceptable battery voltage is reached; after that charging is continued with a fixed charging voltage, and charging is completed when the charging current reaches a preset small value. The multi-stage constant current (MCC) charging method is another well-known fast charging method. Unlike the constant-current charging method, charging current is divided into several levels in the MCC method to reduce the charging time and heat generated inside the battery during charging [8,13]. Generally, the charging current is controlled in a direction where the size of the charging current decreases as the charging time progresses. The constant power (CP) charging method maintains charging power during the charging operation [7]. High charging current is induced at the beginning of charging and it is gradually decreased as the battery voltage increases. The boost charging method is also one of the basic fast charging methods based on the CC-CV charging profile [8,17]. It adopts a high current charging period at the beginning of the CC-CV charging profile to reduce the charging time. Figure 1 shows the various battery charging profiles.

Although various fast charging technologies are being studied through many different approaches to reduce the charging time while minimizing capacity loss, they cannot reflect the characteristics of changes in internal impedance, which depends on changes in state-of-charge (SOC) or SOH. As a result, the battery's life expectancy, which is determined by the available power rating and total remaining capacity, is rapidly deteriorated with use of the fast charging method. Therefore, the conventional CC-CV charging profiles are still widely utilized in applications where the battery price is a significant factor in relation to the overall price and/or the battery's characteristics have a significant impact on the performance of the product.

In this paper, the aging characteristics of the battery with use of the CC-CV method are analyzed and a new charging method that minimizes battery degradation is proposed. The proposed method controls the charging current by considering the difference in internal impedance caused by SOC changes in order to minimize the heating of the battery during charging. To verify the effectiveness of the proposed charging method, the battery aging characteristics when the proposed charging method is applied are compared with those observed with use of the CC-CV method.

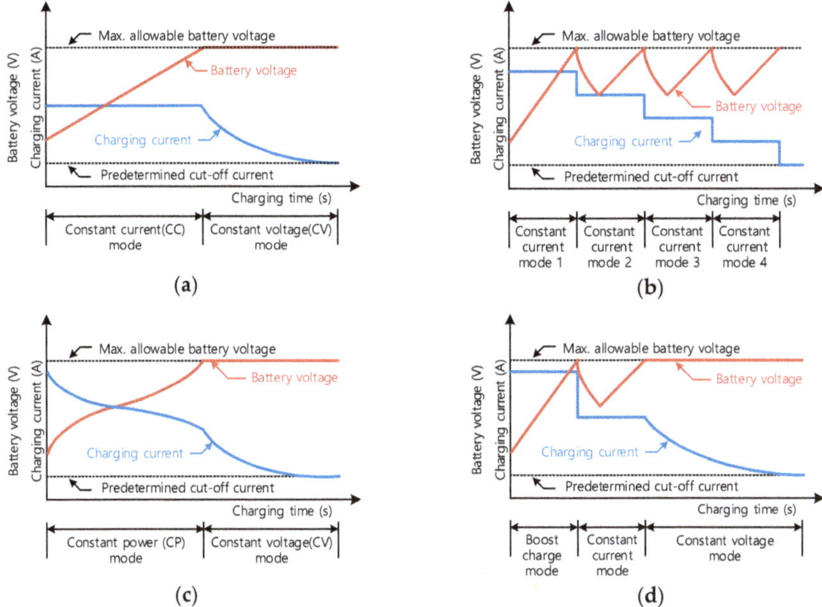

Figure 1. Battery charging profiles: (**a**) Constant current-constant voltage (CC-CV) charging method; (**b**) Multi-stage constant current (MCC) charging method; (**c**) Constant power (CP) charging method; (**d**) Boost charging method.

2. Equivalent Circuit Model for SOH Estimation

Various physical and chemical changes occur within the battery as battery cells age [20–25]. Physically, the structure of electrodes is changed and this reduces the capacity of the battery; chemically, the loss of active material causes capacity reduction, and unwanted chemical reactions within the electrolyte prevent circulation of internal active ions. These aging characteristics typically result in a change in electrical properties, such as reduced capacity of the battery or increased internal resistance ($R_{int.}$), which in turn undermines the output power characteristics of the battery.

Figure 2 shows the conventional equivalent circuit model (ECM), which is widely used in the electrical analysis of batteries. SOH estimation using the ECM offers the advantage of a relatively simple calculation while expressing the change in characteristics of the battery. The characteristics of batteries are explained in the ECM by the open circuit voltage (OCV), which represents the battery capacity, the internal ohmic resistance (R_i), which denotes the resistance of the electrolyte, and the internal diffusion resistance ($R_{diff.}$)–capacitance ($C_{diff.}$) network, which represents the charge diffusion/transfer characteristics. The applied voltages to R_i and $R_{diff.}$-$C_{diff.}$ network are represented as V_{ohmic} and $V_{diff.}$, respectively. V_{BATT} in Figure 2 shows the voltage across the terminals of a battery. The characteristics of batteries that vary with the SOC and SOH can be expressed through changes in the internal resistance, which is determined by the sum of R_i and $R_{diff.}$, and capacitance values in the ECM [20]. Therefore, for accurate prediction of the operating characteristics and SOH of batteries using the equivalent circuit model, it is necessary to clearly identify the variation in the internal resistance parameters according to the condition of the battery and to measure the parameters of the equivalent circuit periodically. Figure 3 illustrates the hybrid pulse power characterization (HPPC) method for measuring the resistances in the ECM. The HPPC method obtains the internal parameters of the ECM by monitoring the change in the internal voltage of the battery while varying the charge current and discharge current within the range of operation of the battery [26,27].

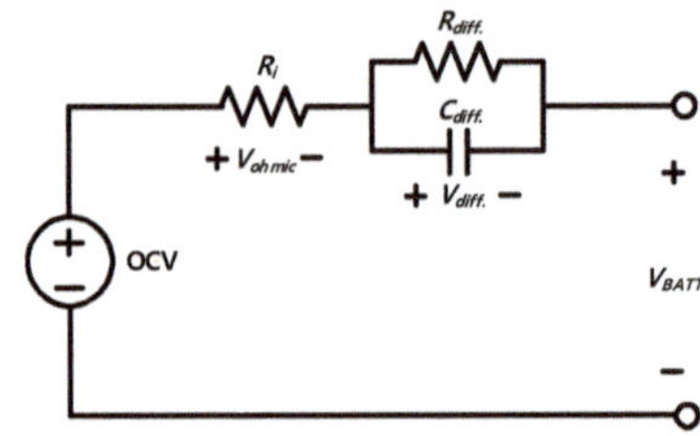

Figure 2. Configuration of the equivalent circuit model.

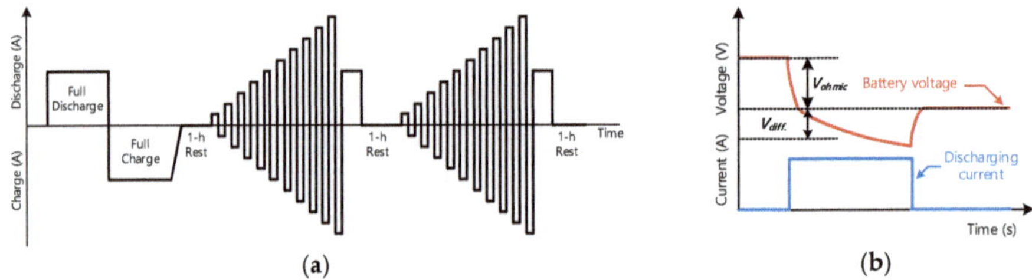

Figure 3. Hybrid pulse power characterization (HPPC) parameter calculation method: (**a**) Charging-discharging profile; (**b**) Parameters estimation method.

3. Effect of Internal Heat on SOH of Batteries

A number of factors affect battery aging including temperature inside/outside the battery, SOC, depth of discharge (DOD), etc., [4]. Among these, temperature, which has a direct effect on the rate of chemical reaction of batteries, is one of the most influential factors in the aging of batteries. According to the Arrhenius Equation (1), which describes a chemical reaction, the rate of chemical reaction and temperature have a linear relationship of logarithmic scale. Therefore, in the case of batteries, which charge and discharge energy through chemical reactions, temperature management is considered to have a significant effect on battery aging.

$$r = A \times e^{\left(\frac{E_a}{kT}\right)}, \tag{1}$$

where r is the reaction rate, k is Boltzmann's constant, A is the frequency factor, E_a is the activation energy, and T is the absolute temperature.

4. Proposed Charging Method for Minimizing Battery Degradation

As mentioned earlier, the life cycle estimation method using equivalent circuit models is widely employed in many applications due to its advantages of being simple and quite accurate in calculation [28]. The internal resistances in the equivalent circuit are measured periodically using the HPPC test method and the measured values of an aged battery cell are compared with those of a fresh cell, and the differences in the parameter values are used to calculate the life of the battery. This paper proposes a variable battery charging current algorithm (VCC) that controls the charging current in a way that minimizes heating inside the battery when charging. This is done using the characteristics of internal resistance change in batteries, which depend on the SOC and the relationship between heating and aging of the battery.

Figure 4 shows the measured internal resistance values of batteries in accordance with the SOC at intervals of 10 cycles using INR 18650 25R cells of Samsung SDI (Yongin-si, Korea). As shown in Figure 4, the internal resistance of a battery varies significantly with the SOC and the life cycle. In particular, the Figure 4 shows that changes in internal resistance do not have a linear relationship with changes in the SOC. The internal resistance shows a proportional decrease in the resistance to the SOC in the range where the SOC decreases from 100% to 40%, whereas below 40% it shows an increase in resistance as opposed to SOC reduction, as can be seen in Figure 4. Given that heating inside the battery is ultimately determined by the value of the current flowing into the resistor, it can be expected that the temperature of the battery will continue to change during the charging process and heat generation will increase in the low and high SOC range, if the CC-CV method, which is the most popular battery charging algorithm, is applied. Therefore, this paper proposes a variable charging current algorithm that changes the amount of charging current according to the SOC, taking into account the internal resistance of the battery in order to slow the aging of battery cells according to the charging power. The proposed charging algorithm is a method to minimize internal thermal variation of the battery in the charging process and thereby prevent fluctuation of the charging power loss (P_{Loss}), which directly affects the heating of the battery while the battery is being charged. Since the internal resistance of a battery generally shows characteristics that vary with the SOC condition, a method of controlling the charging current of the battery according to the changing resistance value is required, and the charging current value (I_{Charge}) is achieved using Equation (2) in the present study. To verify the effectiveness of the proposed charging algorithm, two types of batteries, high-capacity lithium-ion cells (INR 25R) and high-power lithium-ion cells (INR 29E), were compared with use of the conventional CC-CV charging method and the proposed VCC method.

$$P_{Loss}(W) = {I_{Charge}}^2 \times R_{int.},\ I_{Charge}(\mathrm{A}) = \sqrt{P_{Loss}/R_{int.}} \quad (2)$$

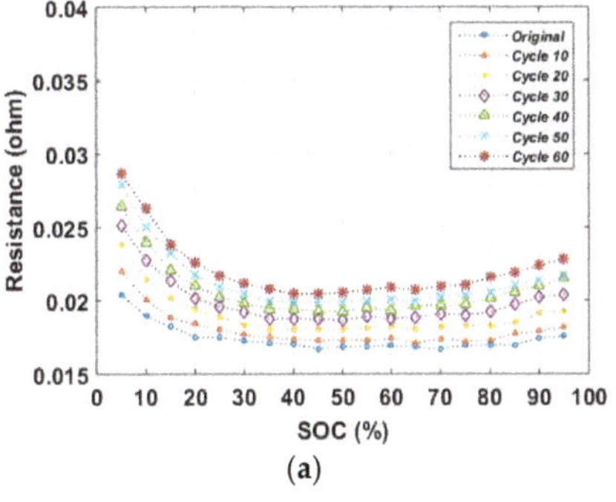

(a)

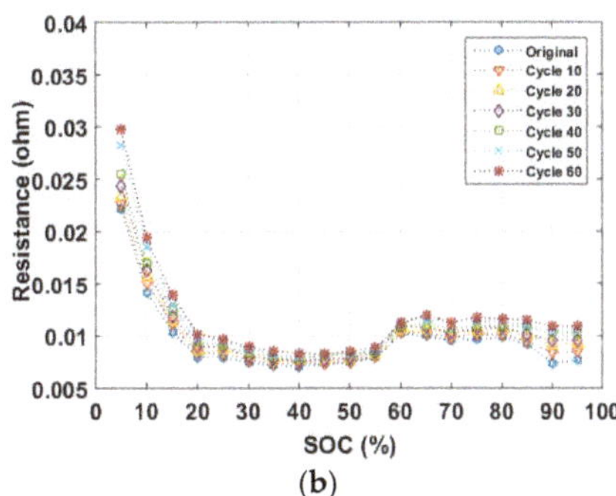

(b)

Figure 4. Measured internal resistance of INR 25R battery as it aged: (**a**) Internal ohmic resistance, R_i; (**b**) Internal diffusion resistance, R_{diff}.

5. Experimental Configuration of the Proposed Algorithm

Two different types of batteries were used in the experiment: a battery with a high output power characteristic, INR 25R, which is composed of nickel cobalt aluminum oxide (NCA) and a battery with a high energy density characteristic, INR 29E, which is composed of nickel manganese cobalt oxide (NMC). Table 1 presents detailed specifications of each battery. The experiment was conducted in a temperature chamber set to be maintained at a temperature of 25 °C and the experiment proceeded as follows.

Table 1. Specifications of batteries used in the experiment.

Characteristics	INR 25R	INR 29E
Rated voltage	3.60 V	3.65 V
Maximum continuous discharge current	20 A	2.750 A
Rated capacity	2500 mAh	2750 mAh
Cut-off voltage	4.20 V (Charge) 2.50 V (Discharge)	4.20 V (Charge) 2.50 V (Discharge)
Materials	C (Negative) $LiCoNiAlO_2$ (NCA)	C (Negative) $LiNiMnCoO_2$ (NMC)

Step 1: Measure the exact capacity of each battery.

To measure the capacity of the battery, first it was fully charged up to 4.2 V using the CC-CV charging method using 0.5 C charging current. A 0.5 C discharge current was then applied to discharge the battery to the cut-off voltage and the capacity of the battery was calculated using the discharge time. After each capacity calculation process, one hour of rest time was given to the battery to complete the internal chemistry reaction.

Step 2: Measure the internal resistance of the battery.

The HPPC test method was applied to measure the internal resistance of the battery using 0.5 C charging/discharging current. The resistance was measured by lowering the SOC by increments of 5% from 100%. Figure 5 shows the experimental procedure used to measure the internal resistance.

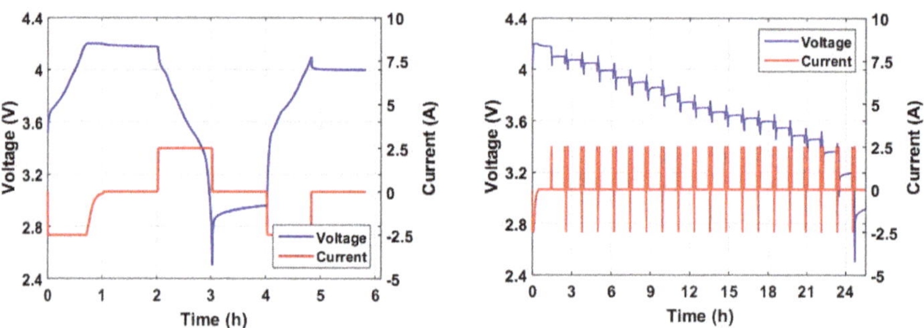

Figure 5. Experimental procedure used to measure the internal resistance.

Step 3: Derive the charging current.

Charging current values for each SOC were derived in this step using the resistance values obtained from the previous step. In order to maintain a constant amount of heat inside the battery generated during the charging process, a constant value of charging loss was set and the charging current was calculated. Considering the tendency of the internal resistance of the battery to increase as the SOC becomes lower, the loss value when 0.5 C charging is applied was taken as the reference value at 50% of the SOC. Table 2 shows the charging current and internal resistance at each SOC. Only the DC internal resistance (DCIR), which indicates the effect of DC current on the battery is considered in calculating the charging current. The resistance is widely used value when comparing the performance of the battery and also has the characteristics of large variation on the SOC [29–31].

Table 2. Measured internal resistance and calculated charging current in the experiment.

SOC (%)	INR 25R (NCA)		INR 29E (NMC)	
	DCIR (ohm)	Current (A)	DCIR (ohm)	Current (A)
100	0.0252	1.2258	0.0349	1.3991
95	0.0252	1.2327	0.0349	1.3991
90	0.0248	1.2431	0.0349	1.3988
85	0.0261	1.2111	0.0364	1.3693
80	0.0268	1.1962	0.0380	1.3412
75	0.0265	1.2015	0.0384	1.3334
70	0.0262	1.2084	0.0377	1.3468
65	0.0268	1.1963	0.0379	1.3429
60	0.0271	1.1893	0.0388	1.3276
55	0.0248	1.2433	0.0384	1.3334
50	0.0242	1.2570	0.0356	1.3861
45	0.0240	1.2632	0.0349	1.3991
40	0.0241	1.2619	0.0344	1.4101
35	0.0243	1.2555	0.0343	1.4102
30	0.0247	1.2464	0.0348	1.4010
25	0.0254	1.2269	0.0351	1.3947
20	0.0255	1.2268	0.0356	1.3859
15	0.0285	1.1589	0.0367	1.3649
10	0.0331	1.0757	0.0400	1.3073
5	0.0425	0.9492	0.0473	1.2012
0	0.0425	0.9439	0.0473	1.2012

Step 4: Aging test using the VCC method.

The battery aging experiment was carried out by repeating the charging cycle using the charging current value obtained in Step 3. The SOH of the battery was checked by measuring the internal resistance of the battery at every 10 charging–discharging cycles.

To compare the aging of batteries according to the charging method, the degree of aging of the batteries with application of the existing CC-CV charging method was measured using the same two types of batteries as applied during the VCC algorithm aging test. The battery aging condition was compared after 60 cycles of the charging test were carried out in different ways to compare the degree of aging according to the charging algorithm.

6. Experimental Results

Figure 6 shows the results of measuring the internal resistance variation of the battery by the SOC according to each charging method by applying the proposed VCC method and the conventional CC-CV charging method. The battery charging with the VCC method shows less variation in internal resistance, and this is more visible in high-power battery cells. The difference in the increase in resistance of a battery resulting from the batteries having different charging algorithms can be seen, similarly to the tendency of capacity reduction due to the aging of the battery. Figure 7 shows the variation of the capacity of each battery as the charging cycle repeats and Table 3 shows the comparison result of the capacity degradation rate on each method. As shown in this Figure 7, there is less capacity change in the battery cell with the proposed charging algorithm. Figure 8 shows the results of the change in charging time following the aging test. In the proposed method, a resistance value at a 50% SOC, which has the lowest internal resistance, and a current value of 0.5 C were used to obtain the charging power loss, and the power loss value is used to calculate the charging current due to the change in resistance. As a result, as the test progresses, the difference in charging time was reduced or reversed depending on the battery type because the charging time is longer with the proposed charging method than with the conventional CC-CV charging method at the beginning of the charging test, but with aging this difference becomes smaller. This is ascribed to a slower aging process and smaller internal resistance characteristics in the proposed charging algorithm when compared to the conventional method of charging [32].

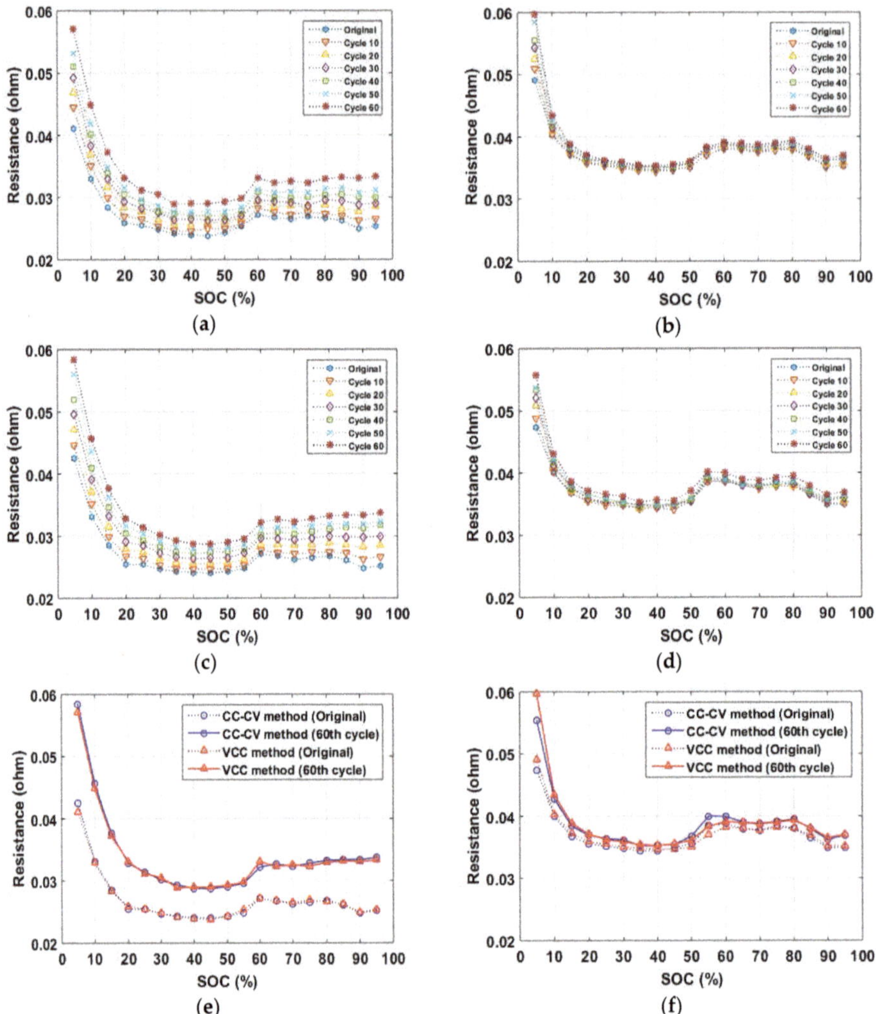

Figure 6. Hybrid pulse power characterization (HPPC) parameter calculation method: (**a**) DCIR of cell with NCA material (CC-CV); (**b**) DCIR of cell with NMC material (CC-CV); (**c**) DCIR of cell with NCA material (VCC); (**d**) DCIR of cell with NMC material (VCC); (**e**) comparison of internal resistance (NCA); (**f**) comparison of internal resistance (NMC).

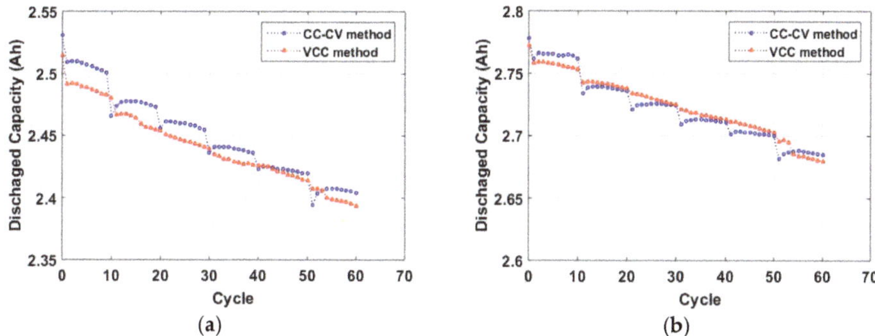

Figure 7. The discharging capacity by cycle number: (**a**) cell with NCA material; (**b**) cell with NMC material.

Table 3. Comparison of the capacity degradation rate.

Battery Types and Cycles		CC-CV Method		VCC Method	
		Capacity (Ah)	Growth Rate (%)	Capacity (Ah)	Growth Rate (%)
25R (NCA)	Original 60th cycle	2.5303 2.4039	−4.993	2.5143 2.3935	−4.804
29E (NMC)	Original 60th cycle	2.7781 2.6850	−3.353	2.7722 2.6798	−3.334

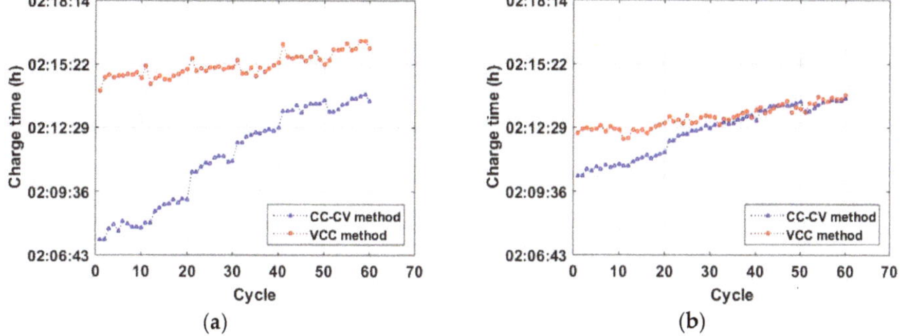

Figure 8. The charge time by cycle number: (**a**) cell with NCA material; (**b**) cell with NMC material.

7. Conclusion

This paper proposes a variable charging current charging algorithm to minimize the aging characteristics of batteries generated by repeated charging/discharging operation. The proposed method is to manage heat inside the battery that occurs when charging the battery to minimize the battery aging caused by repeated use of the battery. For heating management of batteries, the charging current is varied at each SOC level depending on the internal resistance value of the battery to maintain the power loss inside the battery. The effectiveness of the proposed charging algorithm was then compared with that of the conventional CC-CV charging algorithm. By applying the proposed algorithm, a battery with repeated charging would have a slower aging effect compared to a battery with the conventional charging method. Also, the effects of the proposed algorithm vary depending on the chemical composition of the battery and the proposed algorithm was found to be more effective in high-power battery cells. Based on the results of this experiment, the variable charging current battery charging algorithm proposed in this paper is an effective charging method that slows down the aging of the battery, and is expected to achieve better results in applications where the product has a long life cycle and battery replacement is difficult. Further research to compare the results of battery aging through long-term charging cycle testing and to adjust charging current to reflect the changes in capacity due to aging of batteries should be carried out.

Author Contributions: Conceptualization, I.-H.C. and J.-H.K.; methodology, I.-H.C.; validation, I.-H.C., P.-Y.L. and J.-H.K.; formal analysis, I.-H.C., P.-Y.L. and J.-H.K.; investigation, I.-H.C., P.-Y.L. and J.-H.K.; data curation, P.-Y.L.; writing—original draft preparation, I.-H.C.; writing—review and editing, I.-H.C. and J.-H.K.

Funding: This research was supported by a grant from the R&D Program (No. PK1904B1C) of the Korea Railroad Research Institute, Republic of Korea.

Conflicts of Interest: The authors declare no conflict of interest.

References

1. Petzi, M.; Kasper, M.; Danzer, M.A. Lithium plating in a commercial lithium-ion battery—A low-temperature aging study. *J. Power Sources* **2015**, *275*, 799–807. [CrossRef]
2. Vetter, J.; Novak, P.; Wagner, M.R.; Veit, C.; Moller, K.C.; Besenhard, J.O.; Winter, M.; Wohlfahrt-Mehrensc, M.; Voglerc, C.; Hammouched, A.; et al. Ageing mechanisms in lithium-ion batteries. *J. Power Sources* **2005**, *147*, 269–281. [CrossRef]
3. Lee, J.H.; Lee, H.M.; Ahn, S. Battery dimensional changes occurring during charge/discharge cycles—Thin rectangular lithium ion and polymer cells. *J. Power Sources* **2003**, *119–121*, 833–837. [CrossRef]
4. Wu, C.; Zhu, C.; Ge, Y.; Zhao, Y. A review on fault mechanism and diagnosis approach for Li-ion batteries. *J. Nanomater.* **2015**, *2015*, 8. [CrossRef]
5. Belov, D.; Yang, M.H. Failure Mechanism of Li-ion Battery at Overcharge Condition. *J. Solid State Electrochem.* **2008**, *12*, 885–894. [CrossRef]
6. Timmermans, J.M.; Nikolian, A.; De Hoog, J.; Gopalakrishnan, R.; Goutam, S.; Omar, N.; Coosemans, T.; Van Mierlo, J.; Warnecke, A.; Sauer, D.U.; et al. Batteries 2020—Lithium-ion battery first and second life ageing, validated battery models, lifetime modelling and ageing assessment of thermal parameters. In Proceedings of the 2016 18th European Conference on Power Electronics and Applications, Karlsruhe, Germany, 5–9 September 2016.
7. Keil, P.; Jossen, A. Charging protocols for lithium-ion batteries and their impact on cycle life—An experimental study with different 18650 high-power cells. *J. Energy Storage* **2016**, *6*, 125–141. [CrossRef]
8. Zhang, S.S. The effect of the charging protocol on the cycle life of a Li-ion battery. *J. Power Sources* **2006**, *161*, 1385–1391. [CrossRef]
9. Guo, Z.; Liaw, B.Y.; Qiu, X.; Gao, L.; Zhang, C. Optimal charging method for lithium ion batteries using a universal voltage protocol accommodating aging. *J. Power Sources* **2015**, *274*, 957–964. [CrossRef]
10. Zheng, J.; Engelhard, M.H.; Mei, D.; Jiao, S.; Polzin, B.J.; Zhang, J.G.; Xu, W. Electrolyte additive enabled fast charging and stable cycling lithium metal batteries. *Nat. Energy* **2017**, *2*, 17012. [CrossRef]
11. Liu, Y.H.; Luo, Y.F. Search for an optimal rapid-charging pattern for li-ion batteries using the Taguchi approach. *IEEE Trans. Ind. Electron.* **2010**, *57*, 3963–3971. [CrossRef]
12. Botsford, C.; Szczepanek, A. Fast Charging vs. Slow Charging: Pros and cons for the New Age of ElectricVehicles. In Proceedings of the Evs24 International Battery, Hybrid and Fuel Cell Electric Vehicle Symposium, Stavanger, Norway, 13–16 May 2009.
13. Ikeya, T.; Sawada, N.; Murakami, J.I.; Kobayashi, K.; Hattori, M.; Murotani, N.; Ujiie, S.; Kajiyama, K.; Nasu, H.; Narisoko, H.; et al. Multi-step constant-current charging method for an electric vehicle nickel/metal hydride battery with high-energy efficiency and long cycle life. *J. Power Sources* **2002**, *105*, 6–12. [CrossRef]
14. Gol, E.; Heidary, M.; Kojabadi, H.M. Comparison of Different Battery Charging Methods for Sustainable Energy Storing Systems. In Proceedings of the 26th International Power Systems Conference, Tehran, Iran, 31 October 2011.
15. Min, H.; Sun, W.; Li, X.; Guo, D.; Yu, Y.; Zhu, T.; Zhao, Z. Research on the optimal charging strategy for Li-ion batteries based on multi-objective optimization. *Energies* **2017**, *10*, 709. [CrossRef]

16. Yilmaz, M.; Krein, P.T. Review of charging power levels and infrastructure for plug-in electric and hybrid vehicles. In Proceedings of the 2012 IEEE International Electric Vehicle Conference, Greenville, SC, USA, 4–8 March 2012.
17. Shen, W.; Vo, T.T.; Kapoor, A. Charging algorithms of lithium-ion batteries: An overview. In Proceedings of the 2012 7th IEEE Conference on Industrial Electronics and Applications, Singapore, 18–20 July 2012.
18. Shirk, M.; Wishart, J. Effects of Electric Vehicle Fast Charging on Battery Life and Vehicle Performance. *SAE Tech. Paper 2015-01-1190* **2015**. [CrossRef]
19. Kettles, D. *Electric Vehicle Charging Technology Analysis and Standards*; FSEC-CR-1996-15; Florida Solar Energy Center: Cocoa, FL, USA, February 2015.
20. Wang, D.; Bao, Y.; Shi, J. Online Lithium-Ion Battery Internal Resistance Measurement Applicaton in State-of-Charge Estimation Using the Extended Kalman Filter. *Energies* **2017**, *10*, 1284. [CrossRef]
21. Ecker, M.; Gerschler, J.B.; Vogel, J.; Käbitz, S.; Hust, F.; Dechent, P.; Sauer, D.U. Development of a lifetime prediction model for lithium-ion batteries based on extended accelerated aging test data. *J. Power Sources* **2012**, *215*, 248–257. [CrossRef]
22. Fleischhammer, M.; Waldmann, T.; Bisle, G.; Hogg, B.I.; Wohlfahrt-Mehrens, M. Interaction of cyclic ageing at high-rate and low temperatures and safety in lithium-ion batteries. *J. Power Sources* **2015**, *274*, 432–439. [CrossRef]
23. Liaw, B.Y.; Roth, E.P.; Jungst, R.G.; Nagasubramanian, G.; Case, H.L.; Doughty, D.H. Correlation of Arrhenius behaviors on power and capacity fades, impedance, and static heat generation in lithium ion cells. *J. Power Sources* **2003**, *119*, 874–886. [CrossRef]
24. Onda, K.; Ohshima, T.; Nakayama, M.; Fukuda, K.; Araki, T. Thermal behavior of small lithium-ion battery during rapid charge and discharge cycles. *J. Power Sources* **2006**, *158*, 535–542. [CrossRef]
25. Abdollahi, A.; Han, X.; Avvari, G.V.; Raghunathan, N.; Balasingam, B.; Pattipati, K.R.; Bar-Shalom, Y. Optimal battery charging, Part 1: Minimizing time-to-charge, energy loss, and temperature rise for OCV-resistance battery model. *J. Power Sources* **2016**, *303*, 388–398. [CrossRef]
26. FreedomCAR Battery Test Manual for Power-Assist Hybrid Electric Vehicles. Available online: https://avt.inl.gov/sites/default/files/pdf/battery/freedomcar_manual_04_15_03.pdf (accessed on 22 June 2019).
27. Seaman, A.; Dao, T.-S.; McPhee, J. A survey of mathematics-based equivalent-circuit and electrochemical battery models for hybrid and electric vehicle simulation. *J. Power Sources* **2014**, *256*, 410–423. [CrossRef]
28. Lai, X.; Zheng, Y.; Sun, T. A comparative study of different equivalent circuit models for estimating state-of-charge of lithium-ion batteries. *Electrochim. Acta* **2017**, *259*, 566–577. [CrossRef]
29. Qian, K.; Huang, B.; Ran, A.; He, Y.B.; Li, B.; Kang, F. State of health (SOH) evaluation on lithium ion battery by simulating the voltage relaxation curves. *Electrochim. Acta* **2019**, *303*, 183–191. [CrossRef]
30. Kwon, S.J.; Lee, S.E.; Lim, J.H.; Choi, J.H.; Kim, J.H. Performance and Life Degradation Characteristics Analysis of NCM LIB for BESS. *Electronics* **2018**, *7*, 406. [CrossRef]
31. Lu, Z.; Yu, X.L.; Wei, L.C.; Cao, F.; Zhang, L.Y.; Meng, X.Z.; Jin, L.W. A comprehensive experimental study on temperature-dependent performance of lithium ion battery. *Appl. Therm. Eng.* **2019**, *158*, 113800. [CrossRef]
32. Yang, J.; Xia, B.; Huang, W.; Fu, Y.; Mi, C. Online state of health estimation for lithium ion batteries using constant voltage charging current analysis. *Appl. Energy* **2018**, *212*, 1589–1600. [CrossRef]

Article

A Nonlinear-Model-Based Observer for a State-of-Charge Estimation of a Lithium-ion Battery in Electric Vehicles

Woo-Yong Kim [1], Pyeong-Yeon Lee [2], Jonghoon Kim [2,*] and Kyung-Soo Kim [1,*]

1 Department of Mechanical Engineering, Korea Advanced Institute of Science and Technology, Daejeon 291, Korea
2 Department of Electric Engineering, Chungnam National University, Daejeon 99, Korea
* Correspondence: whdgns0422@cnu.ac.kr (J.K.); kyungsookim@kaist.ac.kr (K.-S.K.)

Received: 3 June 2019; Accepted: 28 August 2019; Published: 2 September 2019

Abstract: This paper presents a nonlinear-model-based observer for the state of charge estimation of a lithium-ion battery cell that always exhibits a nonlinear relationship between the state of charge and the open-circuit voltage. The proposed nonlinear model for the battery cell and its observer can estimate the state of charge without the linearization technique commonly adopted by previous studies. The proposed method has the following advantages: (1) The observability condition of the proposed nonlinear-model-based observer is derived regardless of the shape of the open circuit voltage curve, and (2) because the terminal voltage is contained in the state vector, the proposed model and its observer are insensitive to sensor noise. A series of experiments using an INR 18650 25R battery cell are performed, and it is shown that the proposed method produces convincing results for the state of charge estimation compared to conventional SOC estimation methods.

Keywords: nonlinear battery model; state of charge estimation; lithium-ion battery; Lipschitz nonlinear system; Luenberger observer

1. Introduction

Since the first development of hybrid electrical vehicles (HEVs), pure electric vehicles (EVs) have been rapidly commercialized. In contrast to HEVs, the mileage range of EVs is directly affected by the power and energy density of the battery itself and the performance of the battery management system (BMS). Hence, many studies related to lithium-ion (Li-ion) batteries, including the development of new materials and algorithms for inner state estimation, have been conducted by various research groups [1–8]. Increasing the energy and power density via advancements in battery manufacturing technology requires a higher level of monitoring of the battery states to fully and safely use the potential of the battery.

In EV applications, the state of charge (SOC), which represents the amount of charge in the battery, is the most important parameter because it directly relates to the number of miles that an EV can travel. An inaccurate SOC information causes the driver to constantly worried about the EV stopping on the road or the battery being overcharged/overdischarged, causing ignition or explosion. Therefore, it is important to estimate the SOC and accurately determine the dischargeable capacity of the battery to protect the battery itself and help reduce the driver's anxiety [9–11]. However, unlike the voltage and current, there is no way to measure the SOC directly. Hence, advanced algorithms for accurate SOC estimation need to be researched.

Typically, there are two kinds of categories of SOC estimation methods: (1) model-less and (2) model-based methods. The most famous example of a model-less algorithm is the Coulomb counting method [12,13]. This method estimates the SOC by integrating the current through the battery.

Its simplicity and low computational cost make this method valuable in the infancy of the BMS. However, it has obvious limitations: it suffers from an initial condition problem and the accumulation of sensor offset due to the integrator. Artificial neural networks (ANNs) and fuzzy algorithms are also model-less methods that implement intelligent algorithms [14–16]. However, these data-driven methods have inherent problems, such as a long training time and a large number of data sets. In particular, when the type of battery cell is changed, the learning procedure has to be restarted. This is far from a practical concept. On the other hand, model-based methods use an equivalent electrochemical model (EECM) or electrical circuit model (ECM) to represent the current-voltage relationship of the battery. The EECM formulates the key behaviors of the battery cell by deriving a series of differential equations for the chemical reactions inside the battery cell [17,18]. The accuracy of the EECM for SOC estimation is very high, but its practical usefulness is questionable, because a very high complexity leads to a significant memory and computational burden. On the other hand, the ECM represents the current-voltage relationship by using electric components such as resistance, capacitance and a variable voltage source [19–22]. Although the ECM is relatively inaccurate compared with the EECM, the ECM is commonly adopted for real-time SOC estimation because it can be simply implemented and can achieve a high accuracy when the ECM cooperates with a state observer. Therefore, there have been many studies on various state observers and various kinds of ECMs.

The Kalman-filter-based observer, Luenberger observer, sliding-mode observer and proportional-integral observer are widely used for the SOC estimation [23–35]. Previous studies have verified that all methods produce good performance for the online SOC estimation. However, while the battery is a type of nonlinear system due to the nonlinear relationship between the SOC and open circuit voltage (OCV), previous studies focused on linear systems and their observers. Therefore, linearization techniques must be implemented. For example, Kalman-filter-based approaches apply a Taylor series expansion to each operating point at each time step, and the other approaches apply the 'piece-wise' linearization technique, which divides the nonlinear function into multiple linear functions according to each operating region. Linearization techniques are useful for approximating a nonlinear system, but when the operating point changes, the model of the linearized system changes. This means that the performance of the designed observer based on a linearized model at a certain operating point changes with respect to the operating region, and even worse, there can be a critical point where the observer loses its stability. However, most previous studies did not consider time-varying conditions.

This paper proposes a nonlinear model for a battery cell and a nonlinear-model-based observer. This work has two contributions. First, this paper proposes a nonlinear state space representation of a 1st-order Thevenin equivalent circuit model. This allows the system to be time-invariant and the eigenvalues of the designed observer to be fixed in all operating regions. The resulting observability condition of the proposed model, which is generally used as the necessary condition for the design of an observer, is derived regardless of the shape of the open-circuit voltage curve. This means that the observability condition is always satisfied even if there is a voltage plateau on the open-circuit voltage curve [36–38]. The proposed nonlinear model is also insensitive to sensor noise, because the state vector contains the terminal voltage. Second, a nonlinear-model-based Luenberger observer that can address the nonlinear system model is proposed. The stability condition of the proposed observer is strictly derived using nonlinear system theories. The performance of the real-time SOC estimation of the proposed method is evaluated by conducting experiments with INR 18650 25R from SAMSUNG SDI.

2. Nonlinear System Model for a Single Battery Cell

There are many ECMs for battery cells. For an onboard BMS system, there is always a trade-off between the model accuracy and complexity. Therefore, a suitable selection for the ECM must be made according to the application. Generally, for real-time SOC estimation, the 1st-order Thevenin ECM [5] is adopted because it is more suitable for real-time SOC estimation (see Figure 1).

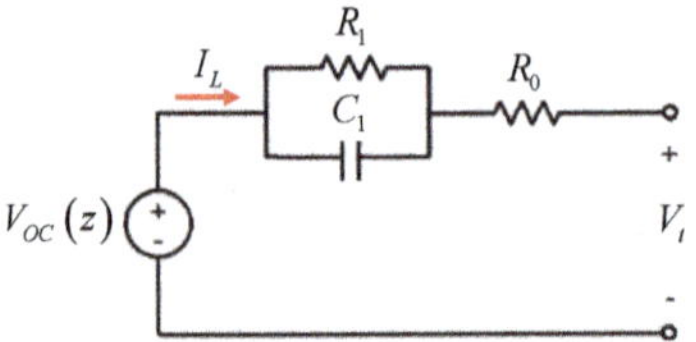

Figure 1. Thevenin's equivalent circuit model with a single RC pair.

The dynamic equations of the selected ECM are derived as follows:

$$V_t = V_{OC}(z) - I_L R_0 - V_1, \tag{1}$$

$$\dot{z} = -\frac{I_L}{C_n}, \tag{2}$$

$$\dot{V}_1 = -\frac{1}{C_1 R_1} V_1 + \frac{1}{C_1} I_L \tag{3}$$

where V_t is the terminal voltage of the battery cell, $V_{OC}(z)$ is the OCV function of z, z represents the SOC, I_L is the load current, R_0 is the equivalent internal resistance, V_1 is the voltage of the RC pair, C_n is the nominal capacity, R_1 is the equivalent resistance of the RC pair and C_1 is the equivalent capacitance of the RC pair. In the case of a Li-ion battery, the nonlinear function $V_{OC}(z)$ representing the relationship between the SOC and OCV always exists. This makes it difficult to build a state space model and design a state observer for the battery system.

2.1. Linearized System Model for a Single Battery Cell

Most previous studies related to observer-based SOC estimation [29,30,32,34,39] linearized $V_{OC}(z)$ by using a piece-wise assumption. The linearized $V_{OC}(z)$ is defined as

$$V_{OC}(z) = k_i z + d_i\,, \quad for\, the\, i^{th}\, SOC\, region \tag{4}$$

where k_i and d_i are the coefficients of each linearized $V_{OC}(z)$ and i is the number of divided sections of the SOC. (1)–(3) can be rewritten as a linear state space representation by using (4).

$$\begin{aligned} \dot{x} &= Ax + Bu, \\ y_i &= C_i x + Du \end{aligned} \tag{5}$$

where $x = \begin{bmatrix} V_1 & z \end{bmatrix}^T$, $y_i = V_t - d_i$, $A = \begin{bmatrix} -\frac{1}{C_1 R_1} & 0 \\ 0 & 0 \end{bmatrix}$, $B = \begin{bmatrix} \frac{1}{C_1} & -\frac{1}{C_n} \end{bmatrix}^T$, $u = I_L$, $C_i = \begin{bmatrix} -1 & k_i \end{bmatrix}$ and $D = R_0$. By utilizing this linearized system model for a single battery cell, it is easy to implement the state observer for an SOC estimation because the theories for linear systems are applicable. However, a major concern of this linearized model is that when the operating point changes, the linearized system model will change. If the change is large, the optimally designed observer will no longer be optimal, and in the worst case, the observer will become unstable. Details about the limitations of the linearized model for the battery system are mentioned in the discussion section.

2.2. Nonlinear System Model for a Single Battery Cell

This paper proposes a nonlinear state space representation for the battery cell. Let us define the nonlinear function $V_{OC}(z)$ as a summation of a linear term and a nonlinear term as

$$\begin{aligned} V_{OC}(z) &= \alpha z + \beta + f(z), \\ f(z) &= \sum_{n=1}^{N} a_n \sin(b_n z + c_n) \end{aligned} \tag{6}$$

where α and β are the coefficients of the linear term, $f(z)$ is a bounded nonlinear function consisting of a sum of sine functions and a_n, b_n, and c_n are the coefficients of the sum of sine functions. Figure 2 conceptually represents (6).

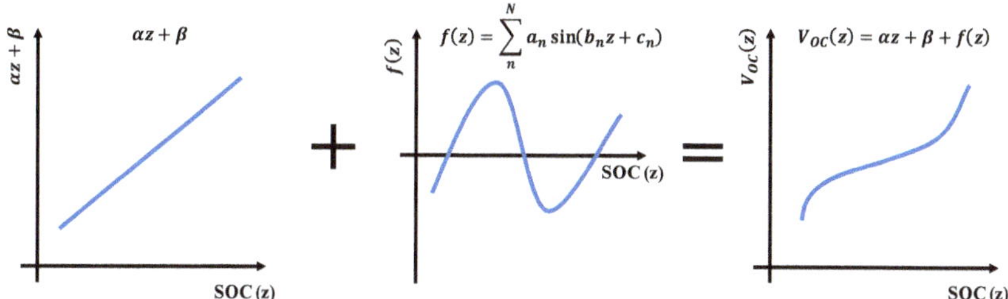

Figure 2. Reformulation of the open-circuit voltage representation.

Equations (1)–(3) can be rewritten as a nonlinear state space representation by using (6). The terminal voltage of the ECM in (1) is

$$V_t = \alpha z + \beta + f(z) - V_1 - I_L R_0. \tag{7}$$

Its time derivative can be calculated as follows:

$$\begin{aligned} \dot{V}_t &= \alpha \dot{z} + \tfrac{d}{dt} f(z) - \dot{V}_1 - \dot{I}_L R_0 = -\alpha \tfrac{1}{C_n} I_L + \tfrac{d}{dz} f(z)\, \dot{z} - \left(-\tfrac{1}{C_1 R_1} V_1 + \tfrac{1}{C_1} I_L\right) \\ &= -\alpha \tfrac{1}{C_n} I_L - \tfrac{d}{dz} f(z) \tfrac{1}{C_n} I_L + \tfrac{1}{C_1 R_1} V_1 - \tfrac{1}{C_1 R_0} (\alpha z + \beta + f(z) - V_1 - V_t) \\ &= \tfrac{1}{C_1 R_0} V_t + \left(\tfrac{1}{C_1 R_1} + \tfrac{1}{C_n R_0}\right) V_1 - \tfrac{\alpha}{C_1 R_0} z + \left\{-\tfrac{1}{C_n} I_L \tfrac{d}{dz} f(z) - \tfrac{1}{C_1 R_0} f(z)\right\} + \left\{-\tfrac{\alpha}{C_n} I_L - \tfrac{\beta}{C_1 R_0}\right\}. \end{aligned} \tag{8}$$

The derivative of the state of charge in (2) is

$$\begin{aligned} \dot{z} &= -\tfrac{1}{C_n R_0} (\alpha z + \beta + f(z) - V_1 - V_t) \\ &= \tfrac{1}{C_n R_0} V_t + \tfrac{1}{C_n R_0} V_1 - \tfrac{\alpha}{C_n R_0} z - \tfrac{1}{C_n R_0} f(z) - \tfrac{1}{C_n R_0} \beta. \end{aligned} \tag{9}$$

Then, the nonlinear state space representation of the given ECM can be obtained as

$$\begin{aligned} &\dot{x} = Ax + F(x, I_L) + G(I_L), \\ &y = Cx, \\ &A = \begin{bmatrix} \frac{1}{R_0 C_1} & \frac{1}{R_0 C_1} + \frac{1}{R_1 C_1} & -\frac{\alpha}{R_0 C_1} \\ 0 & -\frac{1}{R_1 C_1} & 0 \\ \frac{1}{R_0 C_n} & \frac{1}{R_0 C_n} & -\frac{\alpha}{R_0 C_n} \end{bmatrix}, \quad F(x, I_L) = \begin{bmatrix} -\frac{1}{R_0 C_1} f(z) - \frac{1}{C_n} \frac{d}{dz} f(z) I_L \\ 0 \\ -\frac{1}{R_0 C_n} f(z) \end{bmatrix}, \\ &G(I_L) = \begin{bmatrix} -\frac{1}{C_n} \alpha I_L - \frac{1}{R_0 C_1} \beta \\ \frac{I_L}{C_1} \\ -\frac{1}{R_0 C_n} \beta \end{bmatrix}, \quad x = \begin{bmatrix} V_t \\ V_1 \\ z \end{bmatrix}, \quad y = V_t \end{aligned} \tag{10}$$

where A is the state matrix, $F(z, I_L)$ is a nonlinear function with unknown states and $G(I_L)$ is a nonlinear function with known parameters. It is noted that the time derivative of the current I_L can be negligible not only because the sampling time of the algorithm is much faster than the current change [30] but also because its effect is much smaller than that of the other factors. Different from the linearized model in (5), it is easily shown that the proposed nonlinear system model in (10) does not change regardless of the SOC range, and the state vector contains the terminal voltage V_t, which can lead to the model being insensitive to sensor noise when using the measured value directly. However, because of the existence of the nonlinear functions $F(z, I_L)$ and $G(I_L)$, the observers used previously for linear systems are no longer available, and the stability condition for the nonlinear-model-based observer is not determined by considering the eigenvalues of the linear stability matrix $(A - LC)$, where L is the observer gain matrix. Hence, in the next section, a nonlinear-model-based observer is proposed, and its stability condition is verified based on the Lyapunov stability criteria.

3. Nonlinear-Model-Based Observer Design

Theorem 1. *Under the assumptions that the linear observability matrix $(A - L_nC)$ of the given nonlinear system model in (10) has full rank, and the nonlinear function $F(z, I_L)$ can be assumed to be a locally Lipschitz continuous function with a Lipschitz constant χ, which satisfies (11) in the physically feasible range of space X such that*

$$\|F(x_1, I_L) - F(x_2, I_L)\| \leq \chi \|x_1 - x_2\|, \forall x \in X, \tag{11}$$

the observer given in (12) is asymptotically stable if the Luenberger observer gain L_n can be chosen to ensure that the linear stability matrix $(A - L_nC)$ is Hurwitz and the inequality (13) is satisfied.

$$\dot{\hat{x}} = A\hat{x} + F(\hat{x}, I_L) + G(I_L) + L_n(y - C\hat{x}), \tag{12}$$

$$\min_{\omega \in R^+} \sigma_{\min}(A - L_nC - j\omega I) > \chi \tag{13}$$

where χ is the Lipschitz constant in (11).

Proof. Let us prove Theorem 1 by the method of contradiction. According to H_∞ theory, the following well-known condition is satisfied. If the Hamiltonian matrix

$$H = \begin{bmatrix} A & R \\ Q & -A^T \end{bmatrix} \tag{14}$$

has no imaginary eigenvalues; then, there exists a symmetric matrix P satisfying the algebraic Riccati equation

$$A^TP + PA + PRP - Q = 0 \tag{15}$$

(for a proof, see [40]). In the same context, it can said if that the Hamiltonian matrix

$$H = \begin{bmatrix} (A - L_nC) & \chi^2 I \\ -I - \varepsilon I & -(A - L_nC)^T \end{bmatrix} \tag{16}$$

has no imaginary eigenvalues, there exists a symmetric matrix P satisfying the algebraic Riccati equation

$$(A - L_nC)^T P + P(A - L_nC) + P\chi^2 P + I + \varepsilon I = 0. \tag{17}$$

From (13), there exists a finite ω_0 such that

$$\min_{\omega \in R^+} \sigma_{\min}(A - L_nC - j\omega I) > \sigma_{\min}(A - L_nC - j\omega_0 I) = \chi_{\min}. \tag{18}$$

Then, we can say that for all $\omega > \omega_0$, $(A - L_nC - j\omega I)^* (A - L_nC - j\omega I) \geq \chi^2_{\min} I$ is satisfied, where * indicates a Hermitian matrix. Choose ϵ such that

$$(A - L_nC - j\omega I)^* (A - L_nC - j\omega I) \geq \chi^2_{\min} I > \chi^2 (I + \varepsilon I). \tag{19}$$

The eigenvalues of the Hamiltonian matrix (16) are given by [41]

$$\det\left[\chi^2 (I + \varepsilon I) + \{\lambda I + (A - L_nC)^T\}\{\lambda I + (A - L_nC)\}\right] = 0. \tag{20}$$

Without loss of generality, it can assumed that an imaginary axis eigenvalue is represented by $j\omega$. By substituting $\lambda = j\omega$ in (20), we have

$$\det\left[\chi^2 (I + \varepsilon I) + \{-j\omega I + (A - L_nC)^T\}\{-j\omega + (A - L_nC)\}\right] = 0. \tag{21}$$

This means that

$$\{(A - L_nC)^T - j\omega I\}\{(A - L_nC) - j\omega\} = \chi^2 (I + \varepsilon I). \tag{22}$$

This contradicts (19). Hence, the matrix H in (16) cannot have any imaginary eigenvalues if inequality in (13) satisfied.

Let us define the state estimation error vector as $e = x - \hat{x}$ and consider the Lyapunov function candidate $V = e^T Pe$. Then, the derivative of V is

$$\begin{aligned} \dot{V} &= \dot{e}^T Pe + e^T P\dot{e} \\ &= e^T \left[(A - L_nC)^T P + P(A - L_nC)\right] e + 2e^T P\left[F(x, I) - F(\hat{x}, I)\right]. \end{aligned} \tag{23}$$

Using the Lipschitz condition in (11) and the property $e^T P[F(x, I) - F(\hat{x}, I)] \leq \|Pe\| \, \|F(x, I) - F(\hat{x}, I)\|$, the derivative of V can be represented by an inequality as

$$\dot{V} < e^T \left[(A - L_nC)^T P + P(A - L_nC)\right] e + 2\chi \|Pe\| \, \|e\|. \tag{24}$$

Using

$$\chi^2 e^T PPe + e^T e = 2\chi e^T Pe \geq 2\chi \|Pe\| \, \|e\|, \tag{25}$$

and the result of (16)–(22), the upper bound of the Lyapunov candidate can be obtained as

$$\dot{V} \leq e^T \left[(A - L_nC)^T P + P(A - L_nC) + \chi^2 PP + I\right] e = -e^T \varepsilon I e. \tag{26}$$

Hence, the system is asymptotically stable. □

There are two kinds of necessary conditions for Theorem 1: (1) the linear observability matrix $(A - L_nC)$ of a given system has full rank, and (2) the nonlinear function is a local Lipschitz continuous function. The linear observability matrix of the given nonlinear system model in (10) can be obtained as $O_{(A,C)} = \begin{bmatrix} C & CA & CA^2 \end{bmatrix}^T$.

The linear observability matrix of the given system is derived as

$$O_{(A,C)} = \begin{bmatrix} 1 & 0 & 0 \\ p_1 & p_1 + p_2 & -\alpha p_1 \\ p_1^2 - \alpha p_1 p_3 & p_1^2 - p_2^2 - \alpha p_1 p_3 & -\alpha p_1^2 + \alpha^2 p_1 p_3 \end{bmatrix} \tag{27}$$

where $p_1 = \frac{1}{R_0C_1}$, $p_2 = \frac{1}{R_1C_1}$ and $p_3 = \frac{1}{R_0C_n}$. The determinant of the matrix $O_{A,C}$ is

$$\det \left| O_{(A,C)} \right| = \alpha p_3 - p_1 - p_2. \tag{28}$$

It is shown that the given observability condition is independent of the shape of the OCV function. This is a function of the given parameters. Then, if the coefficient α is selected such that

$$\alpha \neq \frac{p_1 + p_2}{p_3}, \tag{29}$$

the given nonlinear battery model in (10) will satisfy the first necessary condition of Theorem 1. The Lipschitz condition in (11) can be rewritten by the following partial differential equation [42] as

$$\frac{|F(x_1, I_L) - F(x_2, I_L)|}{|x_1 - x_2|} = \left. \frac{\partial F(x, I_L)}{\partial x} \right|_{\forall x \in X} \leq \chi. \tag{30}$$

While there are three kinds of states to be considered for the Lipschitz condition, the state V_t is measurable. Therefore, the two unmeasurable states, V_1 and z, must be considered. The Lipschitz conditions for the nonlinearities of V_1 and z can be derived as follows:

$$\frac{\partial F_2(x, I_L)}{\partial x} = 0, \tag{31}$$

$$\frac{\partial F_3(x, I_L)}{\partial x} = -\frac{1}{R_0C_n} \frac{d}{dz} f(z) \tag{32}$$

where F_2 and F_3 are the nonlinearities of V_d and z. As mentioned above, the function $f(z)$ is predefined by the sum of the sinusoidal function, which has a certain boundary. The resulting function in (32) is a function of only one state value z. Therefore, if the function values of (32) in the overall feasible range of z are smaller than the Lipschitz constant χ, the inequality in (11) is satisfied. This means that the second necessary condition of the Theorem 1 is satisfied. The specific parameters of the necessary conditions will be verified in the next section.

4. Experiments

This section inspects the performance of the SOC estimation of the proposed nonlinear battery model and the extended Kalman filter, which has commonly been applied for SOC estimation in previous works in [11,24,38,43] by conducting a series of experiments.

4.1. Experimental Setup

To analyze the performance of the SOC estimation of the proposed observer and previous methods, an experimental battery cell test bench is established. Experiments were conducted using the INR 18650-25R battery cell from SAMSUNG SDI. The test bench for the charge-discharge experiment is shown in Figure 3. The setup consists of a bidirectional DC/DC converter (Maccor 4300 K), a temperature chamber (Jeiotech TH-G-408) and a main PC. The current and terminal voltage of the cell were measured accurately by the Maccor 4300 K converter with a full-scale range (FSR) measurement error below 0.02%. This experiment was implemented with a controlled temperature of 25 °C.

In this paper, all of the charge-discharge experiments were conducted by: (1) charging the battery with CC-CV mode until the battery reaches the charge cutoff voltage of 4.2 V; (2) discharging the battery with CC mode until the SOC reaches the intended initial SOC, where the SOC is calculated by the Coulomb counting method with a precise current sensor; and (3) conducting the target current cycle. As an example, the sequential current and resulting terminal voltage for the UDDS current profile are shown in Figure 4.

To focus on the intended current cycle and target SOC area, the analysis was conducted only for step 3.

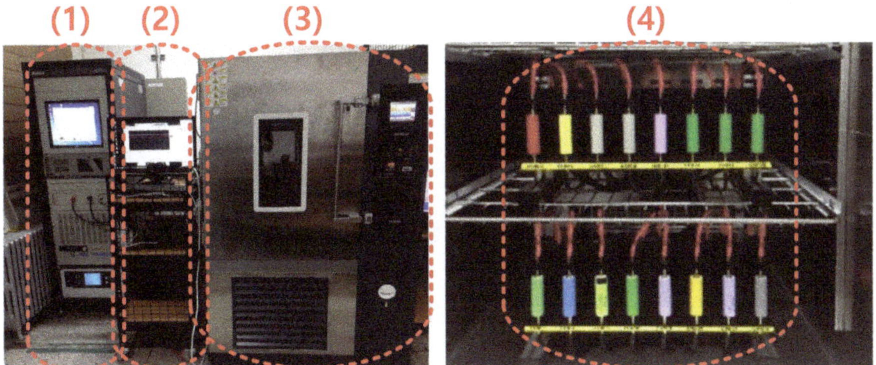

Figure 3. Experimental setup: (1) bidirectional DC/DC converter, (2) personal computer (PC) for data acquisition, (3) temperature chamber, and (4) tied battery cells inside the temperature chamber.

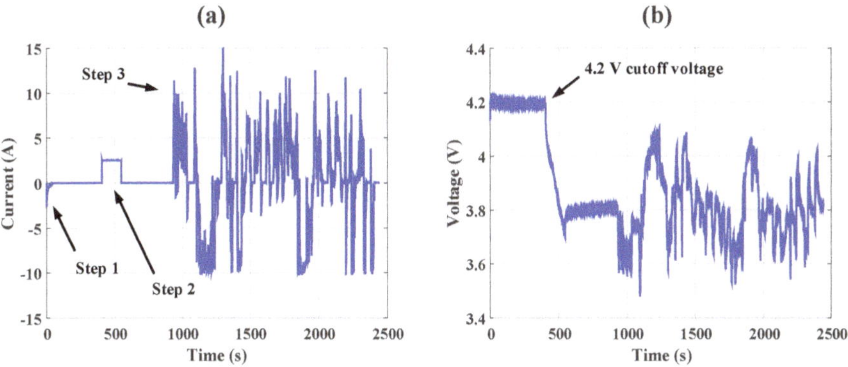

Figure 4. Experimental sequence: (**a**) engaging the current and (**b**) the resulting cell terminal voltage.

4.2. Target Battery Specification and Parameters Extraction

The target battery cell is an INR 18650-25R cylindrical Li-ion battery cell comprising GIC and NMC from SAMSUNG SDI. Before extracting the parameters of the given ECM and SOC-OCV relationship, the pre-cycling procedure including 10 fully charge-discharge cycles was conducted in order to the target battery cell can be warmed-up and ready-to-use state. The equivalent parameters for the 1st-order Thevenin ECM are extracted using an offline hybrid pulse power characterization cycle (HPPC) test at a constant temperature of 25 °C. The current profile and voltage profile of HPPC test are shown in Figure 5.

1 C-rate (2.5A) is chosen for charge-discharge current, and by discharging 30 min with 1 C-rate current, and discharge the battery during 30 min at every cycles so that the SOC level is dropped by 5%. In order to measure the OCV at each SOC level, the battery is rested for an hour. According to the voltage and current data, resulting equivalent parameters and SOC-OCV relationship for each SOC level are shown in Figure 6.

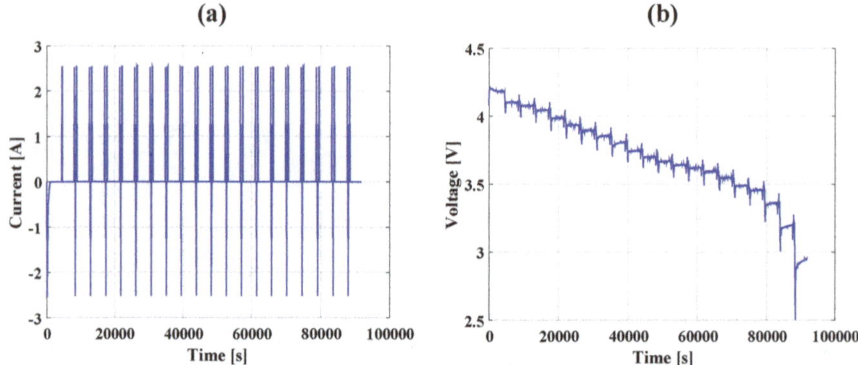

Figure 5. The HPPC test procedure: (**a**) engaged current profile and (**b**) resulting terminal voltage.

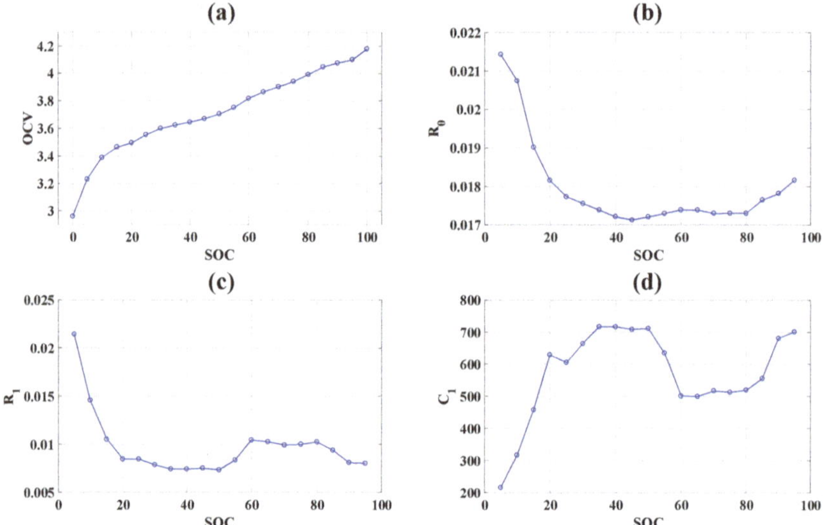

Figure 6. Resulting equivalent parameters: (**a**) SOC-OCV relationship, (**b**) internal resistance (**c**) resistance value of RC-pair and (**d**) capacitance value of RC-pair.

The equivalent parameters are obtained based on the method presented by Kim et al. [44]. The average parameters are listed in Table 1, and these values are used for establishing the state space model and selecting optimal gain of observer.

Table 1. Average values of equivalent parameters of the 1st-order Thevenin ECM.

Parameter	Value
R_0	0.0172 Ω
R_1	0.0097 Ω
C_1	570.86 F
C_n	8972 As

$V_{OC}(z)$ was captured at 5% SOC intervals from 5% SOC to 95% SOC. The linear and nonlinear functions of the proposed OCV representation, which are introduced in Figure 2, are shown in Figure 7.

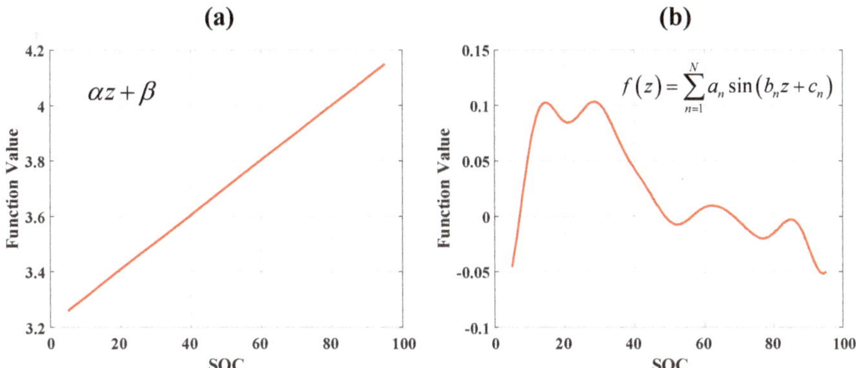

Figure 7. The proposed open-circuit voltage representation: the (**a**) linear term and (**b**) nonlinear term.

The nonlinear term of the OCV function is tuned by utilizing the curve fitting tool in MATLAB (2017a academic version, Mathworks, Natick, MA, USA). The corresponding coefficients are listed in Table 2.

Table 2. Coefficients of the proposed $V_{OC}(z)$ in (6).

Parameter	Value					
α	0.9878					
β	3.2095					
Parameter	**n = 1**	**2**	**3**	**4**	**5**	**6**
a_n	0.07	0.05	0.04	0.02	0.23	0.22
b_n	1.90	0.30	3.39	8.35	10.01	10.10
c_n	−3.30	0.49	−0.98	−1.27	1.74	−1.42

The measured OCV of each 5% SOC and the proposed nonlinear representation of the OCV curve in (6) are shown in Figure 8.

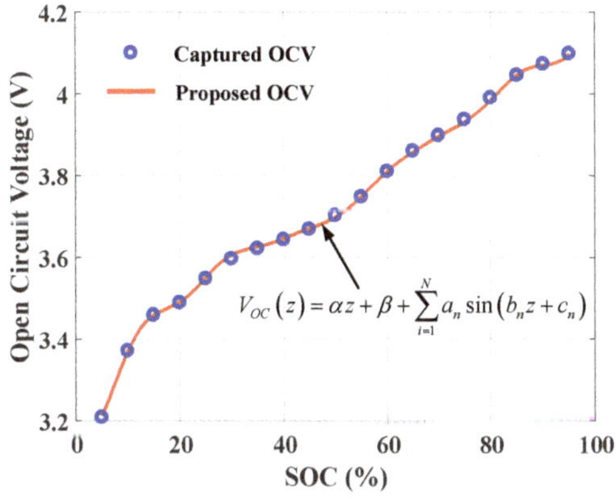

Figure 8. Captured OCV at each SOC point and the SOC-OCV curves fitted using the proposed OCV representation.

Accordingly, the corresponding matrices of the proposed nonlinear battery cell model in (10) are obtained as

$$A = \begin{bmatrix} 0.1018 & 0.2814 & -0.1006 \\ 0 & -0.1796 & 0 \\ 0.0065 & 0.0065 & -0.0064 \end{bmatrix}, C = \begin{bmatrix} 1 & 0 & 0 \end{bmatrix}. \tag{33}$$

From the resulting coefficients, $\alpha = 0.9878$, $p_1 = 0.1018$, $p_2 = 0.1806$, $p_3 = 0.0065$ and the condition in (29), it is known that the linear observability matrix of the given system has full rank. The values of (32) in the overall feasible range of z are shown in Figure 9.

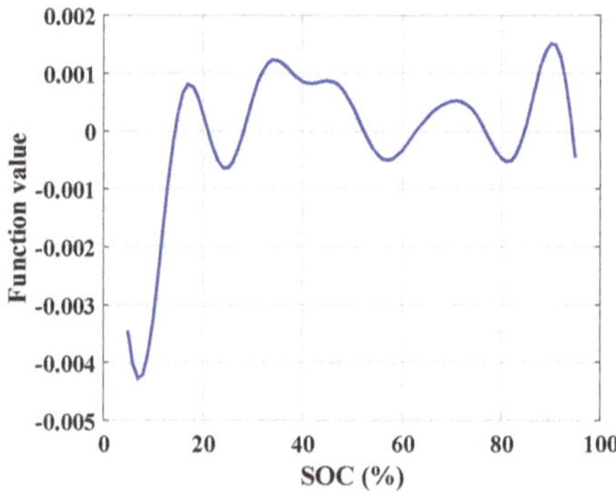

Figure 9. Values of the partial derivative function in (32) for the state in the overall feasible range.

The resulting values are bounded from -0.005 to $+0.002$. This means that the Lipschitz condition in (11) is satisfied if the Lipschitz constant is selected as $\chi > 0.005$. All the necessary conditions of (1) linear observability and the (2) local Lipschitz condition are satisfied. The Luenberger observer gain is selected by utilizing the pole-placement technique so that the eigenvalues of the Luenberger observer satisfy $\lambda\left(A - L_nC\right) = \begin{bmatrix} -0.5 & -0.1 & -0.01 \end{bmatrix}^T$ and the sufficient condition in (13) is satisfied. The minimum singular value of $L_n \min\limits_{\omega \in R^+} \left(A - L_nC - j\omega I\right)$ is 0.0073. This value is larger than the Lipschitz constant $\chi = 0.005$. From the sufficient condition in (13) of Theorem 1, the proposed observer in (12) with the selected observer gain is asymptotically stable and the state estimation error converges to zero as time increases.

4.3. Experimental Results

The UDDS current profile, which is shown in Figure 4, was used for the experiments. To evaluate the performance of the real-time SOC estimation and insensitivity to sensor noise, two types of experiments were conducted. The first working condition is the noiseless condition. Because the experimental setup has a high precise current and the voltage sensors are operated under controlled conditions, it can be assumed that there is no external noise. There are only unknown model uncertainties. To compare the performance of the SOC estimation of the proposed method with that of previous methods, two types of SOC estimation methods are used: (1) a Luenberger observer with a nonlinear model and (2) an extended Kalman filter (EKF) with following form [45]:

$$\begin{aligned} x_{k+1} &= Ax_k + Bu_k + \omega_k, \\ y_{k+1} &= C_kx_{k+1} + Du_{k+1} + v_k, \\ A = \begin{bmatrix} \exp\left(-\frac{\Delta T}{R_1C_1}\right) & 0 \\ 0 & 1 \end{bmatrix}&, B = \begin{bmatrix} R_1\left(1 - \exp\left(-\frac{\Delta T}{R_1C_1}\right)\right) \\ \frac{\Delta T}{C_n} \end{bmatrix}, \\ C_k = \left.\frac{\partial V_t}{\partial x}\right|_{x=\hat{x}_{k-1}^-} &= \left.\begin{bmatrix} -1 & \frac{\partial V_{OC}(z)}{\partial z} \end{bmatrix}\right|_{z=\hat{z}_{k-1}^-}, D = [R_0] \end{aligned} \tag{34}$$

where $x_k = \begin{bmatrix} V_{1,k} & z_k \end{bmatrix}^T$, $y_k = V_{t,k}$, $u_k = I_{L.k}$ and ΔT is the sampling time.

4.3.1. Case 1: Noiseless Condition

The real-time SOC estimation results of the two methods under noiseless condition are shown in Figure 10. The percentage error is calculated by

$$Percent\,error\,(\%) = \frac{True\,value - Estimated\,value}{True\,value} \times 100. \tag{35}$$

At the beginning of the experiment, the initial SOC value is set to be far from the true SOC value. This shows the observer's robustness to the initial state error. As shown in Figure 10, both SOC estimation methods have good performance. Because the EKF is an adaptive and optimal version of the Luenberger observer, it usually shows better performance when the accuracy of the model is sufficiently high and the external noise can be assumed to be Gaussian noise. The EKF also shows a shorter offset compensation time for a well-conditioned experiment than the proposed method. However, after the offset compensation time, the percentage errors of the SOC estimation of the proposed method and the EKF are under ±5%.

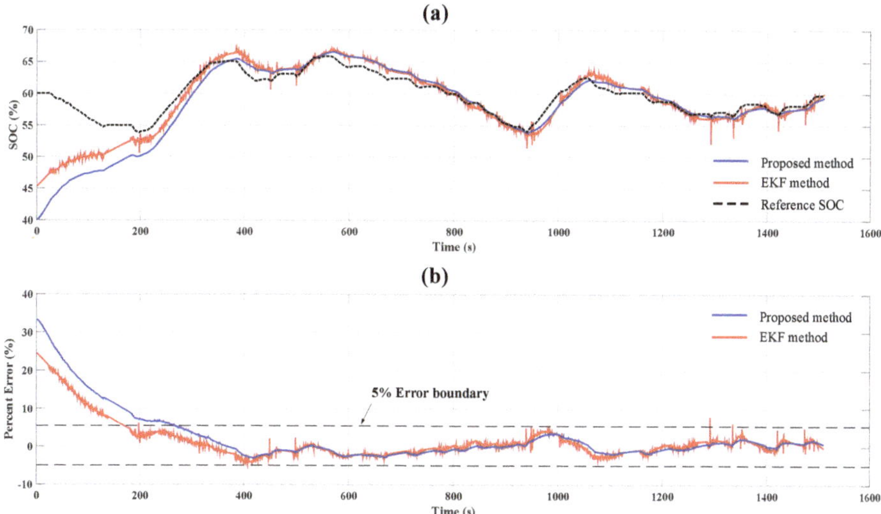

Figure 10. SOC estimation results under noiseless conditions: (**a**) SOC estimation results of the two types of methods and the (**b**) percentage error of each method.

The sensor noise is considered for the other working condition. Two types of sensor noise, voltage sensor noise n_V and current sensor noise n_I, are considered as random noise with zero mean and different peak-to-peak values of $|n_V|_{\max} = 0.02\,\mathrm{V}$ and $|n_I|_{\max} = 2.5\,\mathrm{A}$. Although the given voltage and current sensor noise conditions are quite severe, this level of sensor noise can occur in a real implementation of an onboard BMS as a result of external noise due to an unstable ground, electromagnetic interference (EMI) from electronic equipment or a low sensor resolution, and as a result, the advanced performance of the proposed method can be emphasized. The current and voltage signals with sensor noise are shown in Figure 11.

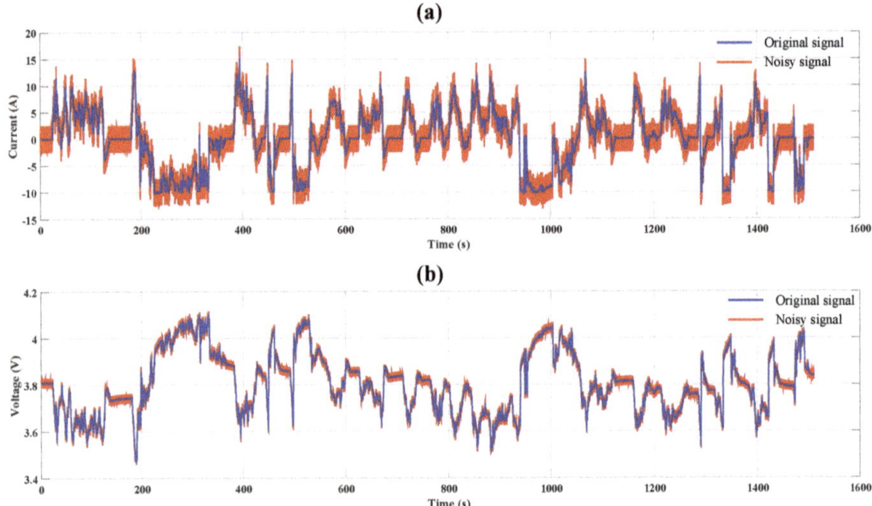

Figure 11. Voltage and current signals: (**a**) original and noisy current signal and (**b**) original and noisy voltage signal.

4.3.2. Case 2: Voltage Sensor Noise Condition

Figure 12 shows the SOC estimation results with only voltage sensor noise.

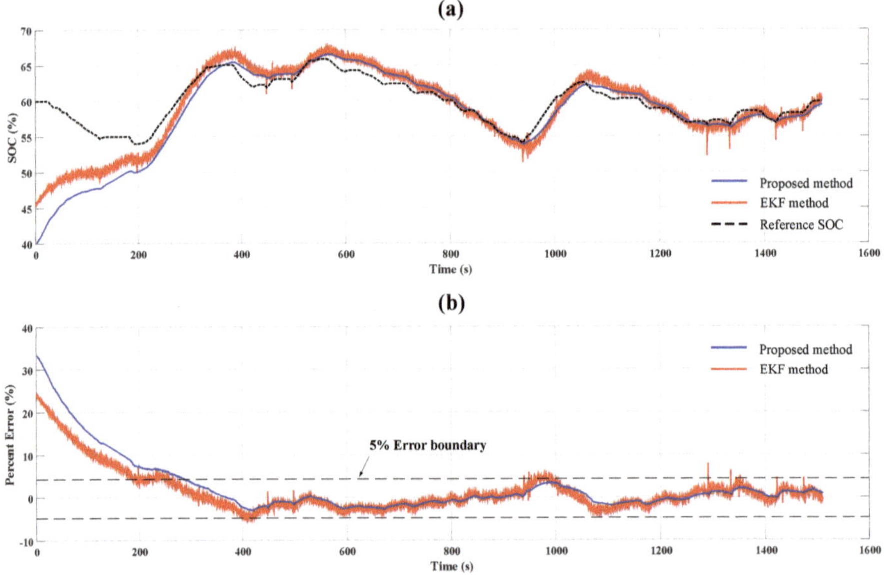

Figure 12. SOC estimation results with voltage sensor noise: (**a**) SOC estimation results of the two types of methods and (**b**) percentage error of each method.

As shown in (34), the output vector y is the measured terminal voltage. Because the EKF is a good estimator whether the measurement noise can be assumed Gaussian noise, EKF shows better SOC estimation performance compared with proposed method by suppressing the voltage noise well.

The percentage error and offset compensating time are slightly increased compared with the results of the noiseless condition.

4.3.3. Case 3: Voltage and Current Sensor Noise Condition

Figure 13 shows the SOC estimation results with voltage and current sensor noise.

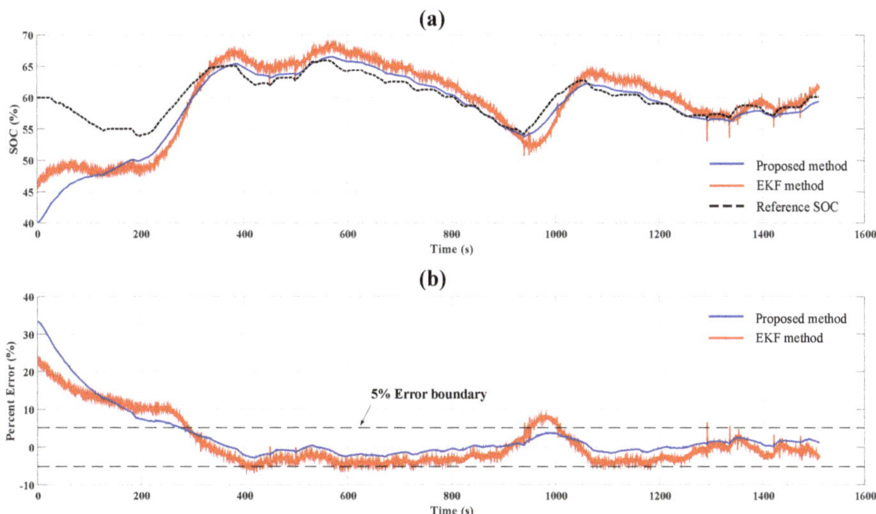

Figure 13. SOC estimation results with voltage and current sensor noise: (**a**) SOC estimation results of the two types of methods and (**b**) percentage error of each method.

Table 3 summarizes the results of the two types of experiments using both methods. Compared with the results of the noiseless experiment, the SOC estimation error is larger with both methods because of the sensor noise. However, in the case of the proposed method, the increases in the mean absolute error (MAE) and the maximum error (after the offset compensation) are relatively smaller than those of the errors of the EKF. The time for compensating the initial offset (when the percentage error is less than 5%) is less affected by the sensor noise.

Table 3. SOC estimation results for different experimental conditions using both methods.

Method	Experiment	Offset Compensation Time (s)	MAE (%)	Absolute Maximum Error (%)
Extended Kalman filter	Noiseless condition	174.59	2.9099	4.1340
	Noise condition	294.62	4.8255	7.8403
Proposed method	Noiseless condition	274.36	3.7413	3.3539
	Noise condition	278.96	3.7646	3.6544

This result occurs because the proposed method includes the terminal voltage in the state vector. Although the terminal voltage of the battery is information that can be measured, the result of adding this information to the state vector is that it is updated when integrating the error between the noisy measurement signal and the estimated value. The block diagram of proposed method is shown in Figure 14.

It is shown that at the last sequence, the state vector is passed through the integrator. This integrator can suppress the zero mean noisy signal in the state vector, which it is similar to the low-pass filter. As a disadvantage, this can decrease the state estimation response. Therefore, it is necessary to set an appropriate gain and achieve a trade-off between these characteristics. On the other hand, the EKF directly updates the state values by using the noisy measurement signal. According to the given model in (34), different from the voltage sensor noise, the current sensor noise is applied to not only the output vector through the matrix D but also the state vector through the matrix B. It means that if there exists current sensor noise, both the measured value $y(k)$ and the estimated states $x(k)$ based on the system model are inaccurate. Therefore, Kalman filter cannot show the convincing performance under this kinds of condition because Kalman filter is designed to estimate the state with a more accurate value between the measured value and the estimated value. This condition occurs neither of these values is accurate. Figure 15 shows the estimated voltage of both the proposed method and the original voltage signal.

It is known that in the case of the EKF, the noisy voltage signal is directly used for updating the SOC, but in the case of the proposed method, although the estimation speed is relatively slow, the noisy signal is filtered out.

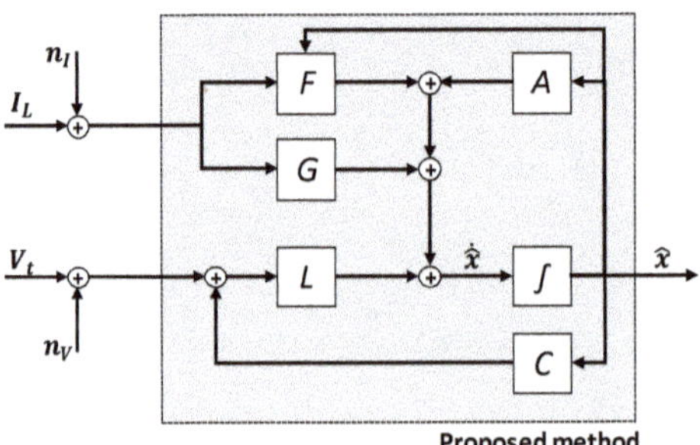

Figure 14. The block diagram of proposed method.

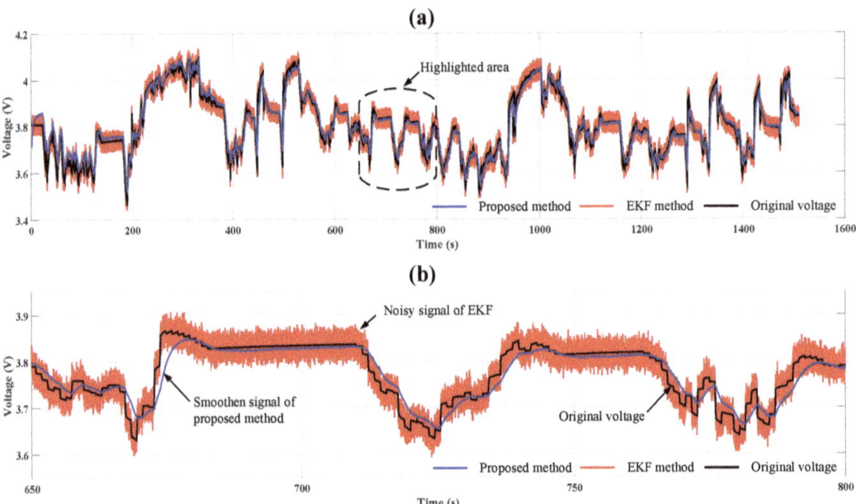

Figure 15. (**a**,**b**) Estimated voltage of both methods and the original voltage signal.

Therefore, the proposed method can robustly estimate the SOC in the presence of sensor noise.

5. Discussion

Although not experimentally verified in this paper, the proposed nonlinear battery cell model can solve the critical limitation of the linearized model in (5). The observability matrix of a given linearized model is calculated as

$$O(A,C) = \begin{bmatrix} C & CA \end{bmatrix}^T = \begin{bmatrix} -1 & k_i \\ \frac{1}{C_1 R_1} & 0 \end{bmatrix}. \tag{36}$$

By calculating the determinant of the observability matrix, the observability condition of the linear system is obtained as

$$\det |O(A,C)| = -k_i \frac{1}{C_1 R_1}. \tag{37}$$

The condition directly states that the given linearized system is observable only if $k_i \neq 0$. That is, if there is a flat voltage region in the open-circuit voltage curve, the linearized model loses its observability. This situation can occur in certain battery types: e.g., $LiFePO_4$ (LFP). The LFP-type battery has voltage plateaus in the SOC-OCV curve [36–38,46]. Therefore, when the SOC range is within such an area, the linearized model cannot estimate the SOC from the OCV curve because there is no state excitation. Previous studies did not consider such problems.

Simulation Study with a Virtual Battery Cell Having Wide Range of Flat OCV Curve

Let assume that there exist a virtual battery cell having flat OCV curve from 20 % to 80 % SOC range. The SOC-OCV relationship of this battery cell is shown in Figure 16, and two kinds of model (1) proposed OCV curve representation in (2) and (2) linearized OCV curve representation in (4) are used for SOC-OCV curve fitting.

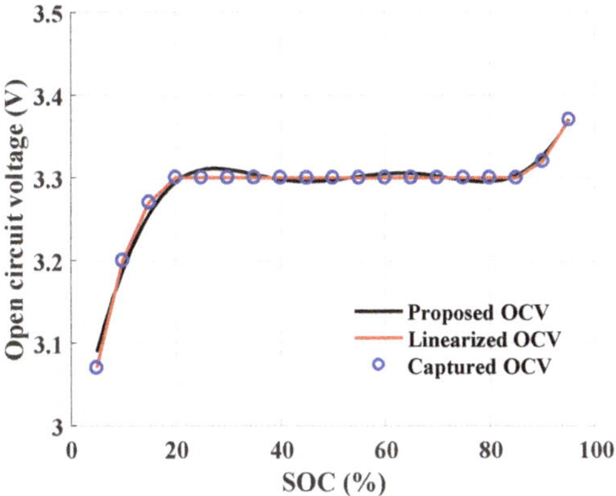

Figure 16. SOC-OCV relationship with wide flat area and two kinds of representation of OCV curve.

The SOC estimation results of proposed nonlinear-model in (10) and linearized model in (5) are shown in Figure 17.

As derived in (37), the linearized model loses its observability when there exists flat area on the OCV curve. Thus, the estimated SOC cannot converge to the true value. However, the observability condition of the proposed nonlinear-model is independent of the form of the OCV curve. It is shown that the estimated SOC based on the nonlinear-model converges to the true value. This result has been obtained empirically, and mathematical validation remains as a further research.

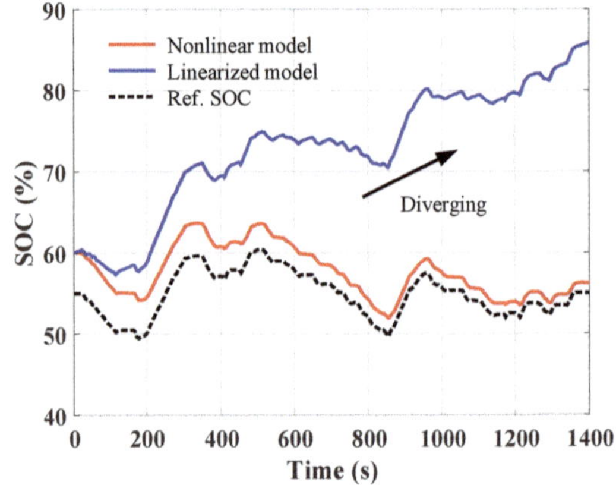

Figure 17. SOC estimation result with nonlinear model and linearized model.

6. Conclusions

This study has proposed a nonlinear state space representation for a Li-ion battery cell and Luenberger observer for a class of nonlinear systems. The proposed nonlinear battery cell model contains the terminal voltage in the state vector, improving the robustness against sensor noise caused by the external operating environment or sensor faults. The proposed method has improved SOC estimation performance in the presence of sensor noise. The improvements of the proposed nonlinear-model-based method are demonstrated with experiments; however, there is room for improvement in this study. Because the proposed observer is a Luenberger observer, an additional performance enhancement can be achieved by adding adaptive observe gain, such as sliding-mode gain and integral gain. However, verifying the stability condition of these observers for nonlinear systems is much more challenging. Thus, this will be considered in a future study.

Author Contributions: This study was carried out successfully with contributions from all authors. The main idea, formulations, simulations and manuscript preparation were contributed by W.-Y.K., P.-Y.L. conducted the experiment and data acquisition. J.K. contributed to the setup and experimental procedures and finalizing the manuscript. K.-S.K. guided the overall flow of the research.

Funding: This research was supported by a grant (17TLRP-C135446-01, Development of Hybrid Electric Vehicle Conversion Kit for Diesel Delivery Trucks and its Commercialization for Parcel Services) from the Transportation & Logistics Research Program (TLRP) funded by the Ministry of Land, Infrastructure and Transportation of the Korean government. This research was also supported by the BK21 Plus Program.

Conflicts of Interest: The authors declare no conflict of interest.

Abbreviations

The following abbreviations are used in this manuscript:

HEV	Hybrid electric vehicle
EV	Electric vehicle
BMS	Battery management system
Li-ion	Lithium-ion
SOC	State of charge
ANN	Artificial neural network
EECM	Equivalent Electrochemical model
ECM	Equivalent circuit model
OCV	Open-circuit voltage
HPPC	Hybrid pulse power characterization
UDDS	Urban dynamometer driving schedule
MAE	Mean absolute error
LFP	$LiFePO_4$; Lithium-ion phosphate battery
EKF	Extended Kalman filter

References

1. Zhou, L.; Zheng, Y.; Ouyang, M.; Lu, L. A study on parameter variation effects on battery packs for electric vehicles. *J. Power Sources* **2017**, *364*, 242–252. [CrossRef]
2. He, H.; Qin, H.; Sun, X.; Shui, Y. Comparison Study on the Battery SoC Estimation with EKF and UKF Algorithms. *Energies* **2013**, *6*, 5088–5100. [CrossRef]
3. Rezvanizaniani, S.M.; Liu, Z.; Chen, Y.; Lee, J. Review and recent advances in battery health monitoring and prognostics technologies for electric vehicle (EV) safety and mobility. *J. Power Sources* **2014**, *256*, 110–124. [CrossRef]
4. Lu, L.; Han, X.; Li, J.; Hua, J.; Ouyang, M. A review on the key issues for lithium-ion battery management in electric vehicles. *J. Power Sources* **2013**, *226*, 272–288. [CrossRef]
5. Hannan, M.A.; Lipu, M.S.H.; Hussain, A.; Mohamed, A. A review of lithium-ion battery state of charge estimation and management system in electric vehicle applications: Challenges and recommendations. *Renew. Sustain. Energy Rev.* **2017**, *78*, 834–854. [CrossRef]
6. Ali, M.U.; Zafar, A.; Nengroo, S.H.; Hussain, S.; Alvi, M.J.; Kim, H.J. Towards a Smarter Battery Management System for Electric Vehicle Applications: A Critical Review of Lithium-Ion Battery State of Charge Estimation. *Energies* **2019**, *12*, 446. [CrossRef]
7. Hoque, M.M.; Hannan, M.A.; Mohamed, A.; Ayob, A. Battery charge equalization controller in electric vehicle applications: A review. *Renew. Sustain. Energy Rev.* **2017**, *75*, 1363–1385. [CrossRef]
8. Nitta, N.; Wu, F.; Lee, J.T.; Yushin, G. Li-ion battery materials: Present and future. *Mater. Today* **2015**, *18*, 252–264. [CrossRef]
9. He, W.; Williard, N.; Chen, C.; Pecht, M. State of charge estimation for electric vehicle batteries using unscented kalman filtering. *Microelectron. Reliab.* **2013**, *53*, 840–847. [CrossRef]
10. Xia, B.; Wang, H.; Tian, Y.; Wang, M.; Sun, W.; Xu, Z. State of Charge Estimation of Lithium-Ion Batteries Using an Adaptive Cubature Kalman Filter. *Energies* **2015**, *8*, 5916–5936. [CrossRef]
11. Fang, Q.; Wei, X.; Dai, H. A Remaining Discharge Energy Prediction Method for Lithium-Ion Battery Pack Considering SOC and Parameter Inconsistency. *Energies* **2019**, *12*, 987. [CrossRef]
12. Yang, N.; Zhang, X.; Li, G. State of charge estimation for pulse discharge of a LiFePO4 battery by a revised Ah counting. *Electrochim. Acta* **2015**, *151*, 63–71. [CrossRef]
13. Ng, K.S.; Moo, C.S.; Chen, Y.P.; Hsieh, Y.C. Enhanced coulomb counting method for estimating state-of-charge and state-of-health of lithium-ion batteries. *Appl. Energy* **2009**, *86*, 1506–1511. [CrossRef]
14. Weigert, T.; Tian, Q.; Lian, K. State-of-charge prediction of batteries and battery–supercapacitor hybrids using artificial neural networks. *J. Power Sources* **2011**, *196*, 4061–4066. [CrossRef]
15. Shen, Y. Adaptive online state-of-charge determination based on neuro-controller and neural network. *Energy Convers. Manag.* **2010**, *51*, 1093–1098. [CrossRef]

16. Lai, X.; Qiao, D.; Zheng, Y.; Zhou, L. A Fuzzy State-of-Charge Estimation Algorithm Combining Ampere-Hour and an Extended Kalman Filter for Li-Ion Batteries Based on Multi-Model Global Identification. *Appl. Sci.* **2018**, *8*, 2028. [CrossRef]
17. Plett, G.L. *Battery Management Systems, Volume I: Battery Modeling*; Artech House: Norwood, MA, USA, 2015.
18. Nikolian, A.; Hoog, J.D.; Fleurbay, K.; Timmermans, J.M.; Noshin, O.; Bossche, P.V.D.; Mierlo, J.V. Classification of Electric modelling and Characterization methods of Lithium-ion Batteries for Vehicle Applications. In Proceedings of the European Electric Vehicle Congress, Brussels, Belgium, 2–5 December 2014; pp. 1–15.
19. Nikolian, A.; Firouz, Y.; Gopalakrishnan, R.; Timmermans, J.M.; Omar, N.; van den Bossche, P.; van Mierlo, J. Lithium Ion Batteries—Development of Advanced Electrical Equivalent Circuit Models for Nickel Manganese Cobalt Lithium-Ion. *Energies* **2016**, *9*, 360. [CrossRef]
20. Lee, S.; Kim, J.; Lee, J.; Cho, B.H. State-of-charge and capacity estimation of lithium-ion battery using a new open-circuit voltage versus state-of-charge. *J. Power Sources* **2008**, *185*, 1367–1373. [CrossRef]
21. Nejad, S.; Gladwin, D.T.; Stone, D.A. A systematic review of lumped-parameter equivalent circuit models for real-time estimation of lithium-ion battery states. *J. Power Sources* **2016**, *316*, 183–196. [CrossRef]
22. He, H.; Xiong, R.; Fan, J. Evaluation of Lithium-Ion Battery Equivalent Circuit Models for State of Charge Estimation by an Experimental Approach. *Energies* **2011**, *4*, 582–598. [CrossRef]
23. Sun, Q.; Zhang, H.; Zhang, J.; Ma, W. Adaptive Unscented Kalman Filter with Correntropy Loss for Robust State of Charge Estimation of Lithium-Ion Battery. *Energies* **2018**, *11*, 3123. [CrossRef]
24. Jung, S.; Jeong, H. Extended Kalman Filter-Based State of Charge and State of Power Estimation Algorithm for Unmanned Aerial Vehicle Li-Po Battery Packs. *Energies* **2017**, *10*, 1237. [CrossRef]
25. Diab, Y.; Auger, F.; Schaeffer, E.; Wahbeh, M. Estimating Lithium-Ion Battery State of Charge and Parameters Using a Continuous-Discrete Extended Kalman Filter. *Energies* **2017**, *10*, 1075. [CrossRef]
26. Yu, Z.; Huai, R.; Xiao, L. State-of-Charge Estimation for Lithium-Ion Batteries Using a Kalman Filter Based on Local Linearization. *Energies* **2015**, *8*, 7854–7873. [CrossRef]
27. He, Z.; Gao, M.; Wang, C.; Wang, L.; Liu, Y. Adaptive State of Charge Estimation for Li-Ion Batteries Based on an Unscented Kalman Filter with an Enhanced Battery Model. *Energies* **2013**, *6*, 4134–4151. [CrossRef]
28. Hu, X.; Sun, F.; Zou, Y. Estimation of State of Charge of a Lithium-Ion Battery Pack for Electric Vehicles Using an Adaptive Luenberger Observer. *Energies* **2010**, *3*, 1586–1603. [CrossRef]
29. Tian, Y.; Chen, C.; Xia, B.; Sun, W.; Xu, Z.; Zheng, W. An Adaptive Gain Nonlinear Observer for State of Charge Estimation of Lithium-Ion Batteries in Electric Vehicles. *Energies* **2014**, *7*, 5995–6012. [CrossRef]
30. Kim, I.S. The novel state of charge estimation method for lithium battery using sliding mode observer. *J. Power Sources* **2006**, *163*, 584–590. [CrossRef]
31. Kim, D.; Koo, K.; Jeong, J.; Goh, T.; Kim, S. Second-Order Discrete-Time Sliding Mode Observer for State of Charge Determination Based on a Dynamic Resistance Li-Ion Battery Model. *Energies* **2013**, *6*, 5538–5551. [CrossRef]
32. Chen, X.; Shen, W.; Cao, Z.; Kapoor, A. A novel approach for state of charge estimation based on adaptive switching gain sliding mode observer in electric vehicles. *J. Power Sources* **2014**, *246*, 667–678. [CrossRef]
33. Jun, X.; Mi, C.C.; Binggang, C.; Junjun, D.; Zheng, C.; Siqi, L. The State of Charge Estimation of Lithium-Ion Batteries Based on a Proportional-Integral Observer. *IEEE Trans. Veh. Technol.* **2014**, *63*, 1614–1621. [CrossRef]
34. Tang, X.; Liu, B.; Lv, Z.; Gao, F. Observer based battery SOC estimation: Using multi-gain-switching approach. *Appl. Energy* **2017**, *204*, 1275–1283. [CrossRef]
35. Klee Barillas, J.; Li, J.; Günther, C.; Danzer, M.A. A comparative study and validation of state estimation algorithms for Li-ion batteries in battery management systems. *Appl. Energy* **2015**, *155*, 455–462. [CrossRef]
36. Zheng, Y.; Ouyang, M.; Lu, L.; Li, J.; Han, X.; Xu, L.; Ma, H.; Dollmeyer, T.A.; Freyermuth, V. Cell state-of-charge inconsistency estimation for LiFePO4 battery pack in hybrid electric vehicles using mean-difference model. *Appl. Energy* **2013**, *111*, 571–580. [CrossRef]
37. Zhang, C.; Jiang, J.; Zhang, L.; Liu, S.; Wang, L.; Loh, P. A Generalized SOC-OCV Model for Lithium-Ion Batteries and the SOC Estimation for LNMCO Battery. *Energies* **2016**, *9*, 900. [CrossRef]
38. Wang, L.; Lu, D.; Liu, Q.; Liu, L.; Zhao, X. State of charge estimation for LiFePO4 battery via dual extended kalman filter and charging voltage curve. *Electrochim. Acta* **2019**, *296*, 1009–1017. [CrossRef]
39. Zou, Z.; Xu, J.; Mi, C.; Cao, B.; Chen, Z. Evaluation of Model Based State of Charge Estimation Methods for Lithium-Ion Batteries. *Energies* **2014**, *7*, 5065–5082. [CrossRef]

40. Francis, B.A. *Course in H_∞ Control Theory. Lectures Notes in Control and Information Sciences*; Springer Verlag: Berlin/Heidelberg, Germany, 1987.
41. Rajamani, R. Observers for Lipschitz Nonlinear Systems. *IEEE Trans. Autom. Control* **1998**, *43*, 397–401. [CrossRef]
42. Khalil, H.K.; Grizzle, J.W. *Nonlinear Systems*; Prentice Hall: Upper Saddle River, NJ, USA, 2002; Volume 3.
43. Chen, Z.; Li, X.; Shen, J.; Yan, W.; Xiao, R. A Novel State of Charge Estimation Algorithm for Lithium-Ion Battery Packs of Electric Vehicles. *Energies* **2016**, *9*, 710. [CrossRef]
44. Kim, J.H.; Lee, S.J.; Lee, J.M.; Cho, B.H. A New Direct Current Internal Resistance and State of Charge Relationship for the Li-Ion Battery Pulse Power Estimation. In Proceedings of the 7th International Conference on Power Electronics 2007, Daegu, Korea, 22–26 October 2007.
45. Plett, G.L. *Battery Management Dystems, Volume II: Equivalent-Circuit Methods*; Artech House: Norwood, MA, USA, 2015.
46. Gerschler, J.B.; Sauer, D.U. Investigation of open-circuit-voltage behaviour of lithium-ion batteries with various cathode materials under special consideration of voltage equalisation phenomena. In Proceedings of the EVS24 International Battery, Hybrid and Fuel Cell Electric Vehicle Symposium, Stavanger, Norway, 13–16 May 2009.

MDPI
St. Alban-Anlage 66
4052 Basel
Switzerland
Tel. +41 61 683 77 34
Fax +41 61 302 89 18
www.mdpi.com

Energies Editorial Office
E-mail: energies@mdpi.com
www.mdpi.com/journal/energies

www.ingramcontent.com/pod-product-compliance
Lightning Source LLC
Chambersburg PA
CBHW051836210326
41597CB00033B/5677

9783039218622